Remedial Treatment of Buildings

To my late father, Stanley A. Richardson

who introduced me to the absorbing science (or is it an art?) of remedial treatment

Remedial Treatment of Buildings

Second edition

Barry A. Richardson
Consulting Scientist
Director of Penarth Research International Limited

Illustrated by the author and Frank Willis,
with occasional assistance from
thelwell.

BUTTERWORTH
HEINEMANN

OXFORD AMSTERDAM BOSTON LONDON NEW YORK PARIS
SAN DIEGO SAN FRANCISCO SINGAPORE SYDNEY TOKYO

Butterworth-Heinemann
An imprint of Elsevier Science
Linacre House, Jordan Hill, Oxford OX2 8DP
225 Wildwood Avenue, Woburn MA 01801-2041

First published by Construction Press Ltd 1980
Second edition 1995
Transferred to digital printing 2002

British Library Cataloguing in Publication Data
A catalogue record for this book is available from the British Library

Library of Congress Cataloguing in Publication Data
A catalogue record for this book is available from the Library of Congress

ISBN 0 7506 2158 3

For information on all Butterworth-Heinemann publications visit our website at www.bh.com

Printed and bound in Great Britain by Antony Rowe Ltd, Eastbourne

Contents

Preface to second edition

When I was working on the first edition of this book about fifteen years ago I realised that the book needed to perform two distinct and different functions: it needed to be readable so that it formed a good introduction to remedial treatments and the problems that they are designed to remedy, but it also needed to be a reference book to which people could turn when they required more information. I achieved these dual aims in part by including Thelwell cartoons in the main text and by providing appendices as a source of more detailed information.

I was evidently successful as I received many flattering comments on the usefulness of the book, and stocks were soon exhausted. We still regularly receive requests for copies but it is not sufficient to simply reprint a book of this type as there have been so many new developments in recent years.

Two changes will be immediately apparent to persons familiar with the first edition of this book. The original text had imperial quantities followed by metric equivalents in brackets but the new text is entirely metric. The current Latin names for fungi have been used, although with the familiar old names included in brackets where appropriate. Both these changes actually caused me considerable personal irritation; as a scientist I understand the advantages of the metric system and the need for taxonomically correct Latin names but it would be much easier if we were able to keep to the systems with which we are familiar. Unfortunately this is no longer possible as we are expected to adopt systems which are internationally understandable.

I am similarly irritated by increasingly stringent health and environmental controls. I recognise the need to ensure safety and to preserve our natural heritage, but I suspect that the present controls are not entirely effective in these respects. The research and development costs for a new safer system are enormous, but when the development is complete the cost of obtaining health and environmental approval is horrendous. The potential market for new remedial treatment products is not very large and these development and approval costs cannot be economically justified. The result in recent years has been a dramatic decrease in research and development on remedial treatment products. The main effect of the more stringent requirements has been to divert money and effort from research on safer products towards political campaigning for the continuing use of established products. The tighter regulations, with their grandiose ambitions, actually discourage the development of safer products and encourage the continuing use of older products.

No new safer remedial treatment chemicals have been developed and introduced since the introduction of mandatory health and safety registration coupled with more stringen requirements. New pròducts are therefore modifications of old-established remedial treatment products. Modified established systems are very satisfactory because they have long service records, not only in terms of performance but also in relation to health and safety. However, the current regulations do not seem to recognise actual performance in service; they prefer to rely on laboratory tests which have no direct relevance to the use of a particlar remedial treatment product, ignoring the simple fact that performance and safety in actual service are the only truly reliable standards for assessing a product!

The situation is immensely frustrating to those of us who are concerned with the development of remedial treatments which are both more effective and safer. The current restrictions completely stifle initiative, the financial implications totally preventing the development of safer new systems. The real question is whether new systems are really required and, in this sense, the conclusion that I must reach is very interesting: we have had effective remedial treatments for many years, and new treatments are only desirable to improve safety, but the development of new materials is prevented by the regulations that have actually been introduced to achieve the same aim!

There have been other dramatic changes in remedial treatments over recent years. The demand for remedial wood preservation treatment has been greatly reduced through the successful use of both pretreatment and remedial treatment wood preservation in the control of, for example, fungal decay in window frames and insect borer infestation in roofs. The demand for remedial thermal insulation treatments has also decreased, both through the increased attention to thermal insulation in the Building Regulations and the success of this section of the remedial treatment industry, partly through the availability of government grants. Several new types of remedial treatment have also developed, based on the reduction of health hazards in buildings such as legionnaires' disease, sick building syndrome and particularly the radiation hazards associated with radon gas, prompting a new chapter in this second edition.

I have been associated all my life with the remedial treatment industries, since 1965 as a consulting scientist serving these and other industries based on materials deterioration and preservation and related environmental and health problems. There have been several times when I have thought that we had exhausted the scope for new developments but new challenges have continuously developed. That is the attraction of this particular industry: it continuously offers new opportunities to a scientist serving the industry, but persons working within the industry also encounter a great variety of problems which demand their interest and attention. However, it is the increasing interest of architects, surveyors and engineers in the diagnosis of defects and the specification of remedial treatments that I find particularly encouraging; the first edition of this book argued strongly for greater professional involvement in remedial treatments and less reliance on specialist contractors alone.

Acknowledgement

The Thelwell cartoons in this book are from *A Place of Your Own* and are reproduced by kind permission of the publishers, Methuen & Co Ltd.

Barry A. Richardson
Winchester, Hampshire

Preface to first edition

To the average householder the term 'remedial treatment' means nothing, as it is a collective description for a variety of specialist operations in buildings; yet most householders will reminisce happily about their encounters with woodworm, dry rot, rising dampness and the many other problems that are today solved by the remedial treatment industry. To the average architect, surveyor or structural engineer the term 'remedial treatment' often has an aura of mystique as it involves a peculiar blend of art and science which suggests to them that these are subjects that are best left to specialists. There are many thoroughly competent remedial treatment contractors operating in the British Isles, some of them members of appropriate technical and trade associations, but there are also firms which offer remedial treatment services without having the knowledge and experience necessary to properly diagnose unusual defects and prepare reliable specifications for treatment.

This book is therefore an attempt at education, principally to make available to professional advisers such as surveyors, architects and engineers the basic but comprehensive information that is necessary to enable them to make decisions on whether treatment is necessary and what treatment should be used. Whilst a few architects and surveyors specialising in conversion or restoration may become relatively experienced themselves in remedial treatments, it is certain that there will be a continuing need for specialist inspections and specifications. As far as the remedial wood treatment industry is concerned, there are a few courses of varying complexity covering inspections and specifications, and the Institute of Wood Science is introducing a certificate of competence for surveyors employed in the remedial wood treatment industry. However, no similar schemes are planned in connection with other remedial treatments, except perhaps in thermal insulation where, in theory, a heating engineering qualification might appear to be appropriate, although of little practical assistance in connection with the actual techniques involved.

Without the incentive of compulsory qualifications, education can be achieved only if it is attractive in some other way. At the moment an experienced and conscientious remedial treatment specialist may have accumulated a collection of several hundred books, booklets and leaflets published by a multitude of independent and commercial organisations, and his reliability when he is faced with an unusual problem may be related to the extent of his recent reading or his knowledge of where to search for further information. My purpose in writing this book has been to simplify the search for these established specialists who are already experienced and knowledgeable, but in addition I have made a particular effort to produce a book that is readable and interesting so that it will be attractive to many others, particularly to surveyors, architects and engineers. I have attempted to achieve this aim in several different ways. The most obvious is the inclusion at intervals of Thelwell cartoons as light-hearted comments on the adjacent text, but I have also included many true anecdotes in order to illustrate some of my comments in a more convincing and interesting manner. Finally, I have arranged the book in two parts, the first a readable description of the situation as a whole and the second as a series of appendices which give the detailed tables, figures and other sources of information that are essential in the reliable diagnosis of defects in buildings that can be remedied by chemical and other specialist treatments.

The term 'remedial treatment' was originally applied to curative or *in situ* wood preservation in buildings. There is no logical reason why its scope should have been confined in this way, as there are many other remedial treatments such as damp-proofing, masonry treatment and thermal insulation, but these and other remedial treatment industries have all developed rather more recently. This book is principally concerned with remedial treatments in the British Isles but the comments are equally applicable to building defects in many other temperate climates. Indeed, comments have been included on the special approach that is necessary to specialist inspections and remedial treatments in other European countries where buildings are constructed in different ways, such as the traditional log buildings in the Nordic countries, or where problems which are unknown in the British Isles may be encountered, such as the termites in southern Europe.

I was born into the remedial treatment industry; my father was the founder of Richardson and Starling Limited, perhaps the first specialist remedial treatment firm. As a child I travelled with him on inspections and visits to sites where work was in progress. Perhaps my earliest memory of remedial treatment was a face full of water-repellent spray caused by a sudden change in the wind, a very uncomfortable and painful experience as the trichloroethylene solvent used at that time was very irritating to the eyes, nose and mouth. In due course I was considered old enough to take an active part in inspections, usually carrying tools or holding torches. Eventually I was permitted to take holiday work as an operative, probably for a period of about nine months spread over several years, an experience that I have never regretted as I served under several very competent foremen who gave me an invaluable insight into the practical problems that are encountered.

It was during this period that I had a second experience involving the solvent trichloroethylene, but on this occasion through its inclusion as a co-solvent in a wood preservative that I was using to eradicate a Common Furniture beetle infestation on an exceptionally hot day. Trichloroethylene is an anaesthetic whose properties can perhaps best be described as schizophrenic, and I clearly recall emerging from a rather unpleasant sub-floor space and then talking to my colleagues, whilst a second 'me' was floating above, persistently enquiring why I was talking such rubbish! A year or two later, when I began to take responsibility for formulation development, I was to remember this incident, promptly eliminating trichloroethylene from all formulations and avoiding the use of similarly unpleasant solvents. I also experienced the problems of attempting to spray exposed external walls on windy days and I made the mistake of using excessive spray pressures when applying rather obnoxious materials such as sodium pentachlorophenate. Once you have experienced such problems they always remain fresh in your mind and I have no doubt that my short experience as an operative accounts largely for many comments that the products which I developed during my years as Research and Development Manager at Richardson and Starling Limited were the most pleasant that were available.

In 1965 I left Richardson and Starling Limited and established the consultancy that is now known as Penarth Research Centre. To the remedial treatment industry we are perhaps best known as investigators into failures, accidents, crimes and civil claims; yet much more of our time is devoted to advising the chemical manufacturers, treatment formulators, specialist contractors, surveyors and architects on remedial treatment problems, usually when the problems are unusual or particularly extensive, and it is this continuing contact with the normal remedial treatment industry that enables us to reliably fulfil our duties as investigators.

Finally, I must acknowledge that little of the information in this book is my own. Instead it is a collection of the information that I have acquired from many other persons over the years. My father, Stanley A. Richardson, introduced me to this absorbing subject and this book is dedicated to him; anyone who has devoted such effort, as we have, to investigating the problems involved in conserving our historic buildings and ancient monuments will appreciate my use of the adjective 'absorbing'.

Whilst my father introduced me to the delights and frustrations of remedial wood preservation and damp-proofing, it was the late Lt-Col. Bertram C. G. Shore, the author of *Stones of Britain* and a similar-minded collaborator in conservation, who introduced me to the entirely different delights and frustrations of natural stone masonry problems. There have been many others who have contributed, sometimes just as sources of information, such as the staff of the Building Research Establishment at both the Garston and Princes Risborough laboratories, including Mr D. B. Honeyborne, Dr C. A. Price, Mr J. M. Baker, Mr J. G. Savory, Dr A. F. Bravery and many others who have always been so consistently cheerful and helpful.

When I am asked to make an inspection I am expected to give advice but I learn just as much myself, particularly from architects and surveyors such as Mr J. A. Ashurst of the Directorate of Ancient Monuments and Historic Buildings and Mr W. Adams of Stenhouse Conservation Centre in Scotland. I cannot start to list the private architects and surveyors who have contributed to a greater or lesser extent to my knowledge and who must therefore be represented by a single name; I have chosen Mr C. W. R. Corfield, known as Roger or Pete to his friends (depending upon their status!), who introduced me about twenty years ago to the magic of the Royal Duchy of Cornwall with its 'pot' granite church towers with the ferns growing inside – towers that were to be waterproofed without killing the ferns as it was said that the villages would die with them! To these various people, to my staff and to so many others I give my thanks, and I hope that they enjoy and appreciate the interchange of information that I have attempted by writing this book.

Barry A. Richardson

1 *Defects in buildings*

1.1 Introduction

Remedial treatments

The term 'remedial treatment' was first applied to *in situ* wood preservation against established wood-destroying borers and fungi. There is no reason why the scope of remedial treatments should be limited in this way and in this book they are considered to cover all repairs of defects which involve unusual problems in diagnosis or unconventional methods or materials for treatment. Most structural materials deteriorate and the problem facing the architect, surveyor or engineer is to select materials and use them in an appropriate way that will ensure good life. Natural durability is not always available or is too costly to be realistic. In some situations where deterioration is inevitable some form of treatment is essential, such as the preservation of wood against wood-destroying fungi. Preserved wood has been used for many years in severe hazard situations where decay is inevitable such as posts, poles and railway sleepers, but it is not widely used in situations where decay is only possible rather than inevitable, as in structural timbers in buildings. In recent years Building Regulations have been introduced but they are not appropriate to all situations and they do not cover all aspects of building construction; they are intended to cover only aspects cncerned with health and safety. The Regulations do not, for example, cover the selection of stone to ensure adequate durability when exposed to the weather, and in the past the Regulations have actually encouraged faults, as in certain aspects of flue design and flat-roof construction.

Remedial treatment is, of course, concerned only with defects that have developed in buildings, or which are anticipated, perhaps due to accidents or 'acts of God' such as fire, storm, flood or earthquake, but more usually to more mundane causes such as faults in design or construction, carelessness in the selection of materials or negligence in maintenance. This book describes the most important remedial treatments which are *in situ* wood preservation, damp-proofing, masonry treatment and thermal insulation, but there are also comments on less important remedial treatments such as fire-proofing, roof repairs and control of radon hazards.

Professional advice

One common feature of all such treatments is that they are not encountered very frequently so that architects, surveyors and general contractors are not usually sufficiently competent, or indeed confident, to deal with such problems and normally seek specialist assistance. Naturally there will be those who will wish to contradict this generalisation, perhaps architects and surveyors who specialise in conservation and renovation work. They will certainly have experience of the defects, but it must be emphasised that few will have the necessary knowledge to cope with the unexpected or unusual defect when it occurs, as such problems or their treatment are not comprehensively covered in either education or current literature. There will also be surveyors and estate agents who are employed, for example, by building societies to make valuations prior to purchase and who will claim that they frequently encounter such faults. They usually forget that they are valuers, not structural surveyors, and they frequently introduce rather generalised comments on dampness and timber decay into their reports, often recommending that a specialist company should be employed to carry our 'necessary' treatment. This system actually results in an enormous amount of entirely unnecessary treatment, but who can really blame the specialist contracting companies? After all, they carry out inspections to obtain business, and if they see a few Furniture beetle holes in a roof most will recommend that treatment should be carried out, perhaps 'as a precaution'; there is no real reason why they should check whether a treatment has been applied in the past and whether there is any continuing activity.

Remedying defects: demolition or renovation?

The various defects requiring remedial treatments are described in detail in later chapters. If defects are present the logical solution is, of course, demolition followed by reconstruction. Such drastic action is usually unrealistic, particularly if the defects are comparatively minor in relation to the structure as a whole. The choice of reconstruction as an alternative to repair and renovation depends on the cost of repair in relation to the value of the structure in both monetary and aesthetic terms, so that repair is usually justified, even if very expensive, for a building with high aesthetic or historic value. In the 30 years to 1975 demolition and reconstruction were widely practised, not usually involving valuable historic buildings but often completely stripping our towns of buildings of secondary value which actually contribute rather more than individual prominent buildings to the 'street scene' or character of an area, so that this reconstruction policy has destroyed the atmosphere and charm of many towns.

Perhaps there is now a more sensitive approach to the scene as a whole rather than to an individual structure, encouraging greater interest in the conservation of entire areas. The economic fluctuations of recent years have also emphasised the need to make maximum use of existing structures and to conserve energy and materials as far as possible. Whilst these developments have together stimulated increasing interest in remedial treatments they do not alone account for the enormous expansion in recent years, which can be largely attributed to a progressive improvement in housing standards that has resulted in an increased awareness of dampness, decay, thermal

"I'm afraid it'll cost more
to knock down than
it's worth."

defects and even radon hazards requiring remedial treatment where in the past these defects were either ignored or concealed.

Professional responsibilities

In the case of limited defects such as ordinary woodworm damage in a roof or damp penetration through a solid wall, a householder will usually approach a specialist contractor and ask for a quote for the necessary remedial work; householders similarly engage decorators, roofers, electricians or plumbers when their specialist services are required. With larger works the situation is more complex; in theory the normal lines of responsibility in the construction industry should apply, involving the appointment of an architect or surveyor who should be responsible for diagnosing the defect, specifying the treatment and supervising the work carried out by the contractor. It has been previously explained that most architects and surveyors usually encounter such problems too rarely to be sufficiently knowledgeable and experienced to either diagnose defects competently or specify appropriate treatment. If the extent of a defect is clearly apparent an architect or surveyor may ask a remedial treatment contractor to propose a scheme for consideration, a perfectly normal procedure adopted in the case of other trades such as heating engineering. However, diagnosis is absolutely crucial in remedial treatment. Often architects and surveyors have suffered civil claims through incorrect diagnosis, or at least have known colleagues who have suffered in this way, and they are naturally keen to seek professional assistance which will bear this responsibility, just as they might employ a structural engineer or a quantity surveyor when appropriate.

Free services

Unfortunately professional assistance with remedial treatment problems is not generally available. Instead architects and surveyors ask remedial treatment contractors to assume these professional responsibilities, expecting the firm to make an inspection and then submit a report and recommendations. As remedial treatment contractors generally offer free surveys, principally to encourage business from householders with comparatively minor problems, they can hardly refuse a request from an architect for a similar service which may, of course, result in a profitable treatment contract. Unfortunately architects and surveyors, and householders of course, are tempted to invite as many as half a dozen different companies to make such inspections since they are offered 'free of charge', completely ignoring the simple economic fact that the final treatment, if it is to be reliable, must therefore incorporate the cost of all these inspections. It is quite clear that a single professional inspection would be both less expensive and more reliable.

Professional services

This book is therefore an attempt to encourage architects and surveyors to inspect and prepare reports, recommendations and specifications, leaving the contractor to submit a competitive estimate which is, of course, the limit of a contractor's normal responsibility. If architects and surveyors consider that they cannot gain the knowledge and experience to enable them to act in this way, perhaps this book will encourage a few of their colleagues, or perhaps scientists or technologists, to become specialists in remedial treatments so that proper professional advisory services are available in the future.

Scope of remedial services

Present professional and commercial attitudes in remedial treatment are often too narrow and this can be very dangrous. For example, the introduction of cavity fill is designed to improve the thermal insulation of external cavity walls but it must be appreciated that unsuitable fill may bridge the cavity and permit rain penetration. Similarly wood-borers are encouraged by fungal development, which is in turn due to dampness. Damp-proofing may involve masonry water-repellent treatment, perhaps with a preliminary biocidal treatment to eliminate moss, lichen or algae. Lichen on roofs may generate acid which will cause perforation of valley gutter lead, leading to dampness and perhaps severe development of fungal decay. Limited ventilation in wall cavities and flat roof spaces is specified in the Building Regulations to improve thermal insulation but it may encourage condensation dampness and perhaps wood decay. Damp-proofing companies sometimes offer only damp-proof courses, entirely ignoring the possibility that the dampness may have some cause which cannot be treated in this way. The possible interrelationships between various treatments are complex and it would clearly be best for all these defects to be considered by a single professional specialist so that all the dangers are fully understood, even if individual remedial treatment contractors prefer to confine their scope to particular types of treatment. Unfortunately it is architects and surveyors who should bear these professional responsibilities, and they are often clearly reluctant to do so.

Remedial treatment

This introduction has mentioned some of the difficulties that arise through divisions of responsibility but, however these problems are resolved, the basic procedure adopted in remedial treatment remains the same. There are three main parts to any remedial treatment: diagnosis, specification and treatment. There are two other aspects of importance: methods and materials, and warranties.

Diagnosis

Diagnosis is, at least in the case of large works, the responsibility of the professional adviser, usually an architect or surveyor, although many will try to persuade specialist remedial treatment contractors to assume these responsibilities. Diagnosis involves an inspection followed perhaps by laboratory tests and the preparation of a report. Many remedial treatment failures result from incorrect diagnosis. Thus the full extent of Dry rot infection is often not appreciated; the diagnosis of a source of dampness is frequently incorrect; and there is seldom any real attempt to decide whether a borer infestation is currently active.

Specification

Specification should detail the methods and materials for treatment and should again be the responsibility of the architect or surveyor, perhaps with specialist contractors submitting schemes for consideration. In fact, it is more usual again for the remedial treatment contractor to be asked to take this professional responsibility but there are few failures that can be attributed to incorrect specification as it seems that, if the diagnosis is correct, the appropriate treatment is usually adopted. The most obvious failures arise through specifying that all borer infestations should be treated by spray alone, whereas Death Watch beetle and House Longhorn beetle must be treated by trimming followed by injection or perhaps bodied-mayonnaise-type emulsion systems which penetrate deeply. There is sometimes a failure to adequately sterilise masonry infected by Dry rot. Remedial damp-proof courses are often specified in solid walls with no precautions against penetrating dampness which may accumulate on top of the new courses, giving all the appearance of the rising dampness that the courses are designed to prevent. There is almost always a failure to adequately specify associated works, but this failure can be largely attributed to the difficult structure of the industry; it must be remembered that, at present, specifications are normally prepared by specialist contractors who are making inspections in a competitive situation and who are thus unwilling to undertake unnecessary additional work.

Treatment
Treatment is the responsibility of the remedial treatment contractor. Some failures occur owing to incompetent or inadequate treatment, again partly due to the structure of the industry as the contractor is often responsible for the professional services and thus supervision of his own treatment. Methods and materials are, of course, the responsibility of the suppliers but there are very few failures that can be attributed to the systems promoted by the principal suppliers, provided that the systems are conscientiously applied by the contractors.

Warranties
Finally, warranties are the responsibility of the treatment contractor, and whether they are valid will depend solely on the status of the firm concerned, a problem that is discussed later in this chapter.

Obviously difficulties arise through the unusual structure of the remedial treatment industry but diagnosis and treatment techniques, with which the rest of this book is concerned, remain the same, whoever may be responsible.

1.2 Inspection and report

Essential equipment
Before attempting an inspection it is necessary to be properly equipped in all respects. The most essential requirement is adequate knowledge and experience in respect of structures in general but defects and remedial treatments in particular. Another essential is adequate time to make the inspection, consider the implications and prepare a report with recommendations for remedial treatment.

Knowledge and experience
Finally, it is necessary to have the ability to write reports and recommendations lucidly. Many architects, surveyors and remedial treatment inspectors will be surprised that sophisticated equipment has not been mentioned but it must be emphasised that diagnostic aids such as moisture meters are no substitute for personal ability, and the essential equipment is quite simple: just a torch, notebook and pencil.

Clothing
An important requirement is suitable clothing. Whilst a professional surveyor may wish to arrive on site neatly dressed, he must be prepared to change so that he can kneel, crouch, squeeze through small gaps and generally perform

"Look at the cellar while you're down there, sir."

the contortions that are necessary to make a thorough inspection of the inaccessible and often very dirty parts of roof and sub-floor spaces. Clothing may become dirty or damaged but this does not mean that a change into old clothes is adequate as it is easy to become trapped when squeezing through small spaces wearing a jacket or even a sweater that will catch on projections. It is always best to use proper overalls but an additional sweater should be available in the winter; there is no excuse for skimping an inspection simply because the property is too dirty or the weather too cold. Similar thought must also be given to suitable clothing for inspecting the exterior of the property; even a simple inspection for woodworm damage should involve at least a brief inspection of the exterior elevations and roof structure. Although the use of an umbrella whilst one is working from scaffolding often causes immense amusement, the laughter gradually dies away as rain begins to trickle down necks or soak notebooks. Certainly an umbrella is ideal protection for a person writing brief notes or taking photographs during rainfall but it is difficult to use when climbing ladders or when making an extensive external inspection for which an anorak, with a hood or separate hat, is essential. Gloves are also essential in frosty weather, and perhaps waterproof trousers in particularly wet weather. Again it must be emphasised that the conditions at the time or a lack of adequate clothing can never be a valid excuse for an inadequate inspection.

Notes and sketches

Taking notes during an inspection can present problems. Even if a ballpoint pen is preferred for normal uses, it is always essential to have a pencil available to use when it is necessary to make exterior notes during rainfall. A normal nib pen should never be used. Some surveyors prefer spiral-bound shorthand notebooks as these can be carried in a jacket pocket, although this does not help when wearing overalls for inspections of roof and sub-floor spaces. Standard A4 pads are really preferable as they give more room for sketches and generally permit all notes regarding a particular area to be made on a single side. Some other surveyors prefer A4 or foolscap hard-back permanent notebooks, excellent in the sense that they provide a permanent record of the inspection notes but not recommended as entries cannot be filed with the final report; if a surveyor leaves a practice it may be impossible to refer back to the original inspection notes. Normal lined paper is adequate, although some surveyors prefer squared or plain paper to simplify sketches; scale plans are not usually required for remedial treatment and special paper is not really necessary.

Figure 1.1 Typical site notes

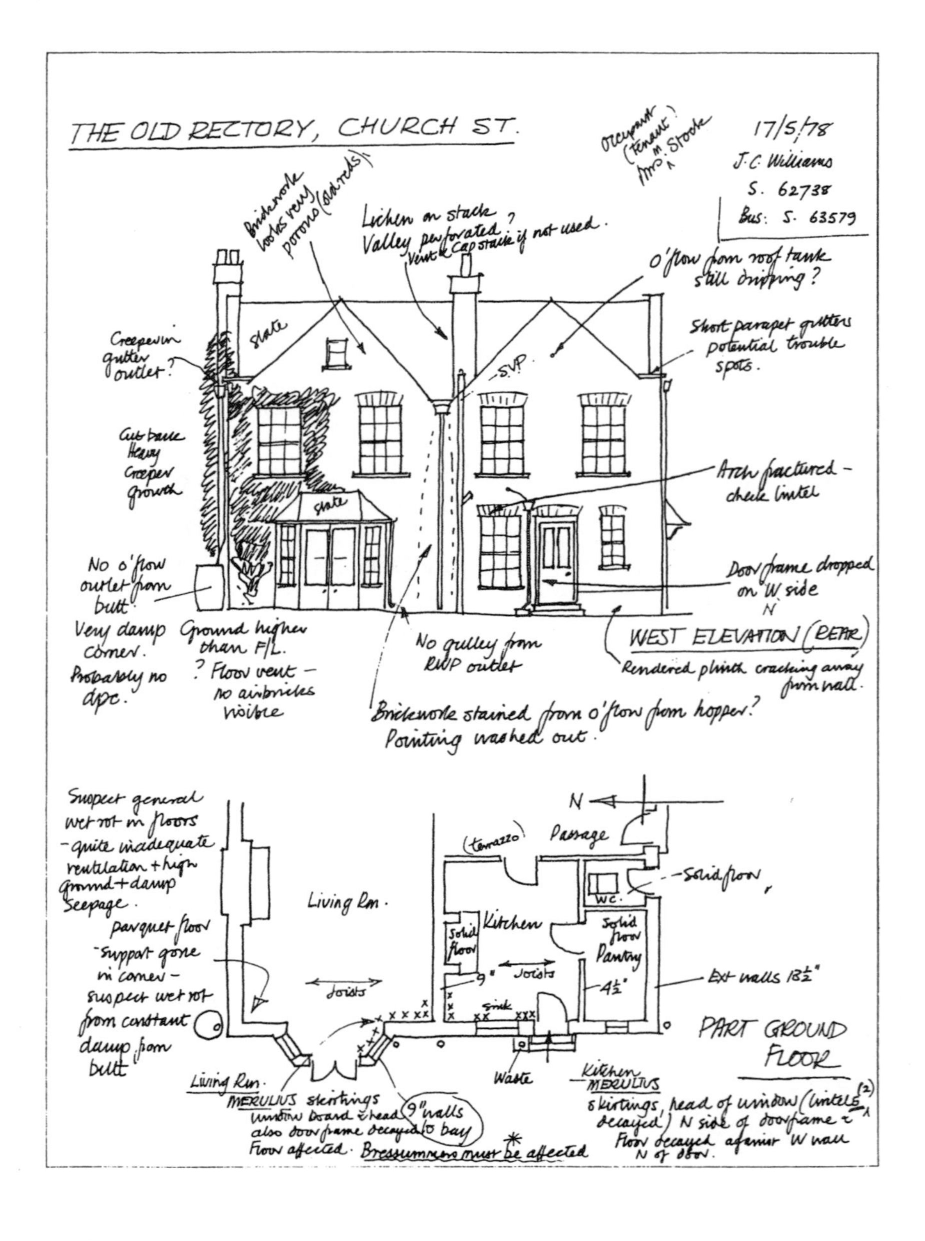

Figure 1.1 continued

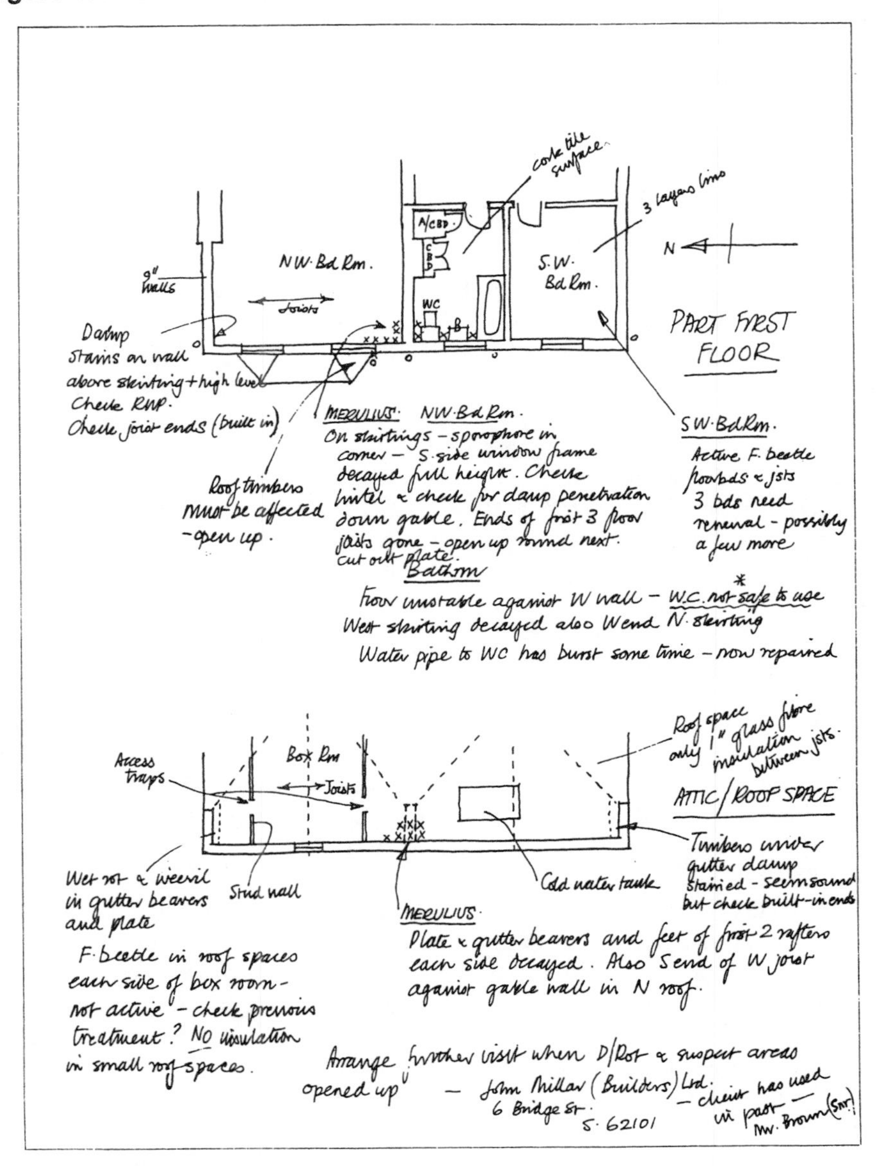

Tape recordings

Taking notes can present serious difficulties when the hands are cold in the winter. Obviously gloves can be worn for exterior work but the necessity to remove a glove to make a brief note can be very irritating. Some surveyors use pocket tape recorders but these cannot be generally recommended as there is really no substitute for a sketch plan annotated with details of the defects as in Figure 1.1, as it is much easier to recall the situation if it becomes necessary to refer back to the original site notes after a period of several years. If a tape recorder is used it is recommended that it should be compatible with office dictation equipment so that the site notes can be readily typed for filing.

Photographs

Photographic records can also be very valuable for recalling an inspection and details of defects. A few general views of the property and close-ups of defects are all that is necessary for prompt recall of details observed during the inspection. The camera should be compact so that it can be carried in a pocket and it should also have a built-in or coupled 'hot shoe' flash system to encourage maximum use. Cameras with automatic exposure are easier to use but automatic focus is not helpful; single-lens reflex cameras are most suitable. The choice of film speed is critical; a high speed will extend the range when using flash but an excessively high speed may cause difficulties with exterior photographs in very bright sunlight. Colour slides are preferred because they show more detail when projected and therefore provide a more useful record, and colour balance is always more accurate with reversal film if it is properly exposed. Negative colour film has the advantage that prints can be obtained very rapidly but reversal prints for inclusion in reports can also be obtained from colour slides. A colour reversal slide film with speed 200 ASA (24 DIN) such as Kodak Ektachrome 200 is suitable for most purposes.

Lighting

A torch is essential for interior inspections, as normal light or even a lamp on a long flex cannot be relied on to illuminate the required areas without dazzle. A normal heavy-duty torch, such as a rubber or plastic torch with two U2 cells, is very popular as it is the largest torch that can be accommodated easily in the pocket when walking round a site writing notes. Some surveyors prefer a hand lamp and this certainly has advantages as the lamp can be stood on a convenient surface and the head adjusted to illuminate a particular area or the notebook. Whatever system is chosen, the most essential point is to ensure that spare batteries and bulbs are available; there is nothing more embarrassing than to find that your torch will not operate when you are conducting an inspection in the presence of a householder or his architect! If rechargeable cells are used it is always advisable to carry a set of alkaline cells as a reserve as they have very long storage life.

One particularly useful item is a very small flat torch which can be used in restricted areas in conjunction with a hand mirror for inspecting concealed ends of beams or wall cavities; a suitable mirror can be readily obtained by rummaging through old handbags! Borescopes provide a more sophisticated alternative and have the advantage that a probe can be inserted through a small hole to inspect, for example, a sub-floor space or a wall cavity without causing excessive damage. The fibre optic systems that are most actively promoted are very cumbersome in use and their high cost can rarely be justified; simple and portable systems such as the KeyMed Checkscope with an 11 mm diameter probe are much more realistic. Another useful item of equipment is a hand lens with 10x or 15x magnification to examine borer flight holes, bore dust and insects. Some surveyors prefer an illuminated magnifier but it is much more bulky than a small folding lens which can be used in conjunction with a normal torch. A small illuminated 'microscope', such as the small Tasco instrument with 30x magnification, can also be useful at times for inspecting bore dust and particularly when checking the identity of natural stones.

Probes

A probe is essential for timber inspections to check the strength of timber and to search for concealed borer galleries, particularly of the House Longhorn beetle. A scar on the author's finger bears witness to the fact that folding jack-knives should never be used as probes, but only a rigid instrument such as a sheath knife or preferably a bradawl or sharpened thin screwdriver. Probes are also sometimes used to locate joists or stud frames concealed by plaster but low-cost electronic instruments are now available which will locate such components without causing any damage, and other instruments will locate concealed wires and pipes.

"Know what you've got *here*, don't you?"

Tools

In addition to these surveyor's 'instruments' it is necessary to have proper tools available so that boards can be lifted and panelling, skirtings or architraves removed. Generally a surveyor will lift only service boards provided by electricians and plumbers. These service boards should be screwed but they are often nailed and must then be lifted with a bolster and crowbar. Occasional difficulties may be encountered where a partition has been constructed over a service board; the technique is then to lift one end of the board and saw over the centre of a joist as near as possible to the partition. If further boards are to be lifted it is generally best to arrange for this with a carpenter who can also ensure proper replacement; tongue-and-grooved boards are difficult to lift and will certainly be damaged if they are secret nailed. Tools must also be available to strip plaster, rendering or screed, although such exposure work is not normally necessary; indeed, damage to the building is usually limited to taking small samples of wood or plaster for analysis.

Taking samples

It will be appreciated that the remaining requirement is therefore for a set of 'house-breaking' tools such as screwdriver, claw hammer, lump hammer, large chisel, bolster, cold chisel and crowbar, as well as extra screws and nails so that boards and panels can be replaced. There are a number of other tools that are helpful, such as the gouge chisel for taking wood shavings for analysis. However, a better method for taking wood for analysis is to use a slow-speed cordless electric drill fitted with a Forstner bit, as described in Chapter 2. The same drill can be used with masonry bits for taking samples for use in diagnosis of dampness, as described in Chapter 3. Envelopes or small plastic bags will be required for the samples; if the drill method is being used it is recommended that normal business envelopes should be used with the opening along the long edge so that the open envelope can be held, pressed against the wood or plaster surface, underneath the drill bit. A few sample bottles are also advisable in case it is necessary to collect insects for identification.

Dampness measurements

For dampness or wood decay inspections an electric moisture meter is very helpful when attempting to diagnose the source of dampness, although it is only accurate for measurement of moisture content in wood; on masonry, brickwork, rendering or plaster it can be used only to give comparative readings which may assist in the identification of dampest and driest areas. The carbide method using a drilled sample, such as the Speedy moisture tester, is far more accurate and suitable for masonry. However, it is not usually helpful to accurately measure moisture contents as materials are too variable for them to have much significance, and the detection of damper or drier areas using an electric moisture meter is usually sufficient to enable sources of moisture to be identified.

Temperature measurements

Temperature measurements may also be desirable in buildings, particularly when attempts are being made to diagnose condensation or determine the thermal properties of the structure; either a special thermometer can be used or a suitable temperature detector in conjunction with the moisture meter.

Metal detector

Some surveyors will also find a metal detector helpful when it is necessary to trace pipes, cables or reinforcement bars beneath plaster, rendering or concrete. A pair of good binoculars is also very helpful when checking the exterior of a building for defects.

Ladder

Finally, it is always advisable to carry a ladder; folding and sectional aluminium ladders are available which can be accommodated in the boot of an average car.

Whilst this may be a reasonably comprehensive guide to the equipment that should be carried by a remedial treatment surveyor, it must be added that he should also carry a first-aid kit. Soap and a towel, perhaps even water, are also very useful as they are not usually available in unoccupied properties and a surveyor can become very dirty if he carries out a proper thorough inspecton of all spaces. Obviously the clothing and equipment are fairly bulky and most easily carried in a car, but it is possible to carry all the recommended items, other than the ladder, in two canvas holdalls if it is necessary to travel by public transport.

Inspection scope

It is often difficult to decide on the scope of the inspection, report and specification. If the instructions are comparatively general, such as to assess defects throughout a building and propose appropriate remedial measures, the main difficulty is to limit the problems that should be considered. Naturally a specialist should confine his investigations to the technical areas in which he is competent, and just mention other defects in passing, perhaps in a covering letter; if the situation has already been generally assessed by a chartered surveyor or architect the report should be made available to the specialist as it is foolish to duplicate work and increase costs.

The main difficulties arise where a householder or professional adviser has observed a localised defect and has given instructions for the specialist to consider that problem alone. If the specialist decides to strictly obey this instruction it is essential to clearly state in the report that the investigation was restricted in this way. For example, an extreme situation might involve a Dry rot outbreak but with access being available only to one side of a wall, despite the fact that the infection may have spread for a considerable distance. The specialist must generally ensure that he has investigated the entire outbreak in a competent and thorough manner, without being rushed owing to his own programme or through the wishes of the occupants. If the scope of an inspection is restricted in any way, through either limited access or difficulties in moving furnishings, lifting floorboards or carrying out any other exposure work, the limitations must be clearly stated in the report as claims for negligence may otherwise result should the report and specification later prove inadequate. These problems must be appreciated at the time of the inspection, as must the purpose of the inspection and report; is it with a view to remedial treatment or is it in connection with a dispute that might involve arbitration, or even civil or criminal litigation?

"The cellar's been walled up since the night poor Alice died."

Whilst an inspection may be limited through circumstances such as limited access or specific restricted instructions, it is essential that it should be adequate. For example, an instruction may be received to investigate rising dampness, or simply to estimate for the insertion of a damp-proof course. This instruction implies that rising dampness is known to be present so that an estimate could be prepared, in theory, from a knowledge of the dimensions of the property and the thickness of the walls alone. However, it may not be apparent from a casual inspection that dampness is confined to external solid walls and that it may be due to bridging of an existing damp-proof course by a high external ground level, or accumulation of penetrating rainwater on top of an existing damp-proof course. Clearly in such circumstances the insertion of a new damp-proof course may serve only to aggravate the situation by damaging the existing course, and other remedies are more appropriate. The moral is clear: always be observant and never keep strictly to restricted instructions, as a failure to observe another defect or failure to correctly diagnose the cause can be a considerable embarrassment later if the recommended remedial measures prove inadequate.

Figure 1.2 Danger points to be checked during an inspection

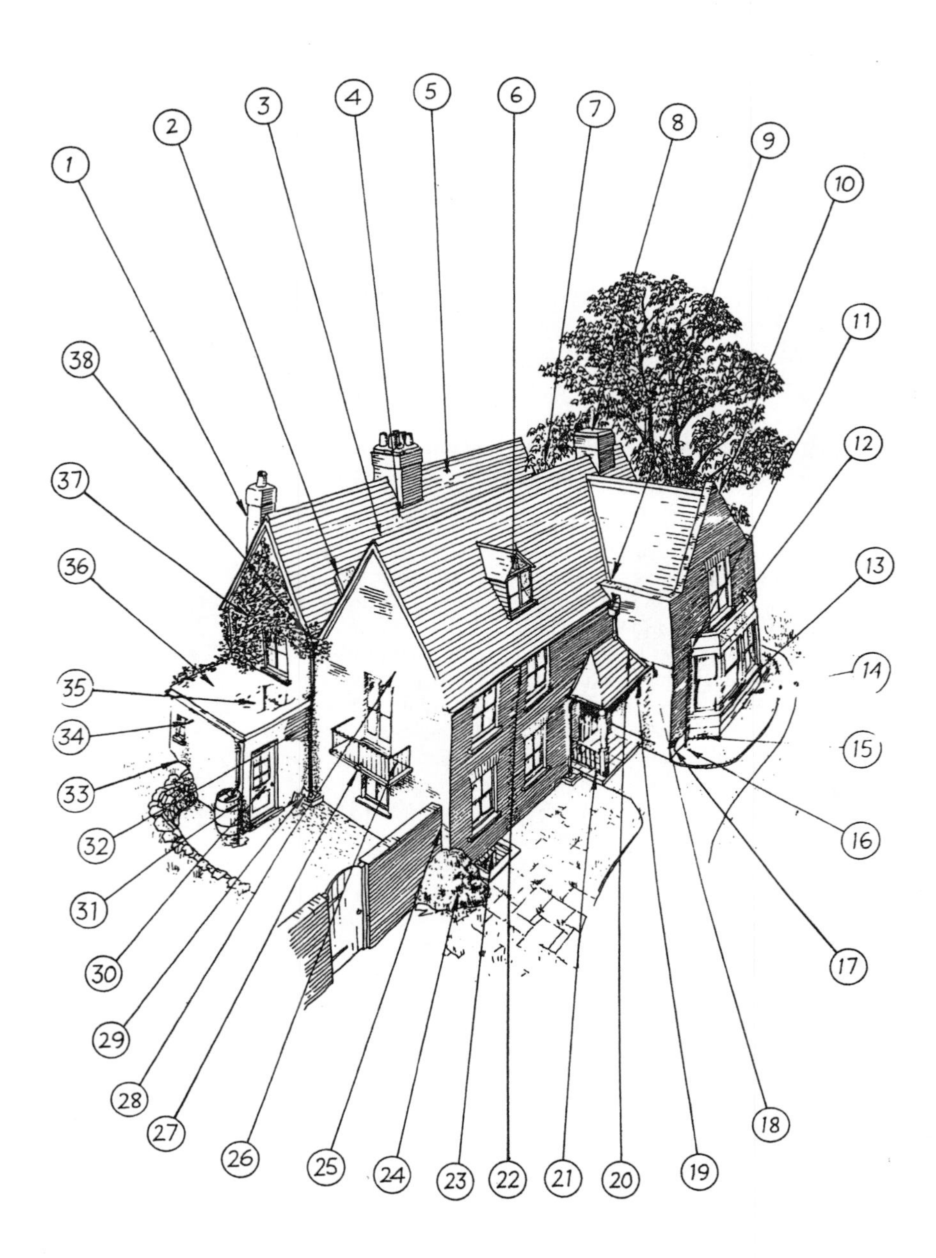

Key to Figure 1.2

1 Sulphate attack in brickwork caused by slow-burning boiler and unlined flue and resulting in distortion of stack
2 Split lead to roof hatch – decay in timber surround
3 Lichen on north slope of roof – lead in valley below deteriorated.
4 Broken slate where hit by falling pot – damage to valley below causing Wet rot in gutter timbers
5 Slipped slates
6 Water penetration of porous render on gable
7 Leaves from overhanging tree blocking gutter – Wet rot in gutter timbers
8 Blocked-off chimney – water penetration, also condensation in flues causing dampness on chimney breasts below
9 Valley and hopper head choked with leaves – Wet rot in roof timbers
10 Water penetration of porous brickwork of gable – Wet rot in ends of purlins
11 Inadequate rainwater outlet from flat roof - roof floods – decay in adjacent timbers
12 Cracked asphalt – water penetration to timbers beneath – Wet rot in bressummers
13 Cracked sill – water penetration into wall below – decay in skirting and window board
14 Plinth bridges DPC causing rising damp
15 Choked airbricks – inadequate sub-floor ventilation – Wet rot in floor timbers
16 No gulley – outflow from rainwater pipe soaks under house – decay in adjacent floor timbers
17 Plinth cracked away trapping water
18 Water penetration of string course – Wet rot in floor timbers
19 Gutter inadequate to take flow from upper pipe – Dry rot in timbers adjacent to damp wall
20 Cement fillet cracked away – water seepage causing decay in adjacent timbers
21 Wet rot in balustrade and foot of post
22 Broken gutter – Dry rot in timbers adjacent to damp wall
23 Wet rot in cellar window frame and lintel from damp wall
24 Heap of sand causing damp wall and decay in floor timbers
25 Damp penetration into house walls from porous contacting garden wall
26 Slab drains towards wall – decay in sill and floor timbers behind and in lintel below
27 Rusting stanchions causing bursting of stone slab
28 Water penetration of porous brickwork - decay of lintel behind
29 Render fragments blocking gulley – rot in floor
30 Wet rot in door – poor quality timber not pretreated – poor maintenance of paintwork and putty.
31 Overflow from water butt soaking wall – Wet rot in sill
32 Render broken away – water penetration of brickwork
33 Water penetration from high ground level
34 Water penetration from overflow – decay in sill
35 Flat roof not draining – water seepage causing Wet rot in timbers
36 Flat roof not ventilated – condensation on timbers causing Wet rot
37 Overgrown creeper keeping wall damp
38 Choked hopper-head overflowing – Dry rot in valley timbers and timbers adjacent to damp wall

Inspection technique

Generally the inspection should commence with a walk around the outside of the property to 'case the joint'! The purpose of this initial inspection is to observe the obvious danger points as illustrated in Figure 1.2. At the roof level damaged tiles may be obvious but features of the construction such as flat roofs, parapet gutters and valley gutters must be remembered so that they can be checked for defects during the internal inspection. Solid walls must naturally be treated with more suspicion than cavity walls, particularly if gutters, fallpipes or drains are damaged or blocked. On dry days persistent green algae may be the only feature available to enable areas of water spillage to be identified. Heavy growth of moss and lichen on roofs must be viewed with suspicion if the roofs discharge into valley or parapet gutters as the acid generated by these growths can cause perforation and severe leakage. Ground levels must also be noted at this stage.

Compass references

This initial inspection is then followed by a systematic tour round the building to make notes regarding each elevation in turn, observing them in a systematic manner commencing at the highest point with the roofs and chimneys, followed by the walls and ground level details. Sketch elevations or photographs are always advisable to refresh the memory, not only during the preparation of the report but also in the future should queries arise. Generally one of the best methods for making notes is to draw an elevation or plan using an entire sheet of paper and then to write very brief notes with arrows pointing to the features concerned. At this stage it is also necessary to decide on the orientation of the property; never use terms such as front, rear, right hand or left hand when referring to rooms, elevations or walls but assume that a particular elevation faces due north, south, east or west.

The internal inspection should then follow in a similar systematic manner, starting at the roof space and working downwards through each floor level to the basement or cellar (Figure 1.3). Some remedial treatment company inspectors use a check list or pro forma report system but this is dangerous as it can never be comprehensive for a particular property; it is far better to adopt a simple systematic approach. Sketch plans are essential and, if they are sufficiently large, they can again incorporate the survey notes, perhaps referring to external features that may be the cause of defects. Photographs are again advisable and can prove very valuable; usually there is only one opportunity for an inspection and, when later preparing a report, a query may arise which can be resolved by projecting colour slides showing the appropriate area. An effort must be made to gain access to all concealed areas by moving furniture as well as lifting floor coverings and boards.

Accessibility

If areas are inaccessible, such as skeelings and spandrels, this should be noted so that it can be recommended in the report that access should be made for inspection, perhaps during remedial work. Naturally the extent of the inspection and the detail that is necessary will depend on the circumstances, particularly the instructions received and the defects discovered. It must be emphasised again that normally only a single inspection is possible

"I'd take that child downstairs ... ... if he's going to jump about."

and this must be completely thorough, perhaps involving taking samples for later laboratory investigations in order to identify a decay organism or to detect contaminating hygroscopic salts in walls or earlier preservative treatments on wood.

Free survey or free estimate?

It is recommended that no report should be prepared following an inspection by a remedial treatment contractor but simply a schedule of observed defects and recommended treatments. It is also recommended that it should be emphasised that the purpose of the inspection was to prepare recommendations for treatment of observed defects and that no assessment has been made of the overall condition of the property; there is no real reason why a contractor should assume any professional responsibility when making an inspection for the purpose of preparing an estimate. Remedial treatment contractors should similarly avoid the words 'free survey' in their advertising and substitute instead 'free estimate' if they wish to avoid these unnecessary responsibilities. It is further recommended that, if a property has been inspected by an architect, surveyor or other professional person, the contractor should clearly refer to this professional report by stating that the schedule of defects and recommendations is prepared in accordance with it.

Figure 1.3 Internal danger points

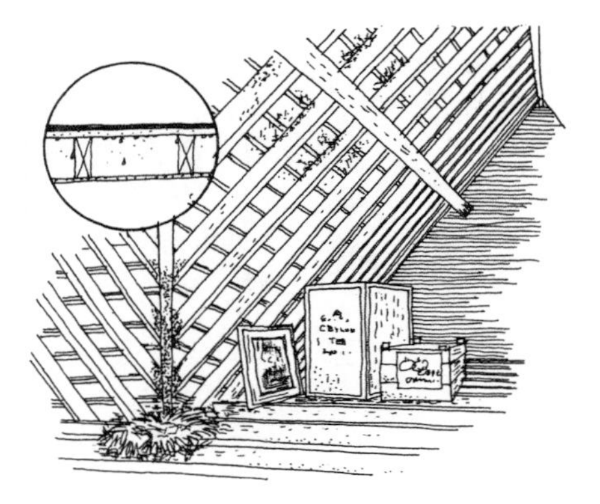

Roof structures. Tiles which are porous or laid with inadequate lap will permit water penetration with staining and salt deposits on supporting rafters, battens and boards. Valleys are always danger points as they may leak if choked by leaves or perforated by acids washed down from moss and lichen growth; decay may result in supporting timbers. Birds' nests show that birds have access, perhaps through gaps that will permit rain or snow penetration. Purlin ends may be decayed if built into solid gables. Stored boxes and furniture may introduce active woodworm infestations, encouraging spread to sapwood in roof members. The inset shows a flat roof in which inadequate ventilation may result in condensation, causing staining on ceilings and decay of joists and boards if the wood has not been adequately preserved.

Staircases and suspended floors. Woodworm (Common Furniture beetle) may be present in sapwood parts of floor plates, joists and boards, as well as in staircase strings, treads and risers; woodworm is more active and can spread into heartwood in damp conditions that encourage incipient or active fungal decay. Hall floor terrazzo may be laid on a wood support, or on cast concrete with wooden shuttering remaining, and there is always a decay danger; linoleum and other impervious floor coverings restrict ventilation and introduce a decay danger unless the floor structure remains completely dry. Blocked fireplaces should always have a ventilator if the chimney remains, as rain entering the pot or penetrating the stack may accumulate to cause dampness, particularly on chimney breasts where flue shelves or bends have caused soot accumulations and hygroscopic salts formed in adjacent mortar which may migrate to spoil decorations. The inset shows a shallow unventilated ground floor space with plates in direct contact with supporting brickwork; there is a severe decay danger.

Figure 1.3 continued

Bathrooms (as well as kitchens and laundries). If ventilation is inadequate humidity from baths (as well as clothes, washing and cooking) may cause condensation on windows (decay danger) and mould growth on walls. Leaks from soil pipe connections and radiators, for example, may cause decay in floor timbers. The inset shows a first-floor joist built into an external solid wall, introducing a danger of fungal decay.

Cellars. The plates supporting the floor joists, and the staircase strings, are in direct contact with the brickwork and floor; there is a severe decay danger. The blocked doorway has no vent, so that the space behind is unventilated, with a severe decay danger to floor plates and joists. In addition the old door frame has been built in, again introducing an unnecessary decay danger. Boxes and other wooden items (perhaps fire logs as well) are stored on the damp cellar floor, introducing yet a further unnecessary decay danger.

Report preparation

As far as a general professional report is concerned, it must clearly be a reliable and impartial assessment of the situation. The report cover page or headings should include the identity of the property and the client. The report itself should then be divided into several sections, basically a general introduction and description of the situation, a report of the inspection and any laboratory tests, a discussion, and finally the conclusions or recommendations, as well as perhaps a separate specification for the recommended remedial treatment and necessary associated work by a general contractor.

Introduction

The introduction should describe the circumstances that have made the report necessary. In many cases this introduction will be very short, simply noting that, following perhaps an inspection by a chartered surveyor, defects were observed and a specialist investigation was recommended. In such circumstances it is normal for the specialist surveyor to have sight of the original report and it is convenient to make reference to it as far as possible so that the specialist report does not contain any unnecessary duplication of information. However, if the inspection forms part of an investigation into a claim that may lead to litigation the introduction may be very extensive, summarising the history of the case and thus explaining why a particular investigation is necessary. The introduction should also define the scope of the report in terms of the property concerned, particularly whether it was restricted in any way through instructions.

Inspection and investigation

The second section of the report should be concerned with the inspection itself and any associated laboratory tests. The date of the inspection should be stated, followed by details of the orientation of the property, for example: 'For convenient reference the elevation to Bridge Street is assumed to face due west.' There should then follow a brief description of the property, for example: 'The property is a Tudor-style eight-bedroomed house with extensive outbuildings, constructed in about 1921 with dummy timber framing set in rendering, apparently on solid 225 mm (9 in) brick external walls.' The report should then continue with sub-headings for each section of the property, such as exterior, roof space, first floor, ground floor and basement. Under each of these sub-headings, minor sub-headings can be inserted for particular rooms, if each justifies at least a paragraph. It is essential to record in this section only what was actually seen during the inspection or the results of subsequent tests on samples taken at that time. The obvious must be stated if it is pertinent but there must be no guesses at the condition of the structure; this is considered in the next section of the report. The present status of defects should be considered carefully: are they past defects which are no longer progressing; are they current defects causing continuing damage; or are there defects that might develop in the future should faults, such as blocked valley gutters, remain unremedied?

Discussion

The next section of the report is the discussion. The previous inspection section of the report described defects as they occurred within the property, but within the discussion section it is necessary to assemble these various defects into 'outbreaks' or 'problems'. Thus Dry rot infection may spread through several floors and may be attributed to a faulty fallpipe against an external wall, or a blocked valley gutter overflowing through the building. It is at this stage that it is necessary to emphasise any doubts regarding the thoroughness of the inspection, perhaps through restricted access or limitations in scope imposed by the original instructions. This section also gives an opportunity to express opinions on the observations that have been reported, perhaps relating defects to faults in design, construction, selection of materials or maintenance, and perhaps commenting on relevant Building Regulations as well as British Standard Specifications and Codes of Practice which represent normal good practice.

Conclusions or recommendations

However, in many cases an extensive discussion will not be justified and this section can be more sensibly included with the conclusions or recommendations. This final section is where defects are summarised and recommended remedial treatment is described, together with associated general contractors' work. It is usually advisable for remedial treatment operatives to expose the extent of a defect but for a general contractor to be responsible for structural work and reinstatement after treatment is complete. In many cases it will be appropriate to attach, in addition, a schedule or specification for the remedial work, together with an estimate if the report has been prepared by a remedial treatment contractor.

Reports must always be comprehensive and thorough, even if they are prepared by a contractor and free of charge. If it is considered uneconomic to prepare such reports then survey or inspection services must not be offered in advertisements or literature; only estimates need be prepared, supported by schedules of observed defects and recommended remedial treatments. The greatest difficulties generally arise through the fact that only limited time can be devoted to an inspection, whether it is professional or carried out by a remedial treatment contractor, and there is rarely an opportunity to return to the site if a particular point is overlooked. If time is available it is always advisable to prepare a draft of the 'inspection' section of the report on site as individual areas can then be checked should any queries

"You'll get my bill in a day or two."

arise. This does not, of course, involve additional cost as it does not add to the total time involved. The causes of defects must be clearly diagnosed as far as possible on site, although it may be necessary to take samples for laboratory examination.

1.3 Legal responsibilities

Let the buyer beware

If a householder is considering engaging a professional adviser or instructing a contracting company in connection with remedial treatment, the general legal principle must be *caveat emptor* or 'let the buyer beware'. It is not intended to imply by this statement that the professional and contracting services associated with remedial treatments are necessarily less competent or less reliable than those concerned with the supply of other goods and services, but simply that this is a situation where adequate knowledge and experience is necessary to ensure reliable diagnosis and recommendations. A householder or any other person contemplating the purchase of remedial treatment services must therefore instruct only firms which appear to be properly qualified, and this usually means relying on advertisements or literature for guidance. Unfortunately descriptions such as 'remedial wood preservation specialists' or 'experts in thermal insulation' do not necessarily mean that the firms concerned have adequate knowledge and experience of these subjects but simply mean that they are the areas in which they wish to attract business. In recent years there have been many attempts to protect consumers against the danger that goods or services are not suitable for the purpose intended, such as the introduction of the Misrepresentation Act 1967, the Trades Descriptions Acts 1968 and 1972, the Fair Trading Act 1973, the Supply of Goods (Implied Terms) Act 1973, the Unfair Contract Terms Act 1977 and the Sale of Goods Act 1979.

Trades Descriptions Acts

It is the Trades Description Acts that probably clarify the situation best by emphasising that a person or firm must not make a false statement recklessly. This means that, if they claim that they are capable of providing a certain service, then they must be capable of doing so. Clearly this implies some measure of standards, often simply a matter of opinion, and it is therefore very difficult to secure prosecutions of persons or firms who are operating with a deliberate intention to defraud the public or which are simply incompetent. There is a danger that a conscientious firm making an innocent error may be prosecuted successfully, whereas a firm operating in a particularly confident or convincing manner may evade prosecution for years despite offering inadequate services.

Perhaps it might be sensible at this stage to give a few examples of prosecutions that have occurred under the Trades Descriptions Acts in recent years. In one case a firm injected a chemical damp-proof course in the outer skin of an external cavity wall, claiming that it would remedy patches of dampness that were apparent at some points internally. The treatment had no effect whatsoever and there were, in fact, effective damp-proof courses in both skins of the external walls. Investigations showed that the damp patches were actually due to bridging of the cavity through mortar droppings accumulating on wall ties during construction, so that failure of the treatment was due to incorrect diagnosis rather than inadequacy of the remedial treatment; the contracting firm had offered a free survey and diagnostic testing service. In another case a firm, claiming to be timber decay treatment specialists, was engaged to carry out Dry rot treatment in a girls' preparatory school. The treatment was restricted to lifting a few floorboards, spraying superficially with a wood preservative, and replacing the boards, packing them up with additional pieces of wood where the supporting joists were sagging through decay. Shortly after the treatment was complete it was noticed that the floor was flexing unreasonably and an investigation showed that it was on the point of collapse through decay.

A rather more sophisticated operation involved the employment of a number of persons to canvass areas for business. The technique involved systematic 'drumming' or door knocking and was aimed particularly at pensioners. Suitable householders were offered free woodworm surveys 'without obligation' and, if they agreed, a surveyor and an assistant, usually drummers from other streets, would appear shortly afterwards. The inspection would normally commence with the surveyor in the roof space shouting comments through the hatch to the assistant standing with the householder beneath. Damage was reported even if absent, and if damage was actually present it was grossly exaggerated, perhaps with comments suggesting that the roof would collapse if it was not treated immediately. Occasionally the surveyor would descend from the roof and produce a small piece of wood thoroughly riddled with woodworm for the benefit of the householder. Floors were never inspected, although occasionally inspections were made under stairs or in garages. Attempts were then made to persuade the householder to agree to treatment immediately, using incentives such as lower rates 'whilst they were working in the area' or 'for pensioners'. If the householder signed an acceptance form, treatment was generally carried out the following day before he had an opportunity to realise what he had done and change his mind. Many householders were unable to afford the treatments and pressure was then applied in various ways to extract money. In almost all cases the treatments were inadequate; it appeared that just sufficient material was sprayed about to give the necessary odour. Vacuum cleaners were often taken into roof spaces but dirt and cobwebs invariably remained undisturbed.

Theft Act

There were, in fact, a number of companies operating in this way, some evidently associated, and a number of prosecutions were successfully brought under the Trades Descriptions Acts, but this was not the end of the story. The practice of saying that woodworm was present when it was entirely absent, or that a roof might collapse if treatment was not carried out promptly, to persuade a householder to pay for treatment, was considered to be theft by deception, an offence under the Theft Act, and several companies were successfully prosecuted by the police.

Such operations were clearly considered by the courts to be deliberate attempts to defraud the public but there are other cases which are far more difficult to decide. In one case a company had injected cavity walls with a urea-formaldehyde foam insulation. Subsequent examination showed that the foam had collapsed, leaving just a small amount of dampness at the bottom of the cavities; indeed, it was the dampness that became apparent and caused the investigation. Use of an inadequate resin concentration is the cause of this type of failure and may be a deliberate attempt to defraud, yet an accidental blockage in the automatic mixing equipment can achieve the same result, so that it is necessary to decide whether the failure was deliberate, an accident or negligence. This illustrates the dangers facing innocent and conscientious firms; Section 14 of the Trades Descriptions Act 1968 represents the most serious problem, and any person who doubts the significance of this section is advised to refer to Ewan Mitchell's *The Merchandiser's Lawyer* (Business Books) for an interpretation that is particularly appropriate to remedial treatment companies.

Duty of care

If firms offer professional or contracting services they have a 'duty of care' to their clients and customers. Prosecutions under the various Acts seldom result in any compensation to the wronged person and it is therefore neccessary to prepare a separate claim which is pursued through the civil courts, often before the Official Referees. This dual system unfortunately results in unnecessary legal costs which may not be recoverable if the defendant firm or person has insufficient assets. However, this is not a problem for the courts alone but must also be considered in relation to the 'guarantees' issued by most remedial treatment firms as these are valueless if they are not covered by adequate reserves; this is another reason why *caveat emptor* has been emphasised as the householder or person requiring treatment must beware that a firm may offer services or guarantees which it cannot properly provide.

Guarantee or warranty?

In this connection it must be emphasised that a guarantee is 'an engagement on the part of a third person to see an agreement, duty or liability fulfilled'. Thus a guarantee cannot, in fact, be issued by a firm to its customer but must be issued by, or at least underwritten by, a third party such as an insurance company. Most so-called guarantees are actually warranties, or 'promises or undertakings from a vendor to a purchaser', and they have value only if the vendor has sufficient reserves to meet all claims that may arise. There are many firms that have been in business for only a few months offering 30-year guarantees for treatments; they have no means of knowing whether their treatment will actually perform for this period and many of them have no reserves to meet any claims in any case. On the other hand, there are many long-established contracting companies which issue valid guarantees in the sense that they are underwritten through an insurance scheme of some sort, either the investment of money to provide a proper reserve or the payment of premiums to an independent insurer.

In all cases a person considering a remedial treatment contract should carefully study any warranty or guarantee that may be offered, checking that it is in addition to rather than a substitute for any statutory or common law liability, perhaps the most important point to check. For example, a guarantee that is 'in substitution for every other condition or warranty, express or implied, common law or statutory' may appear to be attractive if it extends the period of protection to 20 or 30 years compared with the normal limit of 6 years or the extended period of 15 years for latent damage which are allowed under the Limitation Act 1980 and Latent Damage Act 1986. However, the guarantee may be specific for the remedial system involved, stating that various benefits will be available should the system fail to operate satisfactorily. The problem is then often an interpretation of 'satisfactorily' when applied to the system. Thus, a damp-proofing course may be installed in a property but the dampness may persist. Under the terms of the guarantee or warranty the contractor may claim that his system is operating satisfactorily and that he is therefore free of any further responsibility. Under common law it would be necessary to consider the situation in a rather wider context, remembering that the contractor was perhaps instructed to remedy the dampness and was therefore responsible for the selection of an appropriate treatment method as well as carrying out the work, and there is no reason why the contractor in such circumstances should not be responsible should the dampness persist, whatever the cause.

Clearly the law favours the consumer but only really in the sense that it encourages the professional person or contracting firm to ensure that the services provided are adequate and reliable. Clearly a knowledgeable, experienced and conscientious firm has little to fear, provided that it is properly managed. Perhaps the point that must be emphasised most firmly is that contracting companies should avoid assuming professional responsibilities, particularly if they supply such professional services free of charge, since they represent a greater liability than their normal contracting work. This point was emphasised earlier in this chapter when it was pointed out that surveyors and architects should normally assume these professional responsibilities. However, they seldom possess the knowledge and experience that is necessary for the reliable diagnosis of defects and the preparation of specifications for remedial treatment, and it is possible that a separate profession may develop, perhaps through surveyors or architects acquiring the necessary specialist experience of building structures. At the present time it appears that surveyors and architects are reluctant to specialise in this way and it is possible that the scientists and the technologists will assume this responsibility, perhaps through the Institute of Wood Science in the case of remedial wood treatment, so that they are available to assist surveyors, architects or property owners.

Injury through defects

Whilst it would appear that the various Acts favour a property owner rather than a person or firm providing professional or contracting services, it must be emphasised that the property owner also has a number of distinct responsibilities. The most obvious is to ensure that a property remains safe. When stiletto heels are fashionable there is a tendency for them to penetrate the sappy edges of floorboards where these have been heavily infested by Common Furniture beetle, perhaps causing the person to stumble or fall. Edges of stair treads may also collapse through similar attack, but the danger is not confined to the structural parts of the property as furniture may also be affected; in one case a nurse was injured in a hospital changing room when a chair collapsed through such an attack. In another case a widow, living alone in rented accommodation, decided to collect some coal from her cellar. Unfortunately she had not entered the cellar during the previous summer months and had not noticed that the stairs were affected by a severe Dry rot attack. The stairs collapsed under her weight; she broke both her ankles, and it was not until more than a day later that she was able to attract attention by breaking one of the cellar windows. In another case a remedial contractor had already commenced work to eradicate a Dry rot outbreak and a gas leak was caused whilst he was stripping plaster from under a staircase. Two Gas Board employees arrived to attend to the gas leak but, whilst they were working under the stairs, the floor collapsed beneath them. It was alleged in this particular case that the contractor was negligent in failing to advise the Gas Board employees that it was dangerous to stand on this particular piece of floor on account of Dry rot damage. However, subsequent investigation showed that the piece of floor concerned was a later addition spanning a stairwell which consisted of joists supported by skew nails alone; it was not normal to walk on this floor under the stairs, which appeared sound when the boards were in position and failed only when two men were standing together on a single skew-nailed joist.

Value of professional advice

One of the greatest dangers facing a property owner is to ignore professional advice, perhaps because the financial or disturbance costs would seem to be unacceptable. For example, a very substantial semi-detached property was purchased in the Holland Park area of London and plans were prepared for the conversion of the ground and first floors to a substantial maisonette, with the basement, second and third floors converted into six self-contained flats. The cost of the work was considerable and included treatment to eradicate a Dry rot outbreak. Specialist contractors were sent a specification of the areas to be treated and asked to submit estimates. However, the contractor who eventually carried out the treatment submitted a report with the estimate which emphasised the need to expose further areas, refusing to guarantee the work in any way unless this further work was permitted. Unfortunately these further areas had just been restored by the main contractor, including very expensive decoration, and the remedial treatment contractor was therefore instructed to proceed with the specified work alone. After only a few months it became apparent that Dry rot was present in these additional areas and, as the remedial treatment contractor had made the position clear from the start, a dispute arose between the owner and the main contractor concerning the responsibility for ignoring the specialist advice.

Clearly an owner must always ensure that a property is properly maintained and that there is no danger through misuse. One particular financier established a 'hostel' for his 'hostesses' in which each was provided with a suite of rooms consisting of a bedroom, a lavishly equipped bathroom, a sitting room and a small kitchen. After this property had been occupied for about two years, the neighbours complained of Dry rot penetrating through the party walls. Investigations showed that the Dry rot had originated from the flooring timbers beneath the baths, apparently through water seeping down the back of each bath as a result of rather energetic splashing coupled with a defective wall seal between the bath and the elaborate surrounds.

Concealing defects

It is often difficult to decide whether it is reasonable to expect a property owner, general contractor, specialist contractor or professional adviser to understand the significance of a particular defect and the proper manner in which it should be remedied. In one such case a general contractor purchased an old rectory, ostensibly for use as his own residence. However, following extensive restoration, the contractor sold the property to his own solicitor. Only a few months after the purchase the solicitor noticed Dry rot fruiting bodies appearing through the new decorations in various parts of the property. Investigations showed that the Dry rot had been present for many years and it appeared that no remedial treatment had been attempted. The contractor claimed that he did not appreciate the seriousness of the damage that had been concealed and did not realise that the infection would continue and spread extensively. The resulting dispute was taken before the Official Referee and concerned a claim for damages coupled with an allegation of deliberate fraud. It was difficult to decide whether the contractor had concealed the damage to enhance the value of the property or whether he was simply ignorant and incompetent; he was not, of course, a specialist contractor.

Approved contractors

Hopefully a property owner seldom requires the services of remedial treatment contractors but this necessarily means that the property owner has little experience in selecting the most suitable firm for his particular requirements. There are approval schemes in operation, such as membership of the Remedial Treatment Section of the British Wood Preserving and Damp-proofing Association, which imply a measure of competence, experience and financial stability, but there must be serious doubt now regarding even these companies. It is unfortunately a fact that, with the exception of national companies submitting very high estimates, all firms attempting to provide a competent inspection and reliable treatment at reasonable cost are consistently rejected by householders who prefer to accept lower estimates from firms who are willing to skimp inspections and treatment. Many of the old-established companies who were unwilling to compromise simply failed or simply went into voluntary liquidation through trying to offer competent and reliable services at prices competitive with those of irresponsible newer firms. For example, in 1976 there were about 830 remedial treatment firms offering wood preservation and damp-proofing services in the British Isles.

Unreliable companies

A survey sixteen months later in 1977 showed that almost 400 of these firms had ceased trading, amongst them many of the older-established firms, but almost 600 new firms had become established. In such a volatile situation it is clear that few firms are experienced or are likely to be sufficiently stable to honour their so-called guarantees. Even in 1990 the proportion of failures amongst remedial treatment contractors was very high compared with other businesses.

Selecting contractors

There is no reason to suppose that large national companies are more competent or more reliable than small local companies. Indeed, large companies tend to operate as independent branches which often lack the continuity of staff that is a feature of most smaller family businesses, and if branches are inadequately supervised there is clearly a danger that they will give an unreliable service, whatever the national reputation of the company. Obviously these are problems for the companies concerned but such considerations make it very difficult to decide on the best choice of contractor. One method is to select a contractor on the basis of the materials or methods employed; small independent contractors may be linked through manufacturers of materials or may be licensees for particular processes.

Agrément approval

In some cases processes are covered by Agrément Certificates and, where specialist installation is involved as in cavity-fill insulation and injection damp-proof courses, the British Board of Agrément maintains a list of approved contractors.

Building Regulations

Specialist remedial treatments are designed to solve problems that have occurred or are anticipated because a fault can be seen to be developing. The emphasis throughout remedial treatment must be on common sense but many remedial treatments may appear to conflict with the Building Regulations. The Regulations are actually concerned mainly with health and safety, and most of the actual Regulations are common sense. However, earlier Building Regulations for England and Wales contained a number of 'deemed-to-satisfy' requirements which many architects, surveyors, contractors and even local authority inspectors considered to be mandatory. These 'deemed-to-satisfy' requirements were, in fact, included only as examples of systems that could be approved without further consideration, whereas any other system could be used, provided that it complied with the performance requirements defined in the Regulations themselves. This system was abandoned with the introduction of the Building Regulations 1985 which comprise only a brief schedule of requirements, but with a series of supporting Approved Documents which give examples of the ways in which these requirements may be met, but emphasising that alternative approaches are equally acceptable. The earlier 'deemed-to-satisfy' system stifled the introduction of new ideas but, whilst the new system is more flexible and encouraging, there are still some situations in which applications for relaxation of Building Regulation requirements are necessary for approval of a remedial treatment, usually when there is a possible influence on health and safety. However, it should also be appreciated that an application for planning consent is necessary if the treatment is likely to change the external appearance of the building. Thus it is necessary to apply for approval for cavity insulation; cavity walls are a precaution against damp penetration so the use of an unsuitable insulation may lead to severe dampness problems. However, it is unnecessary to apply for approval for ceiling insulation, for wall-surface treatments with water repellents or biocides, or or any form of remedial wood preservation.

Other Acts

This comment on legal responsibilities in relation to the remedial treatment industry is based on the situation in England and Wales. It must be appreciated, of course, that remedial treatment firms have many other responsibilities in common with all other firms, such as the Health and Safety at Work Act 1974 and the various Factories Acts. These Acts are not specific to the remedial treatment industry and they will not be considered further, but remedial treatment companies applying wood preservatives and masonry biocides must be aware of their additional responsibilities under, for example, the Food and Environment Protection Act 1985, the Control of Substances Hazardous to Health Regulations 1988, and the Control of Pesticides Regulations 1986.

2 Wood treatment

2.1 Wood-boring insects

Remedial treatments are, of course, treatments to remedy defects, and treatment methods cannot be fully understood without a proper understanding of the defects as an appropriate remedial treatment cannot be selected until the defect has been reliably diagnosed. These general comments are applicable to all remedial treatments but some of the greatest problems arise in connection with damage caused by wood-boring insects as there is no general treatment applicable to all borer infestations and it is therefore essential to diagnose the damage. It should be possible to identify common wood-borer infestations from the general desciptions in this chapter and thus to decide on appropriate remedial treatments, but diagnosis is sometimes much more difficult, particularly if a rare borer is involved. In addition borer damage in buildings is sometimes attributed to insects which have been found and it is therefore necessary to positively identify these insects and to decide upon the damage that they might cause. The problems of identification of these insects and borer damage are not considered in detail in this chapter but are covered separately in Appendix 1.

"*They're* the little devils
that cause all the trouble."

The Furniture beetles, comprising the sub-family Anobiidae, are the best known wood-borers in temperate areas, probably because damage occurs in furnishings and is thus readily apparent to the householder. This sub-family can be further divided into the Anobiinae, including *Anobium* spp. which produce elongated ovoid or rod-shaped pellets, and the Ernobiinae, including *Ernobius* and *Xestobium* spp. which produce bunshaped pellets, one of the useful diagnostic features describe in Appendix 1. All larvae are curved and the life cycles tend to be long, with a very slow build-up of infestation over many years. All Furniture beetles lay eggs in cracks or open pores and considerable damage may occur within the wood before the infestation becomes apparent through the emergence of beetles through flight holes and associated bore-dust discharge.

Common Furniture beetle, Anobium punctatum

The Common Furniture beetle *Anobium punctatum* is economically the most significant wood-borer, except in areas where there is a risk of damage by the House Longhorn beetle; the Common Furniture beetle does not usually cause as severe structural damage as this other wood-borer but it is very widely distributed and an infestation invariably develops wherever suitable wood is found. The Common Furniture beetle attacks the sapwood of both hardwoods and softwoods, extending into the heartwood in some light-coloured hardwoods or when damp conditions occur. The adult beetle is 2.5–5.0 mm ($^1/_{10}$–$^1/_5$ in) long, the female usually being larger than the male, and reddish- to blackish-brown with a fine covering of short yellow hairs over the thorax and elytra, particularly on freshly emerged insects (the elytra are the wing cases whilst the thorax is the next part of the body hooded over the head which projects downwards in this particular beetle). Longitudinal rows of pits, the punctata, are readily visible on the elytra and these account for the Latin name of this beetle. When viewed from the side the hooded thorax is seen to have a distinct hump, another diagnostic feature mentioned in Appendix 1.

The adult beetles usually emerge from the wood between late May and early August, and can be seen crawling over walls and windows. The beetles are strong fliers on warm days and live for three to four weeks, during which they mate and the female lays about 80 lemon-shaped white eggs, each about 0.3 mm ($^1/_{80}$ in) long, in small groups in cracks, crevices, open joints and old flight holes. An egg hatches after four to five weeks, the larvae breaking through the base of the egg and then tunnelling within the wood in the direction of the grain. The gallery increases in diameter as the larva grows, occasionally running across the grain. The galleries are filled with loosely packed gritty bore dust consisting of granular debris plus oval or cylindrical pellets. When fully grown the curved larva is about 6 mm ($^1/_4$ in) long with five-jointed legs. Eventually the larva forms a pupal chamber near the surface about six to eight weeks before emerging through a flight hole about 1.5 mm ($^1/_6$ in) in diameter. Under optimum conditions the life cycle can be as short as one year but it is usually longer and up to four years.

As infestation is largely confined to sapwood the damage is not usually structurally important, except where individual components in furniture are composed entirely of sapwood or where old types of blood or casein adhesives have been used which encourage infestation. Dampness also encourages infestation, slight fungal or bacterial activity enabling the infestation to extend into normally resistant heartwood. All situations which favour Common Furniture beetle attack result in shorter life cycles; such an

Plate 2.1 Typical Common Furniture beetle damage in hardwood plywood furniture (*Cementone-Beaver Ltd*) and softwood structural timber (*Penarth Research International Ltd*)

acceleration of activity is particularly noticeable in stables and byres which are often extensively damaged. Damage is caused only by the larvae tunnelling within the wood and current activity is entirely invisible, the only indication being a bore dust discharge from a fresh-coloured flight hole suggesting recent emergence, although vibration caused by inspection can also dislodge bore dust. It is therefore often difficult to decide whether an infestation is currently active and whether a remedial treatment is necessary or, if a treatment has been applied in the past, whether it has been effective.

The Common Furniture beetle suffers very high natural mortality, perhaps only 60% reaching the larval stage where they may be further reduced in numbers by the action of predators. The most common predators are two Hymenoptera, the flying-ant-like *Theocolax formiciformis* and *Spathius exarata*, which can often be seen exploring flight holes. Predatory beetles are also sometimes found, such as *Opilio mollis* and *Korynetes coeruleus*, although the latter is more often associated with the Death Watch beetle; these predators are described in Appendix 1.

Remedial treatment

Remedial treatment against Common Furniture beetle is comparatively simple; the special aspects of remedial treatments against other wood-borers are considered later in this chapter. Where timbers are completely exposed,

Plate 2.2 A microemulsion, Wykamol Microtech, being used for spray treatment of a roof (*Cementone-Beaver Ltd*)

as in a roof space, the treatment can consist of spraying all timbers with a suitable eradicant and preservative fluid. The most reliable technique involves the use of comparatively low pressures in conjunction with coarse jets so that fluid can be flooded over the surface of the timber, achieving the maximum loading without wasteful dripping, and avoiding the unnecessary volatilisation of the carrier solvent which occurs if excessive pressures are used; air entrained sprays such as paint sprays must not be used as they cannot achieve the necessary fluid loadings and result in excessive and unpleasant volatilisation. Solvent vapour volatilisation is, in fact, the greatest problem in applying remedial treatments of this type which are normally formulated in petroleum solvents, as conditions can be very unpleasant if solvent vapour accumulations become excessive and there can be severe fire dangers if solvents with comparatively low flash points are used. One technique for reducing unpleasant solvent vapour accumulations is to spray the ceiling joists first, leaving the ridge and upper areas of the roof space until last. Cobweb and dirt accumulations should be removed before spraying as they absorb the treatment fluid, increasing consumption but at the same time reducing the reliability of the treatment by obstructing penetration.

Care must be taken to ensure that fluid does not soak through plaster ceilings and stain the decoration beneath, particularly if ceiling laths are treated; these laths are often sapwood and particularly susceptible to

Common Furniture beetle attack. If floors are to be treated it is necessary to lift floorboards, preferably completely but certainly at intervals of not less than about 600 mm (24 in), so that the area can be properly cleaned and sprayed. Where Common Furniture beetle attack is evident under paint or varnish, or in furniture, superficial spraying should never be used; instead the fluid should be injected into old flight holes. The galleries formed by one larva seldom connect with those formed by others but they are all concentrated within the most susceptible wood so that injection into old flight holes gives a convenient access route to zones of current activity. If activity is particularly intense and widespread, perhaps through incipient or current fungal decay caused by condensation or some other source of dampness, the treatment fluid should obviously contain fungicidal as well as insecticidal components; indeed, the decay may be a far more serious problem than the secondary Common Funiture beetle attack which it may have stimulated.

Treatment fluids

The development of remedial treatment fluids has been described comprehensively in *Wood Preservation* (1993) by the same author. The very effective organochlorine contact insecticides such as DDT, Lindane and Dieldrin were introduced following World War II and Lindane is still in use for remedial wood treatment.

Insecticides

Lindane is volatile and at typical use concentrations of 0.3–0.5% volatile losses from relatively superficial spray treatment result in a fairly rapid decline in effectiveness, despite the extreme activity of this insecticide. Persistence can be improved by formulation with other compounds, and a mixture of Lindane with chloronaphthalene waxes was particularly effective, the waxes acting as persistent preservative insecticides but also absorbing the Lindane and extending the life of its contact action. The chloronaphthalene waxes are not readily soluble in normal petroleum solvents such as kerosene or white spirit, and co-solvents are required in manufacture, generally aromatic solvents. One unexpected advantage of this system is the vapour action of these aromatic solvents as the vapour is able to diffuse within treated wood rather more deeply than the preservative fluid itself. In more recent years many products have consisted simply of mixtures of Lindane and tri-*n*-butyltin oxide, the latter compound usually incorporated as a fungicide but actually giving excellent long-term protection against wood-borer reinfestation.

Organochlorine insecticides were extensively used in agriculture and horticulture but it became apparent that these compounds were very persistent in food chains so that birds eating affected insects would accumulate large concentrations, and predatory birds in turn would be particularly seriously affected. The organochlorine insecticides have now been withdrawn from most uses, although there was never any evidence that remedial wood preservation treatment was responsible for any of these damaging effects, probably because wood-borers were generally killed within the treated wood and did not normally enter natural food chains. These developments caused considerable difficulties in the remedial wood preservation industry as insecticide research concentrated on the development of non-persistent insecticides for agricultural use, whereas the wood preservation industry required persistent insecticides that would give long-term protection. For many years Lindane continued to be the preferred insecticide for use in remedial wood preservatives as reliable and economical alternatives were not available. Today the synthetic pyrethroid insecticides are preferred, particularly Permethrin, as they combine excellent activity against most wood-boring insects with excellent persistence.

There have also been doubts in recent years regarding the safety of tri-*n*-butyltin oxide, apparently prompted by the well-known high toxicities of other organotin compounds. Use of tri-*n*-butyltin compounds is now restricted, although not prohibited, and this has prompted a search for safer alternatives. Chlorophenols have been used in the past but are not popular as they contain dioxin impurities; in fact, whilst some dioxins are extremely toxic, those associated with the tri-, tetra- and pentachlorophenols used in wood preservation are no more toxic than the chlorophenols themselves. Zinc soaps, particularly naphthenates, have been used in remedial wood preservatives for many years but the musty odour of naphthenic acid is not popular. Other acids have been considered and an acid derived from petroleum known as acypetacs is now used, although it seems that it can cause allergic reactions in some individuals. Boron compounds seem to be the most suitable fungicides currently available; they are also insecticidal preservatives and they can be formulated in water or organic solvent systems.

It is believed fairly widely amongst persons who are not involved in the remedial treatment or wood preservation industries that woodworm can be controlled by injecting paraffin into the flight holes. Whilst there is no doubt that paraffin will kill larvae by drowning if applied sufficiently generously, there is no evidence whatsoever that it has any other significant toxic effect and there is certainly no possibility that paraffin treatment will leave a residual preservative effect which will prevent infestation in the future. It is false economy to devote labour and effort to any remedial treatment involving the use of unsuitable materials.

Smokes and fumigants

In this connection mention should be made of the various fumigant and smoke treatments that have been considered and which are still used in some cases at the present time. Fumigants can diffuse deeply and are thus excellent eradicants but they have the distinct disadvantages that they can be used only when a structure can be completely sealed in a gas-tight envelope and they do not give a persistent preservative action, so that treated wood is immediately susceptible to reinfestation. Smoke treatments produce a deposit of insecticide, principally on upper horizontal surfaces, which persists for a short period but, as there is no significant penetration, the existing infestation is controlled solely by the eradication of emerging adult beetles and the prevention of further egg-laying. Smoke treatments must therefore be repeated before every flight season until there is no further possibility that insects remain within the wood. Even then, these multiple treatments will have controlled only the established infestation as the limited persistence of the insecticide deposits will give little protection against new infestation. For these reasons fumigant and smoke treatments are more suitable for use against some wood-borers than against others, and a detailed discussion of the advantages and disadvantages of such treatments and other alternative treatments will be given later in this chapter following the descriptions of the more important wood-borers and the conventional methods by which they are controlled.

Other Anobiid beetles

The Common Furniture beetle *Anobium punctatum* is certainly the most commonly encountered wood-borer in the British Isles and, in fact, throughout all temperate parts of the world, but there are a number of other borers of varying importance.

Ptilinus pectinicornis

The related Anobiid beetle *Ptilinus pectinicornis* is sometimes found causing damage in hardwoods such as beech, sycamore, maple and elm, and it can be a nuisance when it occurs in furniture. The damage is very similar to that caused by the Common Furniture beetle, including the exit holes which are about 1.5 mm ($^1/_{16}$ in) in diameter. Treatment is the same and it is therefore unnecessary to distinguish between damage caused by these two borers. It is important to be able to confirm that adult beetles are wood-borers if they are found in buildings but they are readily idntified by the comb-like antennae of the male and serrated or saw-like antennae of the female, as described in Appendix 1.

Drug store beetle, Stegobium paniceum

Whilst this beetle is relatively easy to identify, another Anobiid, known as the Drug Store, Biscuit or Bread beetle *Stegobium paniceum*, is far more difficult to distinguish from the Common Furniture beetle; it is found in woody natural drugs and cork but most commonly infests biscuits, particularly dog biscuits, and the appearance of adult beetles may generate unnecessary fears of serious woodworm infestation.

Bark borer, Ernobius mollis

Ernobius mollis, sometimes known simply as the Bark borer, is an Anobiid beetle which is only rarely found in buildings, but damage in softwood is often confused with that of the Common Furniture beetle. This borer forms galleries within and beneath softwood bark, sometimes extending the galleries to a depth of perhaps 1 cm ($^1/_2$ in) into sapwood. Only fresh softwood with waney edges (attached bark) is attacked and the infestation dies out naturally after one or two years, but the presence of galleries is often diagnosed as Common Furniture beetle infestation and used as justification for treatment. The concentration of activity beneath the bark and galleries filled with bore dust containing bun-shaped pellets which are either brown or white in colour, depending on whether the larva was feeding on bark or wood, enable the activity of this borer to be readily distinguished from that of the Common Furniture beetle with its deeper galleries and white oval or cylindrical pellets.

Anobium pertinax

Anobium pertinax is much more closely related to the Common Furniture beetle *Anobium punctatum* and is sometimes found in the British Isles, usually in the north of England or Scotland, but it is much more common in Scandinavia. The damage caused by this insect is very similar to that caused by Common Furniture beetle but is always in association with fungal attack, so that severe infestations are normally confined to poorly ventilated cellars or sub-floor spaces, or in timber subject to periodic rainwater or plumbing leaks. These are precisely the same conditions that encourage the Common Furniture beetle *Anobium punctatum* but as damage is so similar and treatment precisely the same it is really unnecessary to distinguish between these species.

Death Watch beetle, Xestobium rufovillosum

Another Anobiid beetle that is associated with incipient or active decay is the Death Watch beetle *Xestobium rufovillosum* which is the largest member of this family. Infestations in the British Isles occur most commonly in oak, probably because this wood was once extensively employed in construction, but infestations can also occur in elm, walnut, chestnut, alder and beech. Sapwood and heartwood can be infested if previously infected by decay, and the infestation can spread into adjacent softwoods; in the Channel Islands Death Watch beetle is often found infesting softwoods but this may be a sub-species. However, infestation is always confined to damp or decayed areas and one common characteristic of Death Watch beetle attack is a brown coloration in the infested wood due to deterioration by Brown rot fungi. It frequently appears that this insect favours churches but this is really a combination of circumstances which results in church timbers being particularly suitable: the roofing frequently consists of sheets of lead or other metals and the periodic heating results in condensation beneath the lead and decay in the supporting boards that encourages infestation. The length of the life cycle depends on the quantity of nitrogen available in the form of fungal attack and can be a single year under optimum conditions but is usually far longer and often as much as ten years. The adults normally emerge between the end of March and the beginning of June but, where longer life cycles are involved, fully developed beetles may be found at other times just below the surface of the wood and apparently ready to emerge; the larva actually pupates in July or August, metamorphosing into an adult after only two or three weeks but remaining in the wood and gradually darkening in colour until it emerges during the normal flight season the following spring.

The adult is not a free flier and tends to mate with other beetles emerging from the same piece of wood, attracting their attention by bending the legs so that the head is struck on the wood surface, producing a series of eight to eleven taps in a period of two seconds. Tapping with the tip of a pencil can generate a similar noise and can stimulate a response. This tapping noise probably accounts for the name Death Watch beetle, perhaps because the sound is apparent in a house which is quiet through a recent death. The tapping should not be confused with the sound produced by the Booklouse *Trogium pulsatorium* which is more like a watch tick than a tap. Although the Death Watch beetle can fly reasonably well when the weather is very warm, it is probable that this pest is spread largely by reuse of old infested wood. The Death Watch beetle has a much more limited distribution than the Common Furniture beetle in the British Isles, being confined to England, Wales, part of southern Ireland and the Channel Islands.

"Have you ever *seen* the old death watch beetle?"

Wood infested by the Death Watch beetle may also be attacked by other insects, such as the Common Furniture beetle and the Wood weevils, with perhaps the rare *Helops coeruleus* in the damper zones of the wood; the identification of borer infestations is described in detail in Appendix 1. Predators may also be present, particularly *Korynetes coeruleus*, a distinctive blue beetle which is an active flier and often the first sign that a Death Watch beetle infestation is present in concealed damp timbers.

Plate 2.3 Typical Death Watch beetle damage in the sapwood edge of an oak timber (*Penarth Research International Ltd*)

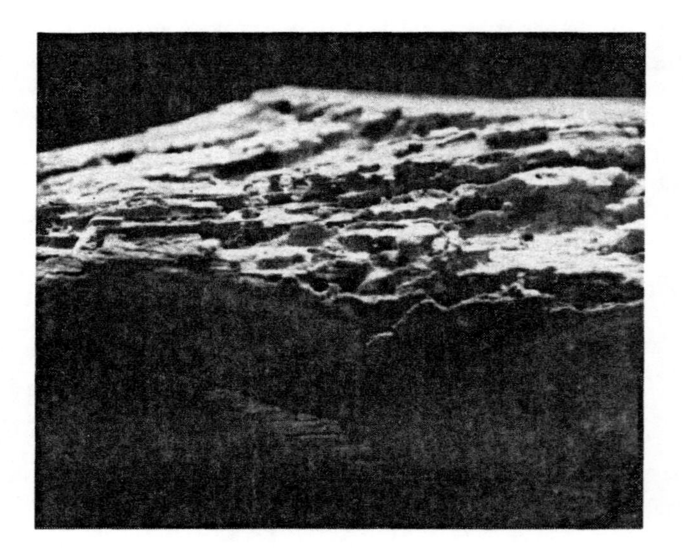

Remedial treatment for a Death Watch beetle infestation is rather more complex than for a Common Furniture beetle attack. Common Furniture beetle damage is normally confined to sapwood, although it may extend to heartwood where activity is encouraged by incipient or active decay. In contrast Death Watch beetle is entirely dependent on decay and is thus able to extend into heartwood, although sapwood is normally attacked first and more heavily. In older buildings the decay and thus the Death Watch beetle infestation may be concentrated within the interior of large beams which are built into damp walls, so that the damage may be largely invisible except perhaps for a few flight holes.

Great care must therefore be taken during inspections to ensure that all possible concealed decay or Death Watch beetle infestation has been identified and attempts should be made, either during the inspection or during subsequent remedial treatment, to expose the ends of beams to check their

Figure 2.1 A moulded plastic injector, used for the treatment of window frames and in other situations in which deep penetration is desirable (*Cementone-Beaver Ltd*)

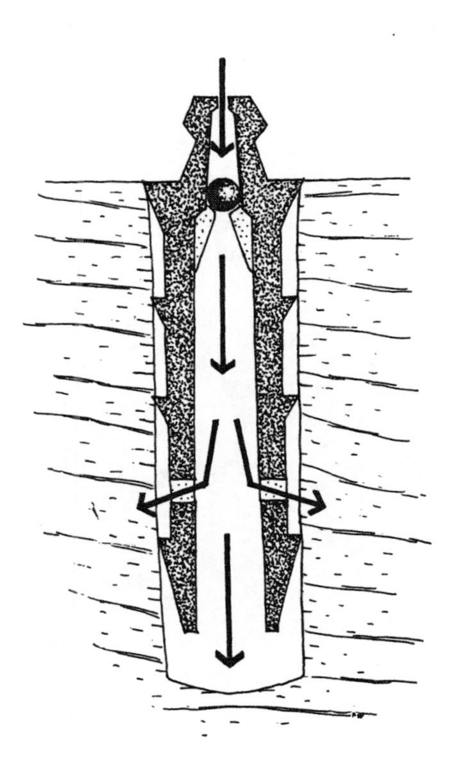

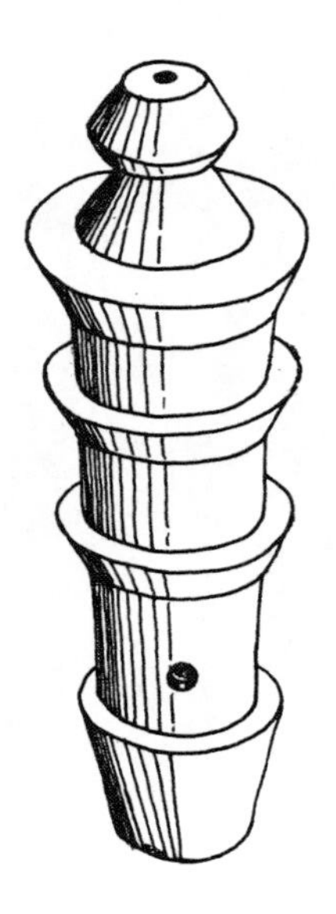

condition. If this is impossible then doubtful beams should be drilled close to their bearing ends so that they can be injected to eradicate any deep-seated infestation. Thorough exposure work to permit reliable diagnosis and then thorough treatment to ensure that the preservative fluid gains access to the infested area are essential features of Death Watch beetle treatments. Even when Death Watch beetle infestation is not concealed in this way it is essential to thoroughly trim all friable frass (damaged wood) to permit a reliable assessment of the remaining sound wood and proper treatment without the preservative being wastefully absorbed by this structurally useless frass.

Plate 2.4 Injecting Death Watch Beetle flight holes using a tapered nozzle (*Cementone-Beaver Ltd*)

As Death Watch beetle is so deep seated in many circumstances, it is essential to always treat by injection rather than rely on superficial spray. If Death Watch beetle flight holes are present, injection is most easily achieved by fitting the spray equipment with a suitable tapered nozzle so that the preservative fluid can be injected into these flight holes. Injection into one flight hole will often result in preservative flowing from others so that this is not such a tedious treatment method as might be anticipated and it normally conveys preservative to the most susceptible zones of the wood. If flight holes are not present but deep damage is suspected, as when a large beam is built into a wall that may be damp, the timber must be drilled to provide access to the danger zone. The first drilling should be about 100 mm (4 in) from one edge, and the second drilling 200 mm (8 in) but also about 300 mm (12 in) along the length of the beam to produce a diagonal series of holes. In the case of a very deep beam a series of holes should be started every 900 mm (36 in) within the zone to be treated, and in this way there is a reasonable chance of finding any deep-seated Death Watch beetle galleries or decayed zones. A small-diameter wood bit or auger should be used with a brace or slow-speed electric drill, although it will be appreciated that it is essential to keep the cutting edges of the bit sharp if drilling is to be reasonably rapid in the hard oak that is normally infested by this beetle. Some remedial treatment companies fit plastic nozzles with non-return valves into larger holes so that fluid can be pressurised and more reliable penetration obtained into the wood aroud the hole (Figure 2.1). Generally holes can be bored from one side to about three-quarters of the depth of a beam or from both sides to half the depth.

Wood weevils, Curculionidae

Wood suffering from incipient or active decay is also attacked by other beetles, particularly the wood-boring weevils. Unlike most of the Anobiid beetles which have been previously described, both the adults and larvae bore into the wood, causing damage which is superficially similar to that made by the Common Furniture beetle, but the holes are smaller, contain fine bore dust without pellets, are often laminar in pattern and are often confined to spring wood and follow the annual rings. The attack is often limited to sapwood but extends into heartwood if fungal decay is more severe, although the fungal attack may not be visible.

In the British Isles *Pentarthrum huttoni* is a native species usually found in buildings in decayed floorboards and panelling, as well as in old casein-glued plywood. *Caulotrupis aeneopiceus*, another native species, is more rarely found in buildings, and then only in association with very decayed wet wood in cellars and subfloor spaces. Several other species may be found but in recent years the weevil that has attracted most attention is *Euophryum confine*, a species that was apparently introduced to Britain from New Zealand in about 1935. It has since spread very widely, apparently because it is able to infest wood which is not significantly decayed and which may have a moisture content as low as 20%. However, it is not really necessary to distinguish individual species or even to clearly distinguish damage from that of Common Furniture beetle, except to note that if the infestation occurs in damp conditions or if active decay is present these features may be of greater structural significance than the borer damage itself. Treatment is therefore similar to that for Common Furniture beetle infestations occurring under the same conditions.

Longhorn beetles, Cerambycidae

Although Common Furniture beetle infestations represent the most common borer damage in the sense that infestation may be encountered in any building within the British Isles, and Death Watch beetle infestation may be best known in the sense that it chooses to attack the most valuable historic buildings, it is the House Longhorn beetle *Hylotrupes bajulus* that represents the most serious borer problem in the areas in which it occurs.

House Longhorn beetle, Hylotrupes bajalus

In the British Isles it is largely confined to southern England (Figure 2.2) where it is sometimes known locally as the Camberley borer. Spread to other parts of England has not occurred, perhaps through climatic restrictions. This is a relatively large beetle which leaves an oval flight hole about 10 mm

Figure 2.2 Distribution of House Longhorn beetle in southern England and local government areas in which preservation of building softwood against this pest is required under the Building Regulations (*Crown copyright*)

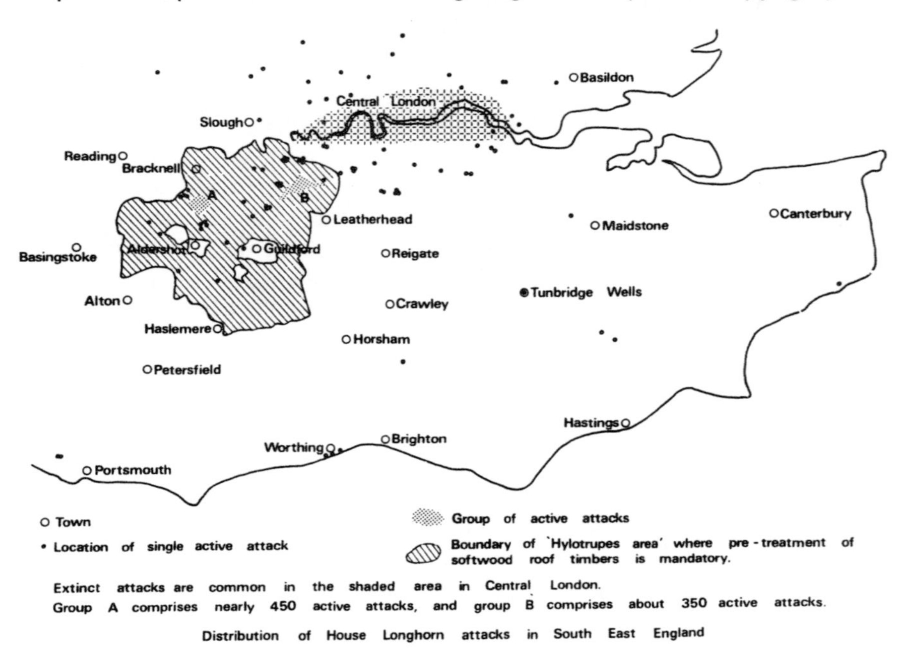

($^{3}/_{8}$ in) across. The appearance of even a single flight hole indicates that severe damage has already occurred and that the condition of the structure should be checked by thorough probing. Because of the seriousness of the damage caused by this beetle the Building Regulations in England now require all structural wood to be preserved in the areas where this pest is established.

House Longhorn beetle attacks the sapwood of dry softwood. It is not as widespread as the Common Furniture beetle but the damage it causes is much more rapid and more severe so that it can be considered the most important wood-borer in temperate areas free from termites. The body of the adult beetle is somewhat flattened, and 10 to 20 mm ($^{3}/_{8}$ in–$^{3}/_{4}$ in) long, the male being smaller than the female. The beetles are brown to black in colour except that they have thick grey hairs on the head and the prothorax (the front section of the thorax); the female has a central black line and a black nodule on either side, whilst the male has white marks in place of the nodules. There are also distinctively shaped white spots on the elytra. The name Longhorn derives from the antennae which are larger than the body.

Plate 2.5 A house Longhorn beetle larva on damaged wood; the galleries have been exposed by removing a thin surface veneer (*Cementone-Beaver Ltd*)

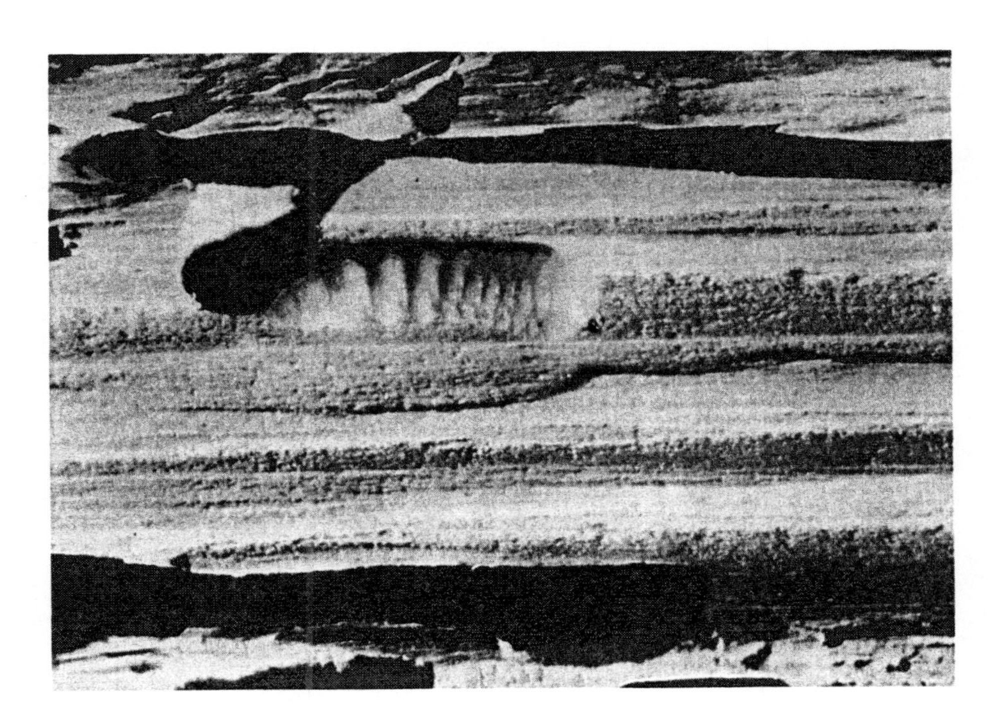

In Europe the beetles emerge in July to September and a single female can lay as many as 200 eggs which hatch within one to three weeks, these eggs being spindle-shaped and 2 mm ($^1/_{12}$ in) long. The House Longhorn beetle attacks the sapwood of softwoods. In a roof structure the larvae from a single clutch of eggs can cause substantial damage within a period of three to eleven years before they pupate and emerge as adults, perhaps entirely destroying the sapwood and leaving only a thin surface veneer which is slightly distorted by the presence of oval galleries beneath. The first signs of damage may therefore be the collapse of a largely sapwood member, though in warm weather the gnawing of the insects can be clearly heard. When fully grown the larva is about 30 mm ($1^1/_4$ in) long, straight-bodied and distinctly segmented with a slight taper and very small legs. Pupation occurs in a chamber just below the surface and is complete in three weeks, the emerging beetle leaving an oval flight hole 10 mm ($^3/_8$ in) across, as noted earlier.

In areas where this beetle is likely to cause damage it is essential to inspect softwood timbers in buildings particularly carefully as considerable damage may be caused by this beetle before the appearance of the first flight hole. If a torch is shone across the surface of a piece of wood at a shallow angle it is possible to detect the slight surface distortion that is caused by the presence of the galleries beneath and, if an infestation is suspected, it is then necessary to probe all sapwood areas to determine the full extent of the damage. In fact, if the purpose of an inspection is simply to confirm that the House Longhorn beetle is present, probing to detect the full extent of the damage can be left until the actual treatment is carried out.

The existence of a characteristic oval flight hole suggests Longhorn beetle attack and, if the hole is in the sapwood of softwood, an extensive infestation by the House Longhorn beetle must be suspected. However, it is important to appreciate that the flight hole may have been caused by a forest Longhorn but, if this is the case, the hole may not be a flight hole but instead a gallery that has been cut through by sawing during conversion. One indication of this form of damage is a hole that is running diagonally to the surface. In most cases such forest Longhorn damage is confined to a few widely dispersed galleries, whereas House Longhorn beetle galleries often extend throughout the sapwood. Active infestations by forest Longhorn beetles in building timbers are very unusual, although a few species may survive for a few years in timber imported from Europe, and Douglas fir and Sitka spruce from North America occasionally contain species which are able to survive, perhaps for 30 years or more in some instances. However, these insects are not very active, causing comparatively slight damage, and are unable to reinfest the timber once they have emerged.

Remedial treatment

Once House Longhorn beetle infestation has been confirmed, perhaps with the assistance of the second section of Appendix 1, remedial treatment is essential to prevent the rapid spread of the infestation to unaffected timbers.

The sapwood areas of all softwood timbers throughout the building must be probed to check for galleries, even if flight holes are absent, as a relatively new infestation may be active but concealed. Where the sapwood area is severely damaged the frass (borer-damaged wood) should be trimmed away to disclose the remaining sound wood so that inadequate structural members can be identified and replaced. Where timber remains relatively sound, holes must be drilled to enable the sapwood areas to be injected to entirely eradicate any larvae in concealed galleries and preserve against reinfestation. Injection can be achieved using a tapered nozzle in a relatively small hole or an inserted plastic nozzle with a non-return valve, as previously described for the treatment of Death Watch beetle infestations. If probing has shown that the infestation is localised then injection treatment can be confined to the infested area, that is, the roof space or floor structure in which the infestation has been discovered, but precautionary spray treatment must be applied to all other softwood timbers in the building as the occurrence of the infestation has clearly shown that they are at risk.

When inspecting timber in buildings it is always essential to decide whether an infestation is active or whether it has been previously treated or died out naturally, perhaps as a result of activity of predators or parasites. It has already been explained that it is difficult to decide whether there is current activity in the case of Common Furniture beetle infestation, but the situation is far easier with House Longhorn beetle. If flight holes are present or surface distortion is observable, and galleries are disclosed by probing, it must always be assumed that the infestation is active as it is certain that an infestation has not been reliably treated if trimming and injection points are not apparent. Longhorn beetle damage in softwood timber in buildings usually results from an infestation originating within the building, but galleries caused by an infestation before installation can always be identified through a careful examination of the surface holes; if the holes run straight into the wood at an angle to the surface they indicate sawing after infestation, whereas a flight hole of an active House Longhorn beetle infestation is perpendicular to the surface but only penetrates for a short distance to the pupal chamber.

Other Cerambycid beetles

There are two other Longhorn beetles that may cause infestations in building timbers. The Two-toothed Longhorn *Ambeodontus tristus* causes serious damage to softwoods in service in New Zealand. It thrives in similar conditions to the House Longhorn beetle and causes similar damage, but the oval exit holes are distinctly smaller and only about 5 mm ($^1/_5$ in) across. In 1974 this insect was found to have caused severe damage to joists in a cellar in Leicestershire, England. The infestation was introduced in the joists which were found to be made from the wood of a *Dacrydium* pine common in New Zealand. Whilst there appears to be no biological reason why this beetle could not become established in England it must be emphasised that it is most unlikely and, if an exit hole of the correct size is discovered, it must first be suspected either that it has been caused by a forest Longhorn beetle or that it is a false flight hole caused by sawing through the existing galleries of an older infestation.

Oak Longhorn beetle, Phymatodes tetaceus

In older buildings constructed from oak the sapwood is sometimes found to be attacked by the Oak Longhorn beetle *Phymatodes testaceus*. Eggs are laid in cracks in the bark and the larvae bore between the bark and the wood, eventually making deeper tunnels to form pupation chambers. Damage is normally caused in the forest in sickly trees but can also occur in stored logs and boards being air seasoned with waney edges. Occasionally this beetle may be found attacking dry oak in buildings but only if the bark remains. As the damage is confined to the zone immediately under the bark it is rarely of structural significance and eradication consists simply of removal of the bark, perhaps accompanied by a spray treatment which will kill any surviving insects in pupation chambers, although the treatment must be principally regarded as a precaution against Common Furniture beetle which is a far greater risk in oak sapwood. The Oak Longhorn beetle can also attack a number of other hardwood species such as beech. If oak and other hardwoods are decayed, however, a number of other Longhorn beetles may be found, such as *Rhagium mordax* and *Leiopus nebulosus*. Whilst these beetles are capable of infesting damp decayed oak in buildings, such infestation is extremely rare unless it has been introduced to the building on the decayed wood involved, the most common circumstances being introduction in fire logs.

Powder Post beetles

There are several other wood-boring beetles that should be mentioned as damaged wood is occasionally found in buildings. The most important are the Powder Post beetles which can be divided in turn into two sub-families, the Bostrychidae and the Lyctidae.

Bostrychid beetles

The European Bostrychidae are all small dark brown or black beetles with the single exception of a species attacking oak, *Apate capucina*, which has brown or red elytra and which is seen occasionally in the sapwood of freshly felled oak. Adult Bostrychids tunnel in bark in order to lay their eggs, producing tunnels which are free from dust. The hatching larvae then bore in the sapwood in their search for starch, producing tunnels which are packed with fine bore dust; this pattern of tunnelling enables these insects to be distinguished from the Lyctid Powder Post beetles. Damage is principally confined to the sapwood of green hardwoods, although softwoods are occasionally found to be attacked, particularly if they have bark adhering. Bostrychid damage is not so common as Lyctid damage, probably because infestation commences with a tunnel bored by the adult which is readily visible so that infested wood is rejected before being used in a building, in contrast with a Lyctus infestation which is initiated by egg-laying and completely invisible.

Lyctid beetles

The Lyctid beetles are all small with a length of only about 4 mm ($^1/_6$ in), elongated but flattened in appearance. From above the head is clearly apparent protruding in front of the thorax. The colour of the various species encountered in Europe varies from mid-brown to black. The beetles are active fliers, particularly on warm nights. After mating the female lays 30 to 50 eggs in large pores on the end-grain of suitable hardwoods containing adequate starch; this means that only freshly converted wood can be attacked. The larvae develop progressively within the wood, feeding on the starch, and pupation occurs in a chamber immediately beneath the surface. The adult beetle normally emerges after one or two years in the spring, summer or autumn, usually between May and early September, leaving a flight hole 0.8 to 1.5 mm ($^1/_{32}$ in to $^1/_{16}$ in) in diameter. The galleries are packed with soft fine bore dust and all the sapwood may be completely destroyed except for a surface veneer, accounting for the name Powder Post beetle.

As the initial attack consists of only an egg laid in an open pore it will be appreciated that the first sign of damage is often collapse or alternatively the appearance of a flight hole, in either case an indication of extensive damage within the wood. Susceptible large-pored hardwoods may be infested soon after conversion during air seasoning or storage and the insect may be introduced to a building within the wood, generally in furniture or decorative woodwork. Eradication treatment is unnecessary as the progressive loss of starch has usually rendered the wood unsuitable for infestation by the time the attack has been discovered. However, it is important to be able to identify Powder Post beetle infestations so that they are not confused with Common Furniture beetle infestations which require eradicant treatment.

Ambrosia beetles

Damage caused by the Ambrosia beetles, the Scolytidae and Platypodidae, is occasionally seen in buildings. The adult beetles bore in the bark of freshly felled green logs, producing galleries beneath the bark in which their eggs are laid. At the same time the adult beetles infect the galleries with the Ambrosia fungus which thrives upon the sap, providing nourishment for the larvae when they hatch. The larvae also burrow, usually forming side tunnels off the original adult gallery and eventually boring deeper in order to pupate. The gallery patterns under the bark are usually characteristic of the species responsible. As Ambrosia beetles are unable to infest wood that is dry or free from bark they represent a problem in the forest rather than in a building or furniture. However, the larvae sometimes bore quite deeply in the sapwood of tropical species, sometimes even penetrating the heartwood in light-coloured woods, creating galleries that are clearly apparent as they are marked by dark staining caused by the Ambrosia fungus spreading along the grain on either side of the hole. Wood damaged in this way may be noticed in buildings or in furniture and may cause unnecessary concern so that it is important to be able to distinguish between this damage and current active infestations caused by other beetles.

Pinholes and shotholes

Galleries formed by small Ambrosia beetles are described as pinholes, and large galleries are usually termed shotholes.

Termites

Although various wood-boring beetle infestations occur throughout the world the damage caused by termites, or white ants, is generally far more serious. About 2000 species of termites have been identified, of which more than 150 are known to damage wood in buildings and other structures. Termites are principally tropical but extend into Australia, New Zealand and North America, although they are rarely seen in Canada. Their introduction into certain parts of France and Germany is clearly related to trade; for example the termite of Saintonge, *Reticulitermes santonensis*, which was established on the west coast of France between the rivers Garonne and Loire, is now found in Paris around the Austerlitz station which serves this region. Similarly *Reticulitermes flavipes* from the United States is concentrated around Hamburg. Neither species can spread widely as they are evidently sensitive to temperature and tend to survive in central heating ducts and other permanently warm areas in buildings. Termites do not appear to be established in the British Isles at the present time but it must be appreciated that similar localised infestations are possible through importation of infested wood and this possibility should be appreciated when investigating wood-borer damage. A more detailed description of termites is given in *Wood Preservation* (1993) by the same author.

Carpenter ants

The Carpenter ants *Camponotus herculeanus* and *C. ligniperda* have been causing increasing damage in buildings in Scandinavia in recent years. In nature these insects tunnel into old trees affected by interior decay to establish their nests. Modern forestry leaves very few suitable trees and stumps, and a search for suitable nesting sites probably explains the increasing incidence of infestations in buildings. Summer homes are much more frequently attacked than permanent homes, probably because they are often situated within or close to forests.

It is usually reported that Carpenter ants will attack only wood which has already decayed but *C. ligniperda* is certainly able to attack and utilise dry wood. Damage is typically internal, being an irregular cavity in soft decayed wood but laminar and following the growth rings in sound wood. *C. vagus* is occasionally found in southern Europe and *C. pennsylvaicus* in North America, together with other species. There is an increasing danger that these insects will be introduced into other countries in wood shipments and there are already rports of isolated infestations in the British Isles, particularly in Ireland.

Predators on wood-borers

There are a number of other insects that are found in buildings and which may be correctly or erroneously related to timber damage. When an insect is encountered which cannot be immediately recognised it should be identified by reference to the identification key in Appendix 1. The most important insects encountered in this way are probably the predators which confim the presence of wood-borer activity in the area, such as the Clerid beetle *Korynetes coeruleus*, and the small fly or ant-like *Eubadizon pallipes, Theocolax formiciformis* and *Spathius exarata*. These insects are all described more fully in Appendix 1, as well as the Spider beetles and Carpet beetles which commonly cause damage in buildings, although not to structural woodwork, and the rather striking Wharf borer which attacks wood in damp situations and which occasionally swarms and may then be found in buildings.

It is assumed throughout this book that remedial treatment surveyors have a basic knowledge of building construction. In fact, difficulties are often encountered in finding the correct term to describe particular components and a glossary of useful structural terms has therefore been included as Appendix 7. The wood infestation surveyor must also have a basic knowledge of wood technology, at least sufficient to know the fundamental differences between hardwoods and softwoods, and to be able to identify the large-pored hardwoods that may be susceptible to Lyctid Powder Post beetle attack. In addition it is essential to have a proper knowledge of normal timber preservation, that is the processes that are used to treat wood prior to installation in buildings and other structures to avoid decay; remedial treatment does not always consist of *in situ* treatment but may involve replacements with adequately preserved materials. It is therefore recommended that any person who claims to be a timber deterioration surveyor should study *Wood in Construction* and *Wood Preservation* by the same author.

Diagnosis, inspection and treatment

Reliable diagnosis is always essential before an appropriate remedial treatment can be considered. In order to assist the timber infestation surveyor as far as possible the illustrations throughout this chapter show the damage that may be observed rather than following the practice, adopted in so many books on this subject, of showing the rarely seen adult beetles of the species that cause the damage. In fact, the reasonably experienced timber infestation surveyor has no need to regularly refer to this chapter as Appendix 1 gives keys for the identification of both the borer damage and the insects that may be encountered in buildings.

The division between inspection and treatment is not always clear. In the case of Death Watch beetle infestation it is important for the surveyor to define the items whch need special attention, such as the exposure of beam ends and the extent of any injection treatment that may be necessary in addition to normal spray treatment. However, in the case of House Longhorn beetle infestation remedial treatment must involve the probing of all timbers, the trimming of severely damaged sapwood and the injection of sapwood where damage is light. Provided an active infestation has been identified by the surveyor it is therefore clear that the full extent of the infestation can be determined during the remedial treatment work, although the surveyor will need to estimate the extent to which timbers require replacement rather than treatment.

Whilst the inspection of exposed timbers in roof spaces, for example, is relatively easy, it must be appreciated that it is equally essential to similarly inspect all other timbers in the building. This will involve the lifting of floorboards here and there but beam ends, plates and lintels may still be concealed so that their condition cannot be reliably assessed before exposure work during actual treatment. In many older buildings great care is necessary to ensure that concealed timbers are not missed; walls may be dry lined, perhaps presenting a severe decay risk if wood battens are untreated and in direct contact with damp solid external walls or affected by condensation.

Where a relatively superficial inspection is sufficient, such as for valuation purposes or where extensive remedial treatment is to be carried out whatever the condition of a particular part of the structure, it is common practice to check the spring of a floor by bouncing, that is by raising oneself on the toes and then allowing the heels to drop and apply a shock to the floor. Clearly the flexing in response to this test will depend on the span of the floor and its structure, quite apart from its condition, and this test should be used only to decide whether the floor seems to be unduly weak and whether the lifting of floorboards during the inspection seems essential to check whether there is severe wood-borer infestation or some other damage causing weakness. It is perhaps worth noting at this stage that this bounce test is really intended to detect joists that have been weakened by wood-borer attack throughout their length whereas fungal decay will tend to be concentrated at the bearing ends and indicated if floorboards are seen to be dropping away from the skirting level. If deterioration is known to be present or even suspected considerable care should be taken to avoid over-stressing the structure as surveyors have been known to fall through ceilings and floors! It should be appreciated that joist-end deterioration is not always easily detectable as old heavy floorboards can transfer the load to remaining sound joists.

Plumber's and electrician's rot

Plumber's and electrican's rot can be very extensive and often far more severe than any natural form of wood deterioration. Employees of secondary trades can be remarkably ignorant and negligent when working in buildings, cutting large notches through joists to accommodate pipes, and perhaps reducing the strength by 50% or more in the process. From the visual point of view the most annoying feature is the damage caused to floorboards, crudely lifted using chisels and bars, and severely damaged in the process. The boards are then refixed using nails whereas screws should be used so that access can be readily obtained to pipe and cable runs in the future. The timber deterioration surveyor should be particularly aware of the dangers that are generated in this way as far as his own position is concerned; if boards are lifted to facilitate the inspection great care must be taken in refixing to ensure that nails do not penetrate the cables and pipes running beneath. Experienced timber deterioration surveyors may be tempted to ignore this warning but beware; the author has made this mistake and the consequences can be very serious. In one case plumbers installing a new central heating system put a nail through their own copper pipe when replacing the boards The nail almost plugged the hole but permitted a very fine spray of water to escape which soon caused the establishment of Wet rot around the hole and the later establishment of Dry rot extending throughout the rest of the same floor.

Remedial treatments

In the case of House Longhorn beetle infestations, remedial treatment should commence with probing to assess the full extent of the damage, followed by trimming of friable damaged sapwood or frass. With Death Watch beetle infestations it may be necessary to drill the ends of beams built into solid external walls in order to check the extent of any concealed interior decay or infestation, the speed of penetration of the drill and the condition of the drillings being used to assess the degree of damage. Structurally unsound wood should be replaced, perhaps using wood pretreated with preservative. The entire area should then be cleaned, an essential part of remedial treatment which is so often ignored by specialist contractors at the present time as it involves rather high labour costs. The purpose of this cleaning is to ensure that all damage has been fully exposed and also to ensure that the wood is freely exposed for treatment so that the preservative penetrates deeply without being wastefully absorbed on covering dirt. In older buildings where laths must be treated as they are usually particularly susceptible to Common Furniture beetle infestation, the removal of dirt reduces the danger of ceiling staining.

Treatment dangers

Treatment should then be applied using low-pressure airless spray through a fairly coarse nozzle; a high-pressure spray through a fine nozzle or an air-entrained spray will cause unnecessary solvent volatilisation, reducing preservative penetration and producing rather unpleasant treatment conditions. Application by brush is entirely unsuitable as adequate treatment loadings cannot be achieved. Generally preservative should be sprayed on all surfaces of existing and replacement timbers, unless the latter have been pretreated, the spray being moved along each timber surface at a relatively slow rate so that the fluid just commences to run down, indicating that the surface has absorbed as much fluid as possible; the permanence of the treatment is substantially reduced if adequate penetration is not achieved. Organic solvent preservatives are preferred as they achieve much greater penetration into the dry wood that is normally treated in buildings, and staining is minimised if the preservative should come in contact with decorations, either through splashing or through penetrating ceiling plaster.

Fire hazard

Unfortunately the light petroleum solvents that are most suitable for remedial treatment preservatives are flammable and there is clearly a severe fire danger. Electrical equipment that is employed during treatment, such as lights and perhaps electric sprays, must not provide a source of ignition. Light bulbs must be protected against damage but a simple cage is not sufficient; if spray accidentally contacts a light bulb there is a danger that the rapid cooling will cause it to burst, igniting the spray. All lead connections should be flameproof. These various precautions mean that installed lighting should never normally be used during treatment, and leads should be plugged into sockets in other parts of the building so that there is no danger that a switch spark may cause ignition. Obviously the fire danger depends on the flammability of the solvents used. Current regulations in Britain require solvents to have flash points in excess of 32°C (90°F) but this is not really sufficient for remedial treatment sprays used in roof spaces where central heating pipes in winter and sunshine in the summer can raise the temperature to 49°C (120°F) or more. Most solvent-based remedial treatment fluids in current use have flash points of about 42°C (105 to 110°F) but slightly more expensive solents are available with flash points over 60°C (140°F) and odourless kerosene, for example, normally has a flash point in excess of 65°C (150°F).

The use of solvents with higher flash points greatly reduces the danger of ignition of solvent vapour but does not significantly affect the influence of the solvent on an actual fire. This is not really a danger during treatment by sensible operatives as, once the danger of vapour ignition has been avoided, a fire can occur only if initiated externally. Plumbers are probably the most common cause of fires in recently treated buildings, usually when they are using blow-lamps to solder pipes close to joists or rafters. The danger of fire is greatly increased by the presence of insulation materials. Obviously insulation should be removed during treatment to fully expose timbers but, if it is relaid too quickly, it will obstruct the ventilation of adjacent timber so that solvent is retained and the fire danger prolonged. It must be emphasised most strongly that treatment must never be applied whilst insulation remains in position as there is then a danger of auto-ignition, or spontaneous combustion. The auto-ignition temperature for the normal solvents used in remedial treatments is about 250°C (482°F) and for high-flash solvents it is about 270°C (520°F). However, when applied to some insulation materials the auto-ignition temperature will be substantially reduced. On mineral wool insulation the auto-ignition temperature for normal remedial treatment products may be as low as 50°C (120°F), a temperature that may occur close to heating pipes or throughout a roof space during hot summer weather. If the preservative contains pro-oxidants such as copper or iron the danger of auto-ignition is further increased as the auto-ignition temperature is reduced.

The auto-ignition dangers are therefore very real when fibrous insulation materials are treated with wood preservatives; the danger arises because the fibrous structure exposes the solvent deposits to an abnormal concentration of atmospheric oxygen, thus increasing the tendency for the solvent to oxidise and ignite. All these fire dangers can be avoided if flammable organic solvents are replaced with water. However, water-based formulations are generally unsuitable for remedial treatment as they do not penetrate readily into dry wood and they introduce a severe danger of staining damage, as previously explained. These factors have been taken into account in the development of some water-in-oil emulsion treatments; the emulsion breaks when applied to wood so that the organic solvent portion penetrates whilst the water evaporates, greatly reducing the danger of fire through vapour accumulations during treatment. Formulations which are simple aqueous solutions or suspensions which do not contain an organic solvent phase are generally far less effective as they do not usually penetrate so deeply.

Electrical defects

One problem that is occasionally encountered is damage to electrical systems following treatment. The most common damage occurs when organic solvent treatments have been applied too generously without sufficient care to avoid cables. The aromatic components in some treatment fluids are able to dissolve the plasticiser from normal PVC cable insulation and in some cases the plasticiser solution will flow down a cable, dripping onto a ceiling at a lighting point or even emerging through wall plaster around a switch, forming a persistent stain. The appearance of this stain implies that the electric cable has been damaged but the electrical properties of the cable are not usually affected, although a test on the system may show a short circuit. The reason for this contradiction arises through the tendency of the solution to follow cables to junction boxes, switches and other fittings which are often manufactured from phenolic insulation material. Some preservative systems can cause electric tracking and carbonisation of the insulation with the development of a short circuit which may cause fuses to blow. In several instances the electrical engineers investigating blown fuses have tested the resistance of the system and reported that the wiring is usually acceptable and it is the phenolic fittings that need replacement. This damage is usually associated with treatments based on odourless kerosene rather than volatile solvents, apparently because this particular solvent evaporates rather slowly and this results in more extensive extraction of the plasticisers from the cables. However, other preservatives have caused this damage but only when applied during very cold winter weather, apparently because the dispersal of the solvent by volatilisation is reduced by the low temperature and because, in view of this low temperature, the property has not been adequately ventilated following treatment. The tracking and short circuit damage caused on electrical fittings with phenolic insulation is generally associated with preservatives containing chlorophenols.

Persistent odours

Another problem that is often associated, perhaps surprisingly, with odourless kerosene rather than other solvents is the development of a powerful paraffin smell for a protracted period following treatment, rather reminiscent of a large airport on a bad day! Investigation usually associates development of this odour with partial burning of solvent vapour. The smell is often particularly severe if a gas cooker is in use, although smell can also be generated with open fires and gas fires. Occasionally smell is generated in electrical heating systems, particularly ducted warm-air systems. Obviously the odour persists only as long as the solvent vapour persists but it appears that the most severe odours are associated with the relatively non-volatile 'tail' of odourless kerosene, and complaints of this type are usually associated with this solvent; some smell for a limited period after treatment is to be anticipated, but a protracted unpleasant smell is unacceptable. Copper and zinc naphthenates were extensively used as remedial treatments in the past but they generate a rather unpleasant persistent musty odour and, as a result, their use for this purpose has declined in recent years. Zinc naphthenate has been replaced in one proprietary range of products by acypetacs zinc which has less odour, although some persons seem to be sensitive to it and find it very unpleasant. It must be added that odour problems are not solely associated with wood preservation treatments but also occur with injection damp-proof courses, again because the smell is rather persistent, apparently because the large quantities of solvent injected into porous brickwork or masonry can disperse only relatively slowly, sometimes into floor spaces which feed air ducts leading to open fires.

Health and safety

Clearly remedial treatment contractors have a duty to ensure the good health and safety of their own operatives as well as the occupiers of treated buildings. It is essential that good ladders and, where necessary, scaffolding should be available to operatives, as well as proper tools for trimming and cleaning, such as axes, chisels, draw-knives and industrial vacuum cleaners. Good light is essential and, where organic solvent treatments are to be employed, the operatives must have their own flame-proof wiring which can be fed from a power socket in a distant part of the building. Electric spray equipment must also be flame-proof. Experience shows that preservative leaks are more likely from electric spray equipment, particularly when long lengths of hose are involved, and hand pressurised equipment is generally preferred. Proper protective clothing must be provided, although it is difficult to ensure that it is worn at all times. For example, during the spring roof spaces may become rather warm in contrast to the winter period and operatives may be tempted to remove clothing, perhaps during spraying. The operative then leaves the roof and perhaps sits in the sun without a shirt or vest and there is then a danger of mild sunburn accompanied by considerable irritation caused by the solvents. Nose bleeding can also occur when some preservative vapours are encountered in high-temperature conditions, but it must be appreciated that some operatives are more sensitive than others and sensitive persons should never be employed for this type of work. Masks and barrier creams are often recommended but neither is really advisable; simple gauze masks tend to absorb treatment fluids, perhaps exposing the user to abnormal concentrations of toxicant vapours, whereas without a mask the operative would probably take more care to avoid unnecessary breathing of preservative spray or vapour. Barrier creams can also give operatives unjustified confidence and it is far better to train operatives to take the necessary care. Obviously all operative gangs must be aware of the health and fire dangers, and must be aware of the action that should be taken in an emergency; a first-aid kit and fire extinguishers should always be available.

There are thus a number of important points that should be borne in mind when carrying out remedial treatment against wood-borer infestations when using normal organic solvent preservatives. Low-pressure sprays should be employed with coarse jets to ensure that the maximum volume can be applied to the timber surface without the excessive volatilisation of solvents that occurs with high pressures and fine jets or air-entrained paint sprays. Whilst it is essential to achieve the maximum loading of preservative on the wood to ensure maximum penetration, dripping of excess fluid must be avoided and care must be taken to ensure that electrical cables are not treated unnecessarily and preservative does not enter junction boxes or other electrical fittings. Treated areas must be freely ventilated to disperse solvent vapour which is a fire hazard and which may affect electrical cables and cause staining around ceiling roses and wall switches. Electrical installations in the treated areas should not be employed during treatment or even for several days afterwards as there is a danger that sparks may ignite solvent vapour. Smoking, naked lights and plumbing activities must be prohibited in the area for seven to fourteen days, depending on the nature of the solvents involved, and notices to this effect should be posted at the entrances of the property and at the entrances to roof spaces and other treated areas. Insulation materials must always be lifted before treatment and replaced later after the solvent has completely dispersed, certainly not less than seven days after treatment in any circumstances, and insulation must never be sprayed with preservative as there is then a real danger of spontaneous combustion.

If there is any doubt regarding the thoroughness of the treatment that has been applied, an inspection will show whether necessary cleaning, trimming or injection has been carried out to check whether treatment has been applied but it may be necessary to take samples of wood for analysis. The normal method is to take shavings with a knife, chisel or gouge, a gouge usually being preferred. However, none of these methods is fully satisfactory as it is not really possible to take a sample of known area and adequate depth from a shaving to enable the loadings of preservative per unit area to be determined and thus the level of treatment to be assessed. The best method that is currently available is to use a special Forstner wood bit which produces a flat-ended hole. It is possible with this bit to take samples to a depth of 5 mm or more on site as this is likely to be more than the normal penetration of the preservative. The area from which the sample is taken is known from the diameter of the bit. The shavings from the sample are usually collected in a long envelope with the flap opening along one of the long edges as this type of envelope, with the flap folded back, is particularly suitable for catching the wood particles as they fall away from a bit. Forstner bits are available with a variety of shanks and can be used in braces or breast-drills, but they are best used for this purpose in slow-speed cordless drills.

Fumigant and smoke treatments

The comments in this chapter on treatment techniques and precautions assume, of course, that conventional remedial treatment techniques are being employed with organic solvent preservatives. Various fumigant and smoke treatments have been considered in the past and are still used in some cases. Amongst the fumigant gases hydrocyanic acid, methyl bromide, methylene chloride, ethylene oxide, ethylene dichloride, ethylene dibromide, acrylonitrile, phosphine and carbon disulphide have all been found effective in the control of insects, particularly in stored foodstuffs. However, the practical problems of applying gaseous fumigants in buildings severely restricts their usefulness in remedial treatments. Some fumigants, particularly methyl bromide, continue to be used in some countries for the eradication of termites but treatment involves the evacuation of the house, followed by complete covering with plastic sheet. When treatment is complete the property must then be freely ventilated for a period sufficient to ensure the total dispersion of the fumigant, particularly with treatments like methyl bromide with high mammalian toxicity. Fumigant treatments have the advantage that they diffuse readily into wood and can therefore eradicate deep infestations but, despite the elaborate treatment involved, there is no persistent preservative action and regular retreatment is necessary as reinfestation can occur.

Plate 2.6 Using a cordless electric drill and a Forstner bit to sample wood for analysis: the drillings are caught in an open envelope held beneath (*Penarth Research International Ltd*)

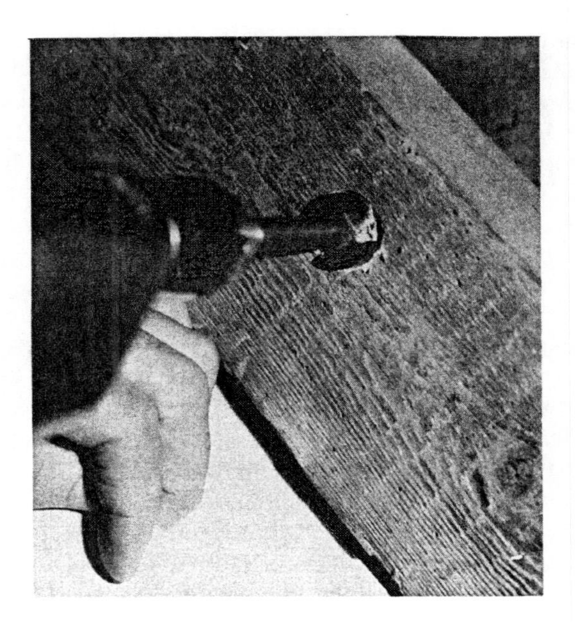

It is much more realistic to incorporate a fumigant as the eradicant insecticide in remedial treatment fluids which contain additional preservative components to prevent reinfestation once the toxic vapour has dispersed. For example, impregnated strips have been used to maintain a low level of Dichlorvos vapour in roof spaces to eradicate Common Furniture beetle infestation. Reasonable success has been achieved but it appears that the system is most likely to be successful in controlling insects entering the roof space during the egg-laying period. However, impregnated strips maintain a toxic vapour concentration only for a limited period and it is necessary to replace them regularly, either on a continuous basis or during the normal emergence and egg-laying periods of each year. Dichlorvos could be incorporated into normal organic solvent remedial treatment fluids, and treatment of timbers with these fluids would, in effect, impregnate the outer layers of wood with Dichlorvos so that toxic vapour would be released slowly in the same way as from proprietary impregnated strips. The treatment fluid would also penetrate readily into old flight holes so that the toxic vapour would also be released deep within the most susceptible wood, a particularly efficient eradicant system. Unfortunately there are doubts regarding the safety of Dichlorvos and other organophosphorus insecticides and it is therefore necessary to await the development of safer vapour insecticides before this concept can be adopted commercially.

Insecticidal smokes are normally applied using a 'candle' which releases a cloud of smoke which then settles on the timber. Obviously the highest loadings are achieved on upper horizontal surfaces and lowest loadings on lower horizontal surfaces, and this method will only effectively eradicate insects that settle on the upper horizontal surfaces whilst the insecticide still remains. Most insecticides are deposited in a very finely divided form on the surface of the wood without any penetration to give protection against loss by volatilisation and the deposits rapidly lose efficacy. With normal insecticides such deposits may not have even sufficient life to span the normal flight season for Common Furniture beetle and, in particularly warm seasons, when a second emergence period may occur in late August or September, a second smoke treatment may be necessary if egg-laying is to be reliably prevented.

Whilst smoke treatments are simple to apply it will be appreciated that treatment must be reapplied to cover each annual flight season for a number of years in order to eradicate an existing infestation by killing emerging adults and preventing further egg-laying. The disadvantage of this system is that further damage is permitted to occur before emergence, an unacceptable situation with, for example, House Longhorn beetle where structurally significant damage may be caused in this way. In addition, no persistent preservative action is achieved and, with a free-flying pest such as the Common Furniture beetle, infestation is likely to occur if annual treatments are eventually abandoned. It is therefore frequently argued that smoke treatments are far more suitable for use against Death Watch beetle which is only free-flying in exceptional circumstances so that there is little danger of reinfestation once an existing infestation in a structure has been eradicated.

Whilst smokes are normally applied using candles which have the advantage that one or two candles can treat an entire space, this method is really unsuitable for the treatment of floor spaces, unless all the boards can be raised when necessary, a completely unacceptable situation for a building in normal use. The alternative is a smoke 'gun', but this considerably increases the labour cost and largely defeats the only real advantage of smoke treatment which is simplicity of application in a suitable large space such as a roof. If regular smoke treatment with a gun is considered acceptable it is far more realistic to attempt eradication and persistent preservation using conventional spray treatments. However, there are always areas which are not readily accessible to a normal spray, such as laths, rafters and sarking boards behind skeelings.

Fogging treatments

The correct method for treatment is to remove the skeelings or roof tiles to expose the timbers, but one alternative is to treat using a mist gun or 'fogger' – equipment that is often used, for example, for the treatment of fruit trees. Some foggers operate using ram jets powered by burning kerosene and are entirely unsuitable for use in buildings, but others operate electrically, using a centrifugal system for forming the fog which is then blown out of the machine by a fan. Some machines of this type can be adapted to feed fog through a wide spout which can then be directed up skeeling linings or into any area that is inaccessible to normal spray. Fog is preferred to smoke as the liquid particles have a greater chance of achieving a degree of penetration, but it will be appreciated that only limited volumes of solution can be applied and a specially designed preservative is essential if treatment is to be efficient. This principle has been used successfully to apply fungicide to protect battens behind valuable plaster work, as described in the following section of this chapter.

Bodied mayonnaise emulsions

It has been explained that preservative may need to be applied by injection to eradicate deep Death Watch beetle attack which may be encountered in the centre of beams built into damp solid walls. One alternative method of treatment is to employ bodied mayonnaise-type emulsion systems such as Woodtreat. The emulsion structure is, in fact, a technique to thicken the preparation so that high loadings can be applied to the surface of wood as a paste. If wood remains continuously in contact with a preservative solution, as in a dipping tank, the absorption and penetration depend on the time involved. The BMT emulsion system is simply a method for placing a supply of relatively non-volatile preservative solution in contact with the wood so that absorption can occur over a protracted period permitting unusually deep penetration. Whilst these systems are much easier to use than injection, it must be appreciated that the act of drilling for injections is actually a testing method to assess the strength of the timber and to check for concealed voids, a particularly important requirement in old buildings where large beams may be built into solid external walls and affected internally by both fungal decay and Death Watch beetle activity. If BMT emulsion systems are employed it is therefore essential to appreciate that drillings should still be made when necessary for normal inspection purposes. It should also be appreciated that BMT emulsion systems may leave an unsightly deposit on the surface and that this is not always readily removed. Such deposits are clearly unacceptable in some circumstances, such as in the treatment of exposed beams and historic log buildings in Norway, and for such purposes injection is to be preferred as injection holes are readily concealed with matched stopping.

Wormy tales

There are many stories about woodworm treatment, or eradication of Common Furniture beetle as it should be correctly described, most of which are repeated here as they are instructive but a few simply because they are amusing. There is the story of the over-enthusiastic representative for a remedial treatment company in south-east England who warned householders that treatment was urgently required as 'a swarm of woodworm was coming over from the Continent'! A more sinister story concerns two 'antique dealers' who persuaded a widow to allow them to take away seven items of furniture on the apparent understanding that she was selling one item and the others were to be treated for woodworm. Whilst some of the items had old flight holes there was no evidence whatsoever of borer attacking one item, no attempt was made to treat them and none of the items were ever returned to her.

There have also been many strange attempts to develop reliable treatments. One serious proposal during the early nineteenth century was the use of snuff; the inventor disliked snuff so intensely that he was sure that it would discourage wood-borers! Another rather obscure method, still occasionally seen in mail-order advertisements, is the use of 'catch' blocks of susceptible wood, usually aspen, which are exposed in roof spaces in the hope that they will attract egg-laying females. The catch blocks are then burnt at the end of each flight season in the hope that they will contain all the eggs that would otherwise have been laid on the roof timbers. The main danger with this system is that seasonal burning is forgotten and, if the catch blocks are really as attractive as claimed, that the development of the infestation will be accelerated by the availability of particularly attractive wood. In fact, there is no reason to suppose that beetles will be discouraged from laying eggs on structural timbers simply because blocks of another suitable wood are present. At one time mail-order advertisements offered kits which represented 'the only completely reliable method for killing woodworm'; the kits consisted of two small blocks of wood with instructions to 'place the woodworm on block A and hit with block B'! Even recorded music has been seriously proposed as a method for discouraging infestation, although it would seem that the method would be more likely to discourage human use of the property protected in this way!

2.2 Wood-decaying fungi

Moulds and staining fungi

If wood becomes damp it will support a wide variety of fungal infections. Many of these infections are superficial and do not affect the strength or integrity of the wood, such as the surface moulds which form greenish or black, occasionally yellow, powdery growths which are easily brushed or planed away. In fresh green wood the residual sap encourages the development of sapstain fungi, frequently known as bluestain as the dark-coloured hyphae running through the wood produce most commonly a greyish-blue coloration. Sapstain fungi do not usually affect the structural strength of the wood but, by attacking the cell contents, staining fungi and invisible bacteria developing in similar conditions can make the wood more porous, allowing it to absorb water more readily and perhaps encouraging the development of other fungal infections in some conditions. If wood remains very wet continuously, Soft rot fungi are likely to develop, causing softening of the surface of the wood to a progressively increasing depth. However, if an intermediate stage of dampness exists in which the wood has an adequate moisture content but is also freely accessible to the oxygen in the air, a much more serious form of fungal infection can develop which represents the most serious wood decay hazard, the group of fungi concerned being known biologically as the Basidiomycetes.

Generally staining and soft rotting fungi are encountered in buildings only when they have been introduced on the infected wood, although there are some unusual situations where Soft rot may develop, such as in boat houses and mills. The superficial moulds do not confine themselves to wood but occur also on damp plaster, wallpaper and carpets; their main significance is that their presence indicates very damp conditions.

Wood-rotting Basidiomycetes

The wood-destroying Basidiomycetes are therefore the only group of fungi that are of special significance in wood deterioration. Most of these fungi are Wet rots, able to develop only when wood is actually wetted by moisture from another source, such as a leaking roof or plumbing, or by absorption of moisture from the soil or from other damp structural materials. In such cases the removal of the source of moisture is sufficient to prevent further deterioration, but replacement of damaged wood may be necessary for structural or aesthetic reasons, and preservation treatment may be desirable as a precaution against rewetting in the future or to prevent the development of a particular Basidiomycete, the Dry rot fungus, which represents a far more severe decay hazard. Once this fungus has become established it is able to generate moisture through digestion of wood so that it can maintain the atmospheric relative humidity under poorly ventilated conditions, permitting the infection to spread actively even in the absence of a source of moisture. The activity of this fungus is therefore often confined to poorly ventilated areas, such as behind skirtings or beneath floors, and the damage becomes apparent only when the visible surfaces become distorted or collapse occurs. In addition, this fungus is able to develop conducting strands which can convey water and nourishment, enabling the fungus to spread across and beneath inhospitable surfaces such as brickwork and masonry in the search for more wood.

It will be appreciated that it is specially important to distinguish between the Dry rot fungus and various Wet rot species. The following general descriptions of the wood-destroying fungi commonly found in buildings will be helpful, but a more detailed description of wood-destroying fungi will be found in *Wood Preservation* by the same author and the identification of fungi found in buildings is considered in more detail in Appendix 2. Whilst fungi have already been divided into Dry and Wet rots, it is also convenient to divide them into Brown and White rots, depending on the manner in which they destroy wood. In a Brown rot the fungal enzymes destroy the cellulose but leave the lignin largely unaltered so that the wood acquires a distinct brown colour and the structural strength is almost entirely lost. As decay progresses the wood appears very dry and shrinkage cracks appear both across and along the grain, the size and shape of the resulting rectangles often being a useful feature in identification. In contrast the White rots destroy both cellulose and lignin, leaving the colour of the wood largely unaltered but producing a soft felty or stringy texture.

"No, no! This is *wet* rot – *that's* dry rot."

Dry rot, Serpula lacrymans

The Dry rot fungus *Serpula (Merulius) lacrymans* is certainly the best known wood-destroying fungus. This species is usually found in buildings in places where ventilation is restricted and it thus tends to develop in completely concealed areas. The conditions for germination and growth are extremely critical, comprising a narrow range of atmospheric relative humidity and wood moisture content. Spore germination occurs most readily in acid conditions and frequently there are infections by other fungi able to germinate and develop in wetter conditions. If wood becomes accidentally wetted one of the Wet rot fungi may develop but as drying progresses conditions may arise in which Wet rot activity is reduced but Dry rot spores may germinate, encouraged by the acid conditions caused by the previous Wet rot. Similarly a source of continuing dampness may support a Wet rot but at a further distance from the source the moisture content may be lower and optimum for Dry rot spore germination, again encouraged by the acidity generated at the fringe of the Wet rot activity.

Dry rot produces white hyphae which are very fine tubes or hollow threads, progressively branching and increasing in length so that they spread in all directions from the initial point of germination, provided that a food source is available. As food is exhausted some hyphae are absorbed whilst others are developed into much larger rhizomorphs or conducting strands which are available to transport food and water. Thus exploring hyphae finding no nourishment are absorbed to form food for growth in more promising directions, giving the fungus the appearance of sensing the direction in which to spread towards a food source. Seasonal changes sometimes inhibit growth which then resumes when suitable conditions return. Hyphae contract in this way on drying to form layers of mycelium, each successive layer indicating a season of growth.

Active growth is indicated by soft hyphal growth like cotton wool, perhaps covered with 'tears' or water drops in unventilated conditions, the system used by the fungus to regulate the atmospheric relative humidity and the explanation for the Latin name *lacrymans*. Rhizomorphs may be up to 6 mm ($^1/_4$ in) in diameter; they are relatively brittle when dry and extend for considerable distances over and through brickwork and masonry and behind plaster, spreading through walls between adjacent buildings and ensuring a residue of infection, even if all decayed wood is removed. Thus treatment of adjacent sound wood and replacement of decayed wood with preserved wood should always be accompanied by sterilisation treatment of infected brickwork and masonry.

Mycelium is greyish and later yellowish with lilac tinges when exposed to light, and often subsequently green in colour through development of mould growth. Sporophores or fruiting bodies generally develop when the fungus is under stress through food or water shortage or temperature increase. Sporophores are shaped like flat plates or brackets and vary from a few centimtres to a metre or more across; they are grey at first wih a surrounding white margin but then the slightly corrugated hymenium or spore-bearing surface develops, covered in millions of rust-red spores which are eventually liberated, coating the surroundings with red dust. As fungal growth in buildings is generally concealed a sporophore appearing through a wall or ceiling may be the first sign of damage, although a characteristic mushroom smell may be noticed if an infected building is closed for several days.

Dry rot can cause severe Brown rot with pronounced cuboidal cracking, the cubes being up to 50 mm (2 in) along and across the grain. Decayed wood crumbles easily between the fingers to a soft powder. Two important features of the decay are the fact that it can be entirely internal and concealed in beams, and that it can spread to dry wood in unventilated conditions as the fungus is able to transport adequate water for decay through the rhizomorphs. A related species *Serpula himantioides* is sometimes found causing similar decay in Scotland and other northern European countries.

Plate 2.7 Typical Dry rot: damaged wood, sporophore, and 'tears' on active growth (*Cementone-Beaver Ltd*)

Cellar rot, Coniophora puteana

The Cellar rot fungus *Coniophora puteana (cerebella)* is the most common cause of Wet rot in buildings and elsewhere where persistently damp conditions arise through, for example, soil moisture or plumbing leaks. Spores germinate readily and this fungus is likely to occur whenever suitable conditions arise. The hyphae are initially white but growth is not as generous as for Dry rot. In addition there is only limited development of mycelium and only thin rhizomorphs are formed. These rhizomorphs, which become brown and eventually black, are not so extensive as those of Dry rot and never extend far from the wood. The sporophore occurs only rarely in buildings and consists of a thin skin covered with small irregular lumps. The hymenium is initially yellow but darkens to olive and then brown as the spores mature. Wood in contact with a source of moisture such as brickwork often consists of a thin surface film concealing extensive internal decay. The rotted wood is dark brown with dominant longitudinal cracks and infrequent cross-grain cracks. The easiest method for controlling this Wet rot in buildings is to isolate wood from the source of dampness.

White Pore fungus, Antrodia sinuosa

The White Pore fungus, *Antrodia sinuosa,* formerly know as *Poria vaporaria,* occasionally occurs in buildings. It requires a higher moisture content than Dry rot but, in contrast to the Cellar rot fungus, it is tolerant to occasional drying and is therefore the normal fungus associated with roof leaks. Growth is generally similar to that of Dry rot but strands remain white, compared with yellow or lilac for Dry rot and brown or black for cellar rot. The rhizomorphs may be up to 3 mm ($^1/_8$ in) in diameter but are not so well developed as those of Dry rot and are flexible when dry. When examined on the surface of a piece of wood or adjacent masonry they appear to be distinctly flattened or to have a flat margin on either side of the strand, and they do not extend far from their source of wood. The sporophore is rare in buildings but it is sometimes observed in greenhouses when severe decay occurs; it is a white irregular plate 1.5 to 12 mm ($^1/_6$ to $^1/_2$ in) thick and covered with distinct pores, sometimes with strands emerging from its margins. The decay damage to wood is similar to that caused by Dry rot but the cubing is smaller and less deep. When decayed wood is crumbled between the fingers it is not so powdery as that attacked by Dry rot but slightly more fibrous or gritty.

Stringy Oak rot, Donkioporia expansa

The Stringy Oak rot *Donkioporia expansa,* formerly known as *Phellinus megaloporus* or *cryptarum,* occurs in oak in conditions in which the Cellar or White Pore fungi are normally found on softwoods, such as in association with roof leaks or masonry dampened by soil moisture, and it is able to resist the relatively high temperatures that frequently occur in roof spaces. As this fungus prefers oak it is largely associated with older buildings constructed with this wood. It is a White rot, causing no distinct colour change in the decayed wood which becomes much softer with a longitudinal fibrous texture but does not powder in the same way as wood decayed by Dry rot or the other Brown rots. Yellow or brown mycelium is sometimes formed on the surface of the wood. The sporophore is a thick tough plate or bracket, fawn coloured but darkening as the spores develop.

Trametes versicolor

Trametes versicolor, formerly known as *Polystictus* or *Coriolus versicolor*, is the commonest cause of White rot in hardwoods, especially in ground contact, but it is usually confined to the sapwood in durable species. The sporophore is rarely seen but consists of a thin bracket of up to 75 mm (3 in) across, grey and brown on top with concentric hairy zones and a cream pore surface underneath from which the spores are released. Infected wood initially suffers white flecking and eventually appears to be lighter and much weaker than sound wood. Although White rots tend to be associated with hardwoods in buildings, this species and less common White rots are sometimes found causing decay in sapwood of softwoods, particularly in external painted joinery such as window and door frames.

Tapinella panuoides

Tapinella (Paxillus) panuoides causes decay similar to that caused by the Cellar rot but in much wetter conditions. The hyphae develop as fine branching strands which are yellowish, never becoming darker, and a rather fibrous and yellowish mycelium, perhaps with lilac tints, may occur. The wood is stained bright yellow in the early stages, but darkens to a deep reddish-brown and shallow cracking occurs. The sporophore has no distinct stalk but is attached at a particular point, tending to curl around the edges and eventually becoming rather irregular in shape. The branching gills on the upper surface radiate from the point of attachment. The colour is dingy yellow but darkens as the spores develop; the texture of the sporophore is soft and fleshy.

Dote

Dote consists of narrow pockets of incipient decay and is sometimes described as Pipe or Pocket rot. The rot in the pocket is brown and cuboidal, and hyphae like cotton wool may be present. If a pocket is absent, except for the appearance of a brown stain, Dote may be confirmed by brashness when the stained area is probed with the point of a knife. Dote is observed usually in softwood imported from North America and is, in fact, associated with heart rot in standing trees and deterioration in logs. The infection occurs in the forest but the fungus will continue to develop and the decay will spread if the wood remains damp, or becomes damp after a period of dryness. Dote sometimes accounts for pockets of decay in wood floors which are regularly washed; it was at one time encountered in large department stores where solid wood strip floors were kept clean by regular washing.

Diagnosis

It will be appreciated from this brief description of wood-destroying fungi encountered in buildings that they are invariably associated with a source of moisture. Visible dampness defects are not usually tolerated and repairs are carried out before fungal infection can develop, so that extensive fungal decay is usually confined to concealed areas where it may be difficult to detect. The Cellar rot is usually encountered in floor plates in direct contact with damp brickwork or masonry, often causing progressive deterioration and slow settlement of the floor. Doorposts and skirtings in contact with damp floors or walls are frequently affected. However, in basements that are subject to periodic flooding the conditions are usually too wet for this fungus and *Tapinella panuoides* may be found instead, particularly in the bases of basement doorposts and the bottom of the stair strings. If continuous dampness is replaced by periodic dampness in these conditions there is a danger that Dry rot will develop if suitable moisture contents arise, and once developed it will spread extensively behind skirtings and plasterwork and through walls.

"Ever fancied home-grown mushrooms?"

In the roof of a building periodic leaks into a relatively well-ventilated roof area will usually result in development of the White Pore fungus or, in oak timbers, the Stringy Oak rot. Dry rot is comparatively rare in roof areas except where unventilated conditions occur, but it is particularly common in suspended floors, panelling, shutter boxes, sash boxes and other parts of window frames which are affected by dampness from, for example, a faulty rainwater downpipe permitting excessive penetration through a solid external wall. Window joinery presents a special problem as movement cracks often occur in the paintwork, permitting rainwater to be absorbed and causing decay to develop; Brown rot may occur in softwood joinery with White rot in both hardwood components and occasionally in softwood sapwood. Another special situation is the boarding supporting a flat roof as condensation may occur which may be sufficient to cause decay. This situation is fairly common in the boarding under metal roofs in churches and other older buildings where incipient Wet rot attack is often sufficient to encourage the development of the Death Watch beetle. However, in less well-ventilated roofs, such as modern flat roofs in which ventilation is deliberately obstructed to improve thermal insulation, there is a danger of condensation which may be sufficient to cause staining on the ceilings beneath. In these conditions there is clearly a severe decay hazard which can be avoided only by designing to avoid condensation and by using reliably preserved wood during construction; condensation risks are discussed n Section 3.4 and in Appendix 3.

Remedial treatment

When fungal decay is encountered in buildings the first priority is to identify and repair the defect that has permitted dampness to occur. Structurally unsound wood should be replaced, and this replacement wood together with sound wood throughout the area at risk should be thoroughly sprayed with wood preservative as a precaution against the decay that will develop should dampness occur in the future. However, if Dry rot should be encountered, usually indicated by rhizomorphs or mycelium spreading across and through brickwork and masonry, it is also necessary to strip plasterwork to at least a metre beyond the last visible sign of growth and then to sterilise the entire exposed area by drilling holes about 16 mm ($^5/_8$ in) diameter at 250 mm (10 in) centres, the holes being downwards at about 45° and penetrating to about two-thirds of the thickness of the wall. The wall should then be sprayed with a suitable fungicide, the drillings being filled several times so that the fungicide can spread deeply within the wall. Dry rot may also be active in the centre of beams built into damp solid walls and, as internal decay in hardwood beams may be supporting Death Watch beetle infestation, it will be appreciated that the same techniques should be employed for drilling during inspection to determine the full extent of the damage and, if the timber is considered adequately sound, for injection to ensure deep treatment.

Fungicidal preservatives

The fungicidal compounds commonly used in wood preservatives are described in detail in *Wood Preservation* by the same author. Fungicidal remedial treatment wood preservatives are formulated and used in a similar way to the insecticidal formulations described in the previous section of this chapter. Thus organic solvent formulations are preferred, usually based on light petroleum solvents. Metal naphthenates were the only fungicides generally used before World War II. Some of these compounds, such as barium naphthenate, relied only on the fungicidal properties of the naphthenic acid which is liberated by slow hydrolysis following treatment. However, this naphthenic acid gives treated wood a characteristic musty odour which is also an indication that the acid is volatile and progressively lost, so that treatments with naphthenates of this type gave only a limited period of protection. However, zinc and copper naphthenates are far more efficient; the naphthenic acid is an excellent eradicant fungicide, partly through its volatility, whilst copper and zinc give excellent persistent preservative action. Unfortunately a low level of treatment resulting in copper or zinc retentions below the toxic threshold for the attacking fungus achieves no real advantage as these preservative components are then progressively detoxified by the fungus. The need for generous treatment to avoid this problem was not generally appreciated in the past and there were many examples of green copper naphthenate treatments where the detoxification process was clearly demonstrated by the loss of green colour as the fungus destroyed the wood. Whilst it is true that zinc and copper naphthenate are efficient when applied sufficiently generously, these past failures and the rather unpleasant musty odour encouraged the industry to search for alternatives. For many years the alternatives were completely different compounds but there have been continuing attempts to replace the naphthenic acid with more pleasant and more effective acids, the most recent development being a petroleum acid which is generally known as acypetacs, the fungicidal compounds being known as acypetacs zinc and copper respectively.

In the years following World War II pentachlorophenol became the preferred fungicidal preservative, although it had the disadvantage that it was very unpleasant to spray and, unless great care was taken in formulation, the fungicide tended to be deposited on the immediate surface of the wood where it was rapidly lost by volatilisation. Orthophenylphenol (*o*-hydroxydiphenyl) was used extensively in more expensive remedial treatment preservatives as an alternative to pentachlorophenol as it had only a relatively mild odour. However, between 1960 and 1980 both these fungicides were largely displaced, at least in the British Isles, by bis(tri-*n*-butyltin) oxide, also known more simply as tributyltin oxide or TBTO. This compound was introduced to the industry by the author, and experience over a period of 30 years has shown that it is generally more reliable than either the phenols or the naphthenates when properly used. In this connection it should be emphasised that it has a relatively narrow spectrum of activity and it was recommended from the start that it should always be employed in conjunction with a low concentration of an acid fungicide, such as one of the phenols or a borate, and where this original advice has been followed few difficulties have been encountered. One advantage of this compound is its ability to preserve against wood-borers as well as fungi, although in remedial treatment it is essential for formulations to have eradicant components in addition to such preservative compounds. Tributyltin oxide has been criticised in recent years on the grounds that it is lost by volatilisation following treatment and that the compound may decompose to ineffective forms. These dangers were well appreciated before tributyltin oxide was considered for this use but there was no evidence at that time that such losses of effectiveness were significant. The present situation appears to be that some formulations and some treatments are affected in this way, and recent investigations suggest that minor components in the formulations and even the moisture content of the treated timber may be important. Unfortunately these limitations and the need for special formulations have not been generally appreciated. Thus tributyltin oxide has become subject to increasingly stringent controls, largely imposed as a result of health and environmental problems encountered in other applications, and use of this compound has been largely discontinued in remedial treatments, although it continues in use in pretreatment preservatives.

Borate esters were also introduced to this industry by the author. Borates are largely water soluble but suitable esters can be incorporated in organic solvent formulations. The advantage of these esters lies in the fact that, if the wood is dry, they penetrate within the oil solution but, if the wood is wet, the esters hydrolyse and the resulting boric acid then diffuses slowly into the wet zone. For this reason borate esters are particularly suitable for injection treatment of window frames which may be mainly dry but may also have patches of dampness. One advantage of borate systems is their insecticidal action; whilst fungicides and insecticides can be formulated separately, it will be appreciated that a fungal infection frequntly encourages an insect infestation so that a combined insecticide and fungicide is advantageous, and a formulation can be most economically prepared if major components serve both functions. Borate esters have replaced most other fungicides in remedial treatments, although acypetacs zinc is still used in certain proprietary products; usually a synthetic pyrethroid such as Permethrin is included in these formulations as a low-toxicity eradicant insecticide.

Organic solvent or water?

Wood in buildings is frequently very dry and difficult to wet with water. Remedial timber treatments are therefore normally formulated in organic solvents as they achieve more reliable penetration, but there are a number of other advantages such as reduced staining when formulations are spilled on decorations, as explained in the previous section of this chapter, although water-based preservatives are being used to an increasing extent as part of a general policy to reduce the use of organic solvents. However, where Dry rot is found to be infecting masonry and brickwork it is generally an indication of a relatively high moisture content and, for this reason, fungicidal solutions in water are preferred. For many years the sodium salts of pentachlorophenol or orthophenylphenol were used, the latter compound being preferred where its higher cost was not considered a serious problem as it is much more pleasant to use. Borates can be used but both boric acid and sodium tetraborate (borax) are insufficiently soluble. However, much higher concentrations can be achieved if a solution is prepared using one part boric acid to 1.54 parts sodium tetraborate decahydrate; this formulation corresponds approximately to disodium octaborate and can be purchased as the proprietary product Polybor. This particular formulation has a very high active content and is therefore very suitable for this purpose, but it should be appreciated that high borate concentrations are required in remedial treatments as there is a tendency for calcium in mortar and limestones to reduce the effectiveness. Borates are not widely used at the present time for this purpose but their low mammalian toxicity and freedom from odour will probably ensure that they are the most popular formulations in the future. Some new zinc borate formulations are now available which are more effective in terms of activity against the Dry rot fungus *Serpula lacrimans* and resistance to leaching in wet conditions.

Quaternary ammonium compounds are widely used as components in disinfectants but they are also effective against wood-destroying fungi and are now widely used for wall sterilisation treatments. They can also be used as solubilising agents to enable tributyltin oxide to be used in water, and this latter combination has been extensively used for wall sterilisation, although there are now increasing restrictions on the use of tributyltin oxide. These formulations are also particularly suitable for the treatment of external masonry surfaces against lichens, mosses and algae as described in Chapter 4.

Conventional fungicidal remedial wood preservation thus consists of remedying defects responsible for dampness and removing decayed wood, then treating the remaining sound wood and replacement wood by spray with a preservative, as well as stripping plaster to well beyond the last sign of any Dry rot infection and carrying out wall injection and spray treatment with a water-based fungicide. Obviously there are some situations where decayed wood must be retained for structural or aesthetic reasons. Where large beams or plates are involved, or the affected wood is concealed by a finish as in window and door joinery, it is essential to inject with preservative to ensure deep penetration throughout the infected areas. It has already been mentioned that, if it is suspected that the wood is damp, as is often the case in external joinery, borate ester treatment is particularly suitable. It can be added that there are a number of other situations in which injection treatments are desirable or essential, one example being the log buildings and stave churches found in Norway. Deep penetration can also be achieved in other ways, such as with the bodied-mayonnaise-type emulsion systems, but these are not generally as reliable as injection and their use has the distinct disadvantage that the condition of the wood concerned remains completely unchecked whereas drilling for injection provides an automatic inspection system. The advantages and disadvantages of these various systems have already been explained in detail in the first part of this chapter.

There are sometimes situations in which none of these treatment systems seem suitable, such as the treatment of Inverary Castle. The Castle was severely damaged by fire one Guy Fawkes night and an enormous volume of water used in the fire-fighting operations. The lower rooms were not damaged by the fire but were affected by the water and there was a severe danger of Dry rot development during the slow drying period. Each floor consisted of massive joists with sound deadening of sawdust, shavings or sometimes ashes supported on trays between the joists. In most cases a false ceiling was provided below each floor on separate lighter framing. Most of the timber and deadening became saturated with water and, in such poorly ventilated conditions, there was a severe risk of decay. The floorboards, deadening and supporting boards were removed as soon as possible to permit ventilation and rapid drying. However, it was found that the plasterwork of the false ceilings was continued as 'false' walls; the walls were plastered on laths supported on battens, providing narrow vertical channels down the walls which had also become saturated by the fire-fighting water. In some places the situation was even more serious as some of the internal walls were also framed with timber beneath the battening. The correct method of treatment would have been the removal of all plasterwork in the staterooms and proper treatment of the concealed wood but, as the plasterwork in the staterooms is also one of the most valuable historic features of the Castle, it was considered necessary to attempt the treatment of the wood *in-situ* without disturbing the plaster. This was achieved by removing the skirtings to provide access to the bottom of each of the ducts formed by the battens. A fogging machine was then used to blow fungicidal fog up each duct until the fungicide was detected at ceiling level. A borate formulation was used which was preferentially absorbed into the damp wood in which the decay hazard was most severe and treatment most necessary.

Thorough inspection

Whilst fungicidal remedial wood preservation treatments are well developed and reliable if conscientiously applied, they are clearly useless if the inspection has failed to detect all the areas that require treatment. It is essential that an inspection must be thorough and all possible areas at risk fully investigated. On the other hand, care must be taken in making an inspection to ensure the safety of the surveyor. In many cases decay is suspected or has been detected, but in some areas severe decay may not be readily apparent. The surveyor may be moving in areas where traffic is not normally encountered such as roof spaces where parapet gutter leaks may have decayed joist ends. The surveyor must always be aware that a structure may be unsound and should never take unnecessary risks.

Fungal decay in building timbers is as old as buildings themselves and even the properties of Dry rot were fully understood several thousand years ago. The Latin name for Dry rot is *Serpula lacrymans*, the word *lacrymans* being derived from the Latin word *lacrima*, a tear, as Dry rot is the weeping fungus and fresh growth can often be observed covered with drops of water.

Fretting leprosy

It is this weeping which enables it to be identified as the 'fretting leprosy of the house' described in the Old Testament book of Leviticus. Until comparatively recent times the priest was the person who was consulted on any kind of trouble or pestilence and consequently Leviticus contains full instructions on how to deal with the plague of Dry rot or leprosy of the house. The priest, when carrying out his inspection, was to look for 'hollow strakes, greenish or reddish, on the walls' (i.e. rhizomorphs). If they were present the house was to be shut up for seven days and if after that the plague had spread he was to 'command that they take away the stones in which the plague is, and . . . cast them into an unclean place without the city . . . the house to be scraped within round about, and . . . pour out the dust . . . without the city into an unclean place'. This may appear rather ruthless but even today an affected area must be stripped to the bare masonry to ensure the successful application of fungicide. When Leviticus was written there was no such fungicide. If after this treatment, 'the plague be spread in the house, it is a fretting leprosy in the house' (i.e. Dry rot as opposed to Wet rot) and there was only one solution, to 'break down the house, the stones of it, and the timber thereof, and all the mortar of the house; and . . . carry them forth out of the city into an unclean place'. Drastic treatment indeed, but anyone who has had serious Dry rot will know how determined it can be. It is in these verses of Leviticus, which also describe leprosy in man, that we may read of early ideas of contagion. People entering the infected house were required to wash themselves and their garments thoroughly on leaving. If the priest found that the fungus had not developed after replastering the affected area he was to apply final 'fungicidal' treatment: 'He shall take to cleanse the house two birds, and cedar wood, and scarlet, and hyssop.' He was instructed to sacrifice one bird and sprinkle the house several times with the blood. The other bird, after being dipped in the blood, was freed and flew away, presumably taking the infection with it.

Wet rot and particularly Dry rot became an increasing problem as buildings and wooden ships became more sophisticated and more valuable. In 1784 the Royal Society of Arts offered a gold medal 'for the discovery of the various causes of Dry rot in timber and the certain method of its prevention'. It was awarded in 1794 to Mr Batson who, in 1788, had treated an outbreak of Dry rot in a house by removing sub-floor soil and replacing it with 'anchor-smith's ashes'. Whilst it is possible that the sulphurous fumes from the ashes may have had some fungicidal action, it is probable that the removal of the sub-floor soil improved ventilation and perhaps improved resistance to moisture absorption.

Duty of care

Great care is always necessary when inspecting buildings for timber defects; inspections are often requested if defects are known or suspected, but in other cases inspections are required to assess an unknown situation. A rather old public building was in continuous use as offices and it was not suspected that some of the supporting beams were on the point of collapse until damage was disclosed during a routine inspection. Perhaps the worst danger is to approach an inspection with a preconceived notion of what may be found. For example, remedial treatment companies are frequently asked to inspect properties and prepare estimates for 'necessary woodworm treatent' following a valuation survey made in connection with a mortgage application. This type of instruction tends to be automatic in the case of building societies, largely as a precaution to protect the professional reputation of the surveyor. The most prudent surveyor will, however, request an inspection for 'timber defects' but, whatever the detailed instructions, the timber decay inspector is offering specialist inspection services and has a duty of care to detail all defects that require attention.

In one such case a specialist remedial treatment contractor carried out an inspection prior to purchase, noting scattered woodworm but failing to detect Dry rot deterioration in a ground floor room, despite the fact that the skirtings and floorboards were distorted and weakened so that the area of infection was readily detectable. The report was accepted by the owner of the property and treatment against woodworm was carried out; the operatives, meant to be specialists in this field, again failed to report the extensive decay, even though they should have lifted floorboards as part of the woodworm treatment. The owner of the property had observed the defect and, as it had not been remedied, complained to the specialist contractors, who stated that they had only carried out an inspection and treatment for woodworm. It must always be appreciated that a layman may not understand the difference between 'woodworm' and 'decay', and this defence for negligence must be considered totally unreasonable.

In another case a specialist remedial treatment company had been employed to inspect, report on and estimate for treatment of timber defects in a property. Its recommendations were accepted and the treatment was carried out. Shortly afterwards the property was sold and the purchaser instructed a surveyor from the author's consultancy to check that the treatment had been carried out in a proper and competent manner. It was found that, whilst preservative had been applied reasonably generously, there had been no attempt to clean parts of the property before treatment and in places this had caused a failure of the preservative to penetrate adequately into the wood. However, the surveyor also discovered that part of the ground floor of this old property was laid direct onto the soil with no protection against absorption of dampness, and extensive Wet rot decay had already occurred which had not been mentioned in the original specialist report. Surprisingly the owner challenged the invoice for the inspection, stating that the purpose was simply to check whether Death Watch beetle treatment had been reliably carried out and that the inspection should not therefore have extended to cover the other parts of the property affected by Wet rot. It is certain that, if the author's surveyor had neglected to report this additional decay but it had been detected by some other person at a later date, the owner would have been equally energetic in pursuing a claim for negligence! Happily such clients are rare but it is unfortunately true that surveyors must carefully guard against unfair and dishonest clients, just as clients must guard against unfair and dishonest surveyors.

Many housing associations have been formed in recent years to enable government funds to be employed for the purchase and renovation of substandard properties. Quite often considerable numbers of properties are purchased but, whilst it is theoretically necessary for each to be inspected and a plan for renovation prepared, it will be appreciated that a fairly uniform approach is normally applied. Whilst competitive tenders are no doubt desirable this uniform approach appears to extend in some cases to the employment of preferred specialist contractors for tasks such as remedial damp-proofing and timber treatment. An investigation of one of these housing schemes disclosed that about 12% of the completed properties could not be occupied because of failure of the specialist treatments. In many cases these failures involved inadequate inspection and treatment of Dry rot which then extended throughout the completed area, causing complete destruction of floors and internal joinery in some instances, accompanied by a uniform covering of rust-red spores. This high incidence of failures clearly demonstrates the need to employ only experienced surveyors to inspect properties and experienced operatives to supervise treatment; the large number of properties requiring treatment apparently caused the firms concerned to expand their staffs and, in such a competitive situation, it was necessary to devote as little effort and cost as possible to both inspections and treatments.

An estimate submitted by a specialist contractor for the remedial treatment of Dry rot in a Scottish castle was beyond the financial means of the owner who therefore applied for grant aid. The castle was a fine example of its period and contained particularly valuable plaster ceilings, and it was therefore decided to carry out a more extensive investigation, surveyors from the author's practice being instructed to assess the Dry rot situation. Their inspection confirmed that there was current Dry rot activity that required prompt remedial action but much of the estimate prepared by the specialist contractor was concerned with the treatment of very extensive but very old Dry rot infection in the roof area. The damage was largely confined to the feet of the rafters and the plates, but it was noted that the damage had been repaired by fixing spurs to the rafters. Although the area was lined internally with lath and plaster on battens, these linings were not affected by the attack and it was therefore suspected that the infection may have been old and inactive for many years. A study of historical records showed that the roof had been constructed originally with a parapet gutter which had leaked and caused the damage to the rafter feet and plates but, about 130 years before, the parapet had been removed and the spurs fitted to the rafters to enable the roof slope to overshoot the wall in the form of conventional eaves. The internal linings had been provided as part of the complete restoration carried out at that time and it was therefore evident that the decay had not been active for many years. Whilst it is always a wise precaution to treat an old inactive infection as there is clearly a danger that it may become reactivated in the future, treatment is clearly not essential if the budget is limited and the funds are required more urgently elsewhere. As the top floor had been abandoned for many years and was unlikely to be required in the forseeable future, it was recommended that the linings should be stripped out to provide improved ventilation and to enable the area to be easily inspected at regular intervals as a precaution. The savings achieved in this way were enormous, reducing the cost of the treatment to about one-third of the original estimate, a figure that was realistic and acceptable to the body providing the grant aid.

2.3 Fire

It is well known that wood is combustible and the natural reaction to a fire disaster in a building is to demand that only non-combustible material be used in construction in the mistaken belief that this will ensure fire safety. In fact, the first-class reputation of non-combustible structural materials originates from observations on the behaviour of, for example, traditional masonry, but the same excellent fire performance is not necessarily achieved by other non-combustible materials. In a building, fire is usually concerned principally with the contents, and the contribution from wood structural components is limited owing to their size and position within the structure. Even in a completely wood-framed house there is a total of only one-third more wood than in traditional mixed construction. It can therefore be appreciated that the combustibility of structural members is of limited importance compared with their ability to withstand fire originating in furnishings and other building contents.

The rate of combustion is particularly important. Combustion cannot occur in the absence of oxygen so that large-section wood components burn only slowly at their surface which then progressively chars. As the temperature rises, first the wood releases the volatile components which flame on the surface, and then the residue chars as the temperature increases. The thermal conductivity of wood is low, the same order as for cork, lightweight gypsum plaster and other insulation materials. Indeed balsa, a very low-density wood, is used as an insulation material. The natural thermal insulating properties of the wood therefore limit the rate at which heat can be transferred inwards from the burning external surface and these insulating properties improve steadily as moisture is lost from the wood and charring progresses; charcoal is an even better insulator than wood. Eventually the heat transfer from the surface is insufficient to release volatile components from the interior and surface flaming ceases. The charring rate also progressively slows, until heat is contributed only from surrounding burning materials; the significance of the contents of the structure is again apparent.

Charring

The charring rate is considered to be the rate of loss of dimensions, whilst the burn-through rate is considered to be the weight loss that occurs. In large structural components the rate of charring is important as the rate of change of dimension naturally controls the ability of components to continue to support the structural load. The rate of charring varies with wood properties such as thermal conductivity and density which must be considered in the construction of a wooden barrier to achieve fire resistance, although detailed design is of even greater significance; for example, there must be no gaps around doors or windows through which fire can penetrate, or thin areas where early penetration can occur. Despite combustibility wood possesses excellent fire resistance, largely because of its low thermal conductivity, and this is not greatly affected by any form of treatment.

Ignition

The ignition point, usually about 270°C (518°F) for wood, is the temperature at which the rate of heating exceeds the rate of supply of heat, and the fire becomes self-supporting with perhaps visible flame or glow. There is usually no ignition, even on the superficial surface, if the moisture content of deeper areas of wood remains above about 15%. A pilot source of flame is necessary to cause ignition as spontaneous ignition of wood is possible only at exceptionally high temperatures. Large pieces of wood do not ignite easily, and ignition of smaller pieces is achieved only because of the rapid rate at which they reach the ignition temperature.

Flame spread

Flame spread across the surface of a piece of wood is really a series of ignitions, one area on fire acting as pilot ignition for the adjacent area. Thus flame spread is influenced by moisture content, as is normal ignition, but it also depends on the density and chemical nature of the wood. Chemical treatments can achieve considerable success in resisting ignition and thus preventing flame spread. Indeed, the propagation of a fire by flame spread is the most serious criticism of the use of wood as a structural material, yet this is the one property that can be readily modified by comparatively inexpensive treatment.

Smoke

Smoke generation represents the most serious hazard to human life during fires in buildings. The most harmful smokes arise from the plastics and synthetic fibres which are often used in furnishings, whilst the smoke from wood is comparatively innocuous. It is thus important to ensure that dangerous smokes are not released by any chemical treatments applied to wood in attempts to limit flame spread and fire propagation.

Despite its combustibility wood thus possesses distinct advantages when used as a structural material. The contribution of the structural woodwork to a fire is minimal, at least in the early stages; wood components possess excellent resistance to fire penetration and suffer neither significant distortion nor rapid loss of strength. In comparison steel collapses when it reaches its yield temperature whilst concrete shatters or spalls, particularly if it encloses a steel frame. Where reinforcement bars are present in concrete they may be stressed at high tension and a rise in temperature will result in yielding with distortion of the structure, even if complete collapse is avoided. Even if a fire is controlled before the yield temperatures of steel are reached there is still a danger that considerable expansion will rupture the building structure.

Fire resistance

It will be appreciated from these remarks that fire resistance is really a matter of providing an adequate thickness of wood coupled with freedom from gaps through which fire can penetrate. In new buildings these are problems of design and workmanship during construction, and are, in fact, requirements under the Building Regulations. As far as existing buildings are concerned, they may also be subject to fire precautions, particularly hotels, where it is necessary to ensure the control of fire spread and the provision of adequate means of escape. These are really problems for a fire protection engineer, the control of fire spread being achieved largely by the provision of barriers at intervals to compartmentalise the building, perhaps by using fire doors that close automatically when a fire is detected, coupled with sprinkler systems that may be able to control the fire in the affected area. Unfortunately fire protection engineers usually have little architectural appreciation and many fine buildings have been entirely ruined by insensitive fire protection work.

Fire retardants

Whilst the provision of satisfactory fire resistance is principally a matter of design, charring rate varies with properties such as density, and wood selection is therefore a significant factor, although charring rate cannot be greatly influenced by treatment. In contrast, fire-retardant treatments can be highly effective in reducing surface spread of flame. An efficient fire retardant acts in several different ways. When the combustible surface is exposed to fire the treatment should prevent flaming and preferably should reduce the rate of degrade charring. It should reduce ignitability and prevent after-glow, the sustained combustion when heat is removed. If it achieves these requirements it must also be inexpensive, readily applied and permanent, and have no unacceptable defects such as excessive toxicity.

There are two principal fire-retardant systems in use at the present time. Intumescent coatings act on exposure to high temperatures by foaming into an insulating layer, thus reducing the rate of temperature rise of the protected wood. This foaming action, which must be completed below wood charring temperature, generally follows a sequence consisting of film softening, bubble formation and ultimate setting into a rigid foam with good adhesion to the surface. Ideally the bubble should contain inert or preferably fire-retardant gases. Intumescent coatings are really paints and can, of course, achieve a decorative function. They can also be applied as a remedial treatment to an existing structure. The fire resistance of partitions often depends on the ability of a door to seal a doorway, and one solution to this problem is to insert a special strip in the edges of the doors or the door frames. This strip is composed of an intumescent material which swells when subjected to fire and thus seals the gaps. Non-intumescent fire-retardant coatings are also available. In principle they consist of heavy insulating coatings containing components designed to inhibit flaming, perhaps by the generation of fire-retardant gases when exposed to heat.

Plate 2.8 Interdens intumescent strip rebated into a door edge and adjacent frame in order to improve fire resistance by sealing the gap should fire occur (*Dufaylite Developments Ltd*)

Impregnation treatments are widely used as alternatives to these coating systems. In principle these impregnation treatments are similar in action to coatings as they are designed to inhibit flaming as well as insulating the wood. Whilst their ability to inhibit flaming is as good as or better than that of coating systems, the insulating action is achieved by sacrificing the surface layers of wood, allowing them to char to a limited extent to provide the insulation required. Unlike the surface coatings which have no further effect once the surface has been physically destroyed, the deep impregnation treatments continue to control the rate of charring and prevent flaming and after-glow, even after prolonged exposure to severe fire conditions. However, these properties can be achieved only provided that sufficient loadings of fire retardants are achieved to an adequate depth; this is a virtually impossible requirement for a remedial treatment, but wood impregnated with fire retardant can be used in remedial fire protection engineering.

It will be appreciated that the application of intumescent coatings and the installation of intumescent strips around doors are the only realistic fire protection treatments that can be applied by remedial treatment companies. As these processes can represent only a small part of any fire protection scheme it must be emphasised that remedial treatment companies should never offer fire protection or even flame-retardant treatments, but should instead co-operate with fire protection architects and engineers if this is a service that they would like to offer.

Smoke and water damage

In passing it can be mentioned that there is another far more extensive remedial treatment business associated with fires. In many cases the part of a building affected by a fire is comparatively limited as the fire is usually extinguished provided it is reported in time. However, a much greater part of the building may be affected by smoke or water from fire-fighting. The fire-fighting water introduces a serious danger of fungal decay, particularly of Dry rot attack, and this is clearly an area of suitable activity for remedial treatment companies. In principle the affected area must be opened up as rapidly as possible to permit drying, followed by precautionary treatment; an example of such treatment is given in the previous section of this chapter. Treatment of a building and its contents following smoke damage is, of course, a rather different activity which cannot be generally recommended to remedial treatment companies. However, it is of interest to note that such work is often associated with remedial treatment in some other countries such as Norway where remedial timber treatment is often carried out in connection with timber deterioration insurance schemes, the insurance company also offering fire insurance. Restoration following fire damage and remedial treatment following borer or fungal attack are often both performed by the same company.

2.4 Other wood problems

Painted joinery

Reference has already been made to the use of injection methods for remedial treatment of external window and door joinery. As far as rain is concerned it is normal to rely on paint or varnish to protect these exposed wood surfaces. Paint manufacturers often claim that paint functions as a wood preservative. In fact, fluctuations in atmospheric temperature, particularly differences in temperature between the interior and exterior of a building, cause redistribution of the water within the wood by evaporation and condensation so that this water becomes concentrated immediately beneath a cooler paint or varnish coat. Water introduced as rain penetration through a small damaged area of paint or absorbed through contact between an inadequately painted surface and adjacent porous brickwork or masonry will contribute to the moisture within the wood until eventually the moisture content reaches a level at which an appropriate fungal attack can develop.

Whilst it is true that an intact film prevents rain penetration, it is equally true that paint traps moisture within the wood. Random surveys carried out in England have shown that after a few years most window frames have moisture contents of perhaps 20% to 30% around the joints with the sills, and there is a very real danger of decay. This dampness can usually be attributed to the failure of the paint at joints where it is unable to tolerate the differential movement between the side-grain and end-grain. A crack develops in this way, allowing water to penetrate, and the remaining paint inhibits evaporation so that moisture accumulates and decay ultimately occurs. Coating systems may thus actively encourage decay and in some countries it has become normal practice in recent years to require all window and door frames to be adequately preserved. Although preservation prevents deterioration of the wood, it is not the complete answer to the problem; water accumulations still occur and cause adhesion damage between the surface coating and the wood, a failure known as preferential wetting. This problem can be avoided only by adopting a primer system that will permanently bond with the wood even in the presence of moisture, a bond that is lacking in all normal commercial systems. The use of a water-repellent treatment prior to painting can be helpful but only if it is able to establish this bond. The fact remains that a considerable amount of external joinery becomes infected by decay and remedial treatment is therefore necessary; the use of injection methods with borate ester formulations has already been described previously and will not be repeated here.

Staining fungi

Moisture accumulations under paint or varnish coatings can also cause the development of stain fungi. One of the most important staining organisms is

Plate 2.9 Stain on western red cedar cladding is confined to areas exposed to rainfall and is thus less severe under a window-sill (*Penarth Research International Ltd*)

Aureobasidium (Pullularia) pullulans, which may develop in this way but can then attack the surface coating medium, eventually causing small unsightly black pustules to form on the coating surface. The formation of these pustules is associated with the development of a hole through the coating which allows further rainwater penetration, perhaps resulting eventually in the preferential wetting failure that has been previously described. However, this species is also able to develop on the surface of the coating, boring through it into the wood beneath.

This is a relatively new problem, at least in decorative paint systems in the British Isles, apparently because it was usual in the past for primers and undercoats to contain lead pigments which were resistant to this form of damage. Attempts have been made to treat wood and also to add toxicants to priming oil systems but generally these have been relatively ineffective, apparently owing to their limited persistence. It is probable that toxicants or toxic pigments must also be added to the paint system if comprehensive control of stain development is to be achieved. These are all problems that must be solved by modification of the decorative systems for both new and maintenance work, and are thus of limited interest in remedial treatment. However, these problems will certainly be encountered during surveys, particularly where joinery defects are involved, and the remedial treatment specialist should therefore be aware of them.

Decorative preservatives

Surface coatings such as paint and varnish are not essential to the decorative use of wood. Bare wood can be very attractive and perfectly satisfactory in service provided a species is selected which possesses natural durability to give resistance to decay and low movement to give resistance to distortion and splitting. Alternatively another species might be used if appropriate preservation treatment is applied to achieve these requirements. However, the wood will suffer from loss of colour, principally through leaching of extractives and the development of surface staining fungi, these factors generally combining to give the grey shades that develop when wood is naturally weathered. This natural coloration tends to be patchy, largely through the degree of exposure to rainfall and the local extent of leaching and staining. This patchiness can be avoided for the most part by the use of a water-repellent treatment containing a toxicant that is effective against staining fungi, but this treatment must be resistant to weathering if it is to achieve a reasonable life. There are few components that can achieve the desired resistance to weathering and it is more normal to use a formulation which also contains pigments and binders which give the desired persistent coloration but leave the natural grain of the wood visible. Whilst the pigment and coating build-up on the surface of the wood is extremely low compared with that of conventional coating systems, a high degree of permanence is achieved through the deep penetration of the system.

Although these systems were originally designed as a means for introducing an artificial pigment system to maintain the 'natural' colour of the wood, a complete range of alternative colours is now available and the systems have become generally known as architectural finishes. As with normal surface coatings, the life of these decorative treatments depends largely on their resistance to preferential wetting, achieved either by using media which establish permanent bonding to the wood or by deep penetration. Colour retention depends on the stability of the pigments but also on the presence of toxicants which will resist stain development. These systems are now extensively employed for maintaining cladding; in the British Isles Western Red Cedar is usually used for cladding and is maintained using brown pigmented formulations, but in other countries, particularly Benelux and Scandinavia, a much wider range of colours is employed.

Cladding maintenance

Western Red Cedar cladding is sometime used unprotected but the attractive warm brown coloration is soon lost and replaced by a grey coloration which progressively darkens, except where it is sheltered and where a patchy brown coloration may persist. Remedial treatment is often necessary to achieve more attractive appearance. Generally normal pigmented systems are applied but they are not very satisfactory as they usually produce a rather dark and patchy appearance. It is far better to prepare the surface first using a fungicidal cleaning agent, such as a quaternary ammonium system; a 2.5% solution (i.e. a 5% solution of the 50% concentrates that are usually supplied) of a suitable quaternary ammonium compound, such as benzalkonium chloride, can be prepared in water and sprayed on the surface. After the treatment has been left for several hours, perhaps the time taken to spray round a large building, a more dilute solution can be used to scrub the surface clean. This cleaning will be found to be far easier than the dry wire brushing or scraping that is normally employed, but care should be taken to ensure that the compound is not unnecessarily absorbed into the skin by using gloves whilst scrubbing. The treated surface must be washed with clean water and then allowed to dry thoroughly for several weeks before a normal pigmented system is applied. The same treatment can be used on cladding previously treated with pigmented systems and again the maintenance will be found to be considerably simplified and the final appearance greatly improved. However, a word of warning is necessary at this point. Whilst most proprietary pigment systems contain fungicides, they are usually compounds such as pentachlorophenol or tributyltin oxide which are added because they are well known to be effective wood preservative fungicides. Unfortunately they have little effect against the staining fungi that cause the progressive darkening of decorative treatments such as *Aureobasidium pullulans*, and care must therefore be taken to use only formulations that have these necessary properties. The design of such formulations is discussed in more detail in *Wood Preservation* by the same author.

Movement defects

Changes in the moisture content of wood in buildings can cause a number of defects. Wood fibre is hygroscopic and will absorb water vapour from the atmosphere; the moisture content depends on the atmospheric relative humidity, a saturated atmosphere corresponding to fibre saturation point within the wood. However, further liquid water can be absorbed into the pores within the wood, increasing the moisture content above this fibre saturation point. Any increase in the moisture content of wood will tend to make it more susceptible to fungal decay, as described earlier in this chapter, but changes in moisture content below fibre saturation point invariably involve movement, that is shrinkage with drying and swelling with wetting.

Although it is normal to dry wood to a moisture content equivalent to the average atmospheric relative humidity anticipated in use, it is common to encounter movement problems. Faults such as gaps appearing between floor blocks or boards are due to the wood drying after installation, either through inadequate kilning or perhaps through rewetting between kilning and installation, whereas outwards displacement of floor blocks or boards, a less common problem, is due to very dry wood absorbing moisture after installation in a building. A door or drawer jammed in humid weather may be exceedingly slack under drier conditions. Frames which introduce an end-grain surface in contact with side grain will inevitably result in cracking of any surface coating system as the cross-sectional movement is so much greater than the longitudinal movement; this is the defect that causes splitting of paint along joints on external joinery such as window frames and which causes the internal decay that has been previously described. In other situations the cross-sectional movement may become apparent as warping through twisted grain effects. The obvious solution to all these problems is to use only wood with natural low movement but this is not always realistic. Indeed, remedial treatment is also entirely unrealistic, except by the replacement of the components concerned with wood of lower movement, and the only reason for describing these defects here is to explain their cause, should they be encountered in buildings during remedial treatment inspections; again these problems are discussed more fully in *Wood Preservation* by the same author.

3 Damp-proofing

3.1 Causes of dampness

Damp-proofing is certainly the most challenging branch of remedial treatment in buildings as diagnosis of the cause of dampness is particularly difficult and treatments are often unreliable. There is a very high level of complaints, compared with other remedial treatments, and even long-established and thoroughly conscientious firms admit that treatments are sometimes unsuccessful. The property owner must therefore take particular care in selecting a specialist contractor; some are unable to reliably diagnose causes of dampness and some offer inadequate treatment processes, perhaps advertising themselves as rising damp specialists and offering only damp-proof courses, whatever the actual cause of dampness.

Dampness

Dampness normally causes a material to darken and staining may develop, perhaps due to movement of staining substances within the dampness or through biological activity encouraged by the dampness; wood decay resulting from fungal activity has already been considered in Chapter 2, whilst surface moulds and algae on masonry will be described in Chapter 4. Such biological activity frequently results in a characteristic musty or mushroomy odour. Salts that may be responsible for staining may also be hygroscopic, absorbing moisture from the atmosphere and preventing drying, except on the rare occasions when the air has a particularly low humidity. The evaporation of water from a damp surface also results in cooling, as explained in Chapter 5, but cool surfaces also encourage condensation dampness.

When inspecting a property it is not therefore sufficient to observe with the eyes; it is necessary to take equal account of smell and temperature. If dampness is not apparent to the senses in these respects it can be considered insignificant with no justification for treatment. At the present time there are some firms who advertise 'free damp surveys without obligation', thus encouraging householders to have the state of their properties investigated as a precaution, even in the absence of any obvious signs of dampness. Unfortunately dampness from one cause or another is bound to be detected by sensitive instruments and, whilst the dampness may be insignificant or even normal, treatment may be carried out quite unnecessarily. Conversely, recent decoration work can conceal dampness which will later become apparent, and these examples alone are sufficient to demonstrate the considerable care that is necessary in assessing dampness during inspections.

There are four principal sources of dampness in buildings: rainwater

"I don't care what *you* say –
I say it's damp."

descending through roof defects, rainwater penetrating through the walls, rising dampness absorbed from the soil, and condensation. Precautions are normally taken during design and construction to ensure freedom from these causes of dampness, and it would therefore appear to be relatively simple to detect a source of dampness and the defect responsible, and thus to recommend and carry out appropriate repairs. In fact, errors in diagnosis frequently occur, even if dampness is derived from these simple sources, but there are also other sources of dampness that must be considered.

Water spillage and leaks

Water spillage can occur in a variety of ways. Roof penetration has already been mentioned but rainwater disposal systems such as valley gutters and fallpipes may be overflowing or leaking. Waste-water disposal systems from toilets, baths, basins and sinks may be defective; in fact the entire plumbing system of the building must always be suspected. Defective packings in taps and valves are a frequent source of dampness, particularly radiator and other valves involved in central heating systems. To a certain extent these various leaks from plumbing systems are to be expected but there are also other less obvious defects that sometimes occur. The service pipes concealed within walls or under floors can sometimes freeze and split, often causing completely unexplained dampness. In this connection it is perhaps worth mentioning that compression joints frequently start to leak after freezing, even if the pipes remain intact, and such joints should be avoided whenever possible. Another unexpected source of dampness can arise through nails penetrating service and heating pipes. These pipes usually run under floorboards, and PVC or copper pipes can be easily penetrated by nails when floorboards are refixed. Plumbers are themselves often responsible for this damage. The nail typically blocks the hole that is caused and, as a result, a distinct leak does not always occur but simply an oozing of water that may remain entirely undetected in sub-floor spaces, perhaps eventually resulting in dampness elsewhere or development of timber decay as the nail corrodes and the leak becomes progressively more severe, spraying the surrounding timber. Damage following dampness resulting from bath splashes and fire-fighting has been described in previous chapters; whilst such sources of dampness might appear to be obvious, they are easily missed during an inspection.

Condensation

Condensation is a source of dampness that causes considerable difficulty in diagnosis. Normal condensation results from the occurrence of a high atmospheric relative humidity followed by cooling. The kitchen, laundry and bathroom are the normal sources of humidity but it must also be appreciated that every breath we breathe is charged with moisture, and humidity from normal life processes can be a significant cause of condensation in properties with limited ventilation. Condensation typically takes two forms: dew on cold surfaces and mist through cooling of the atmosphere. If a bathroom is unheated the use of the bath will usually result in dew on all surfaces but, even if it is heated, dew may still occur on windows if the external temperature is low and the window insulation is poor. Mist may also occur in the bathroom air, simply because it remains cold and cools the warmer humid air rising from the hot water. If the door is open the humid air may pass into other parts of the building, increasing the atmospheric relative humidity and thus increasing the tendency to condensation. Condensation often occurs in this way in colder corridor areas adjacent to bathrooms, kitchens and laundries, but humidity generated in kitchens often results in almost continuous condensation in adjoining cold larders and perhaps the development of mould on surfaces which, in addition to being unsightly, may encourage infestations of small plaster beetles.

It must also be appreciated that condensation may be occurring invisibly. For example, interstitial condensation occurs when humid air diffuses into the porous building structure, condensing as it approaches a cold exterior surface and perhaps accumulating to a sufficient extent to cause appreciable dampness. One interesting point is the use of a vapour barrier within a wall, not perhaps in brick, concrete or masonry, but certainly in wood construction. This vapour barrier prevents diffusion to the exterior and can prevent interstitial condensation if it is positioned at the warm internal surface, but if it is at the cold external surface it causes condensation to accumulate as it cannot evaporate to the exterior. A less obvious source of condensation occurs in air-conditioned buildings during the summer months when high external temperatures may result in condensation. One obvious example is a flat roof with an impervious metal, felt or hot bitumen covering; any source of high relative humidity may result in condensation beneath the impervious covering. A less obvious source of condensation occurs in air-conditioned buildings during the summer months when high external temperatures may result in condensation through contact with the cool interior of the property, and if the condensation occurs in contact with wood there is a severe danger of fungal decay; this is a serious problem in tropical areas but equally a serious problem in temperate areas with refrigerated stores.

There is one final source of condensation which causes perhaps more trouble in diagnosis than any other single source of dampness: condensation through the presence of hygroscopic salts. These salt deposits may be introduced in the original construction materials, such as the use of beach sand contaminated with sea salt or bricks containing sodium sulphate. Similar salts may be introduced in dampness, particularly dampness rising from the soil. Finally, hygroscopic salts can be generated in buildings by the reaction of atmospheric pollution or flue gases with normal building materials. The difficulties arise through the need to identify both the presence of these salts and their sources.

Roof defects

It is usually easy to discover that dampness is originating from the roof but often not so easy to explain the cause. In common with most defects, roof faults can be attributed to errors in design, construction, selection of materials or neglect of maintenance. Flat or shallow pitch impervious systems are particularly at risk, not roofs alone but also valley and parapet gutters. Whatever their construction, there is serious danger of physical damage caused by work on the surface, such as damage caused by walking, erection of ladders or scaffolding, or the dropping of building materials. Even in the absence of such accidental damage, physical defects may still develop through stresses caused by movement of the building, perhaps settlement following construction or seasonal thermal changes. Movement in a steel framework may introduce thermal stresses but metal roof coverings may damage themselves if thermal movements are ignored in design or construction; each section should not exceed 2.4 m (8 ft) in length, longer sections frequently leading to damage, particularly in lead coverings.

Metal roofs are also particularly susceptible to acid damage, either through atmospheric pollution gases absorbed in rainwater or from oxalic and other biological acids which arise when moss and lichen growth is present on, for example, tile roofs that discharge onto metal-covered flat roofs or valley gutters. This biological acid attack is extremely serious and can cause major leaks. Whenever moss and lichen occur there is a clear danger of this damage, and this is usually confirmed by examining the line of discharge onto the metal covering where a noticeably cleaner area may be observed, particularly where discharged water is concentrated such as under hip valleys.

Felt and bitumen roofs

Laminated felt and hot bitumen roof coverings are durable if they have adequate aggregate protection, but working on a flat roof may cause aggregate to puncture the covering. There is also a tendency for the bitumen or tar components to volatilise, causing the covering to become rather brittle and leading to splits. Generally water penetrating due to accidental damage or these cracks can be identified as dampness is normally relatively localised, but there is a danger in all roofs with impervious coverings that condensation will occur. Generally warm humid air from the interior of the building will diffuse through the roof support, usually tongued-and-grooved boards, chipboard or wood-wool slabs, to condense on the underside of the impervious roof covering (Figure 3.1). Clearly this introduces a danger of decay in the supporting material which must always be preserved.

Figure 3.1 Roof condensation

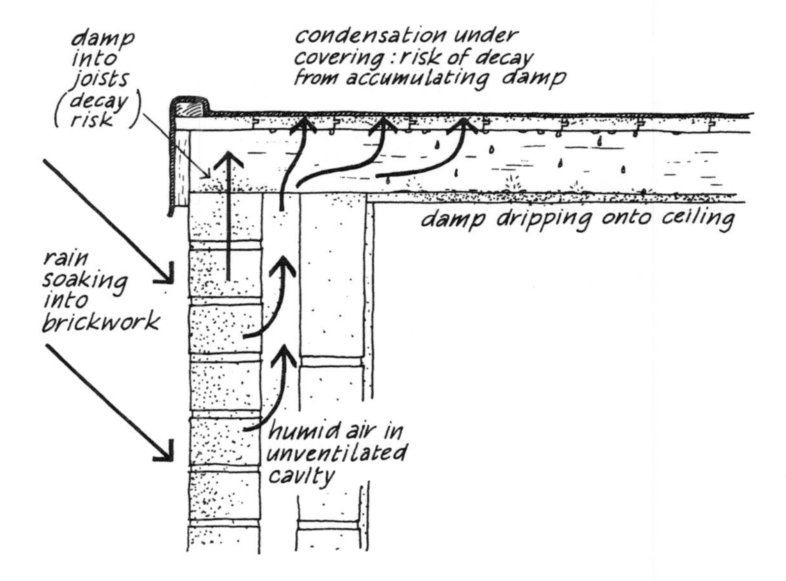

This danger has been considerably reduced in recent years by the use of foil-backed plasterboard for ceilings, largely because the introduction of the foil improves the thermal insulation but also because it limits water vapour transfer. Unfortunately the use of this foil-backed plasterboard does not completely remove the danger of condensation as there are clearly other sources of humid air and, in recent years, an entirely new danger has occurred. It has become the practice to reduce or even eliminate ventilation from wall cavities to improve thermal insulation, yet water penetration still occurs into the external skin and the air within the cavity becomes progressively more humid as a result. This humid air then diffuses into the roof space and condenses under the impervious covering, often leading to condensation to a sufficient extent to cause dampness penetration through ceilings.

Pitched roofs

Some pitched roofs are covered with impervious felt or metal and are subject to many of the problems previous described for flat roofs. Most pitched roofs are, however, protected by large or small sheet materials. Amongst the large sheets, corrugated iron was at one time widely used. In polluted atmospheres the protective zinc galvanising fails relatively rapidly, exposing the steel beneath which then commences to corrode. Corrugated asbestos-cement sheet may be damaged by atmospheric pollutants which can cause the development of sulphate attack, whilst on stables or byres the sheet may be damaged by the formation of nitrates; both these processes are described in more detail in Chapter 4. Although corrugated glass-reinforced plastic was introduced as a permanently durable material, it is common to find that the surface gel coat has completely disappeared after a few years, exposing the glass-fibre reinforcement. Several clear plastic sheets are also used, polycarbonate being most durable, but all tend to deteriorate progressively. Whilst protection can be applied to corrugated iron and asbestos-cement roofs, and perhaps to some glass reinforced plastic sheets, the clear sheets must normally be replaced when they have deteriorated to an unacceptable extent. However, if dampness appears to be originating from roof structures of these types progressive deterioration must never be suspected initially; it is far more likely that penetration is localised and occurring through inadequate side or head laps or through placing fixings at the bottom of corrugations instead of at the top, or if the dampness is uniform then condensation must be suspected.

Slate, tile and shingle roofs

In slate, tile and shingle roofs it is often said that damp penetration can occur as the roof materials become more porous. In fact, changes in porosity are normally insignificant. Indeed the most reliable clay tiles are normally the most porous types; it will be explained in Chapter 4 that resistance to frost damage is greater for a tile with large pores than for one with smaller pores, so that tiles that show spalling due to frost damage are generally relatively impermeable. When damage can be attributed to roof penetration, it is more often due to physical damage such as broken tiles or slates, or alternatively to defects in design and construction causing inadequate pitch or lap. In some cases it will be found that dampness is concentrated at the eaves areas, perhaps resulting in fungal decay damage to the rafter feet and plates, and perhaps explained because the chosen lap was adequate for the main pitch of the roof but inadequate for the shallower pitch over the eaves sprockets. In other cases damp penetration may be concentrated around the hip valley rafters and attributable to defective construction of the valley such as inadequate side lap on a lead gutter.

In tile, slate and shingle roofs the side lap cannot exceed half the width so that with narrow width the side lap may give inadequate protection. Standard tile and slate dimensions are therefore designed to ensure adequate side lap. In plain tiling, clay tiles are traditionally used with a double lap so that the top edge of a tile is covered by two tiles above, the length of tile covered in this way being known as the lap. The required lap must be increased with increased exposure or shallower pitch. The exposure depends on the situation, both the geographical position and local factors such as the height above sea level and surrounding features such as tall buildings, as explained in Appendix 5. Pitch is another name for the slope of a roof. The rafter pitch must be sufficient to ensure that the effective pitch of the tiles is not too low; the effective pitch is the angle of the tile surface and is rather less than rafter pitch, as shown in Figure 3.2. Gauge is the length of the exposed tile, or the spacing between the battens, so that it can be measured both inside or outside the roof. In order to ensure the two tiles cover above the lap length, gauge must be half the difference between tile length and lap.

Figure 3.2 Tile roofs

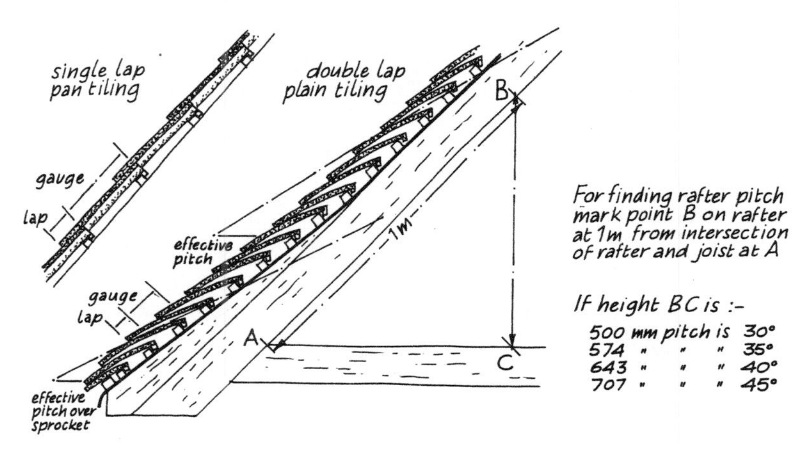

Double-lap plain tiles

With normal clay tiles, plain tiling must be used at a minimum pitch of 40° assuming a lap of 63 mm (2½ in) for a standard 265 x 165 mm (10½ x 6½in) tile. In fact, this lap is sufficient only if the minimum 40° pitch is being used in particularly sheltered conditions, and the lap should be increased to 70 mm (2¾ in) for a normal exposure or to 75 mm (3 in) or even 90 mm (3½ in) for severe exposure conditions. Valleys are particularly at risk as the valley rafter has a relatively shallower pitch and rainwater naturally concentrates in the valleys. Valleys can be constructed in swept, laced, mitred or purpose-made tiles, or a sheet metal gutter can be provided. Metal gutters are most efficient, provided an adequate side lap is allowed, and should always be employed for shallow pitch roofs, although purpose-made tiles are perfectly adequate for normal pitches. Secret gutters involve closing the tiles at the valley across a relatively deep gutter formed from metal but they tend to become blocked with leaves and must never be used in conjunction with sprockets; fungal decay at the feet of hip valley rafters can often be attributed to secret gutters overflowing through leaf blockages at the turn onto the sprockets. Gable roofs are best built out and finished with barge boards, rather than forming verges bedded onto the walls. If verges are considered essential they should be formed so that the tiles project and are bedded to give a slope into the roof so that rainwater discharge down the gable is avoided. Eaves should be formed using proper eaves undertiles, which are short tiles designed so that the ends coincide with those of the course above and the effective slope of the roof is maintained in the bottom two courses; the tiles should overhang the gutter by 38–50 mm (1$^1/_2$-2 in). Generally eaves sprockets should not reduce the roof slope by more than 15°, and the sprockets' slope should still be a minimum of 40° so that sprockets should be used only on steep pitch roofs.

Single-lap interlocking tiles

Special tiles that are close-fitting at their head and side laps can be used with a single lap, that is with only one thickness of tile above the head of the tile, as shown in Figure 3.2. Pantiles are generally used with single lap in this way but there are a variety of tiles which interlock at their heads and sides to reduce rain penetration. Generally single-lap clay tiles can be used at rafter pitches in excess of 35° and concete tiles at pitches in excess of 30°, but it will be appreciated that proprietary interlocking designs vary in their ability to prevent capillary absorption or blown rain penetration and they may therefore be used at greater or shallow pitches. As a general rule steeper pitches must be used in exposed situations such as coastal sites or where it is necessary to prevent excessive snow accumulations. Sprockets must never be used with single-lap tiles. The head lap should always be greater than 75 mm except with interlocking designs, the gauge being the length less the lap.

Slates

In fact, both clay and concrete tiles are available for both single-lap and double-lap construction. It is often said that concrete tiles can be used at a shallower pitch than clay tiles because of their lower permeability. Slates are suitable for use at shallow pitches; a lap of 75 mm (3 in) permits slates to be used down to 35°, and an increase in lap of 6 mm ($^1/_4$ in) enables a reduction of 5° in minimum pitch. However, slates should not generally be used below a pitch of 25° and, if used below 30°, the width should be increased to more than half the length. Plain asbestos tiles can be used in the same way as slates but they tend to increase in porosity with time and must be considered less durable.

Shingles

Almost all the shingles in use in the British Isles are obtained from British Columbia and are sawn from western red cedar. The standard nominal length is 400 mm (16 in) and they are designed to be laid to a gauge of 125 mm (5 in) if used at a minimum rafter pitch of 30°. If the gauge is reduced to 95 mm ($3^3/_4$ in) shingles can be laid down to a rafter pitch of 20°, but there is then a severe danger that water will be retained between adjacent shingles and severe fungal decay may develop, despite the good natural durability of western red cedar.

Thatch

Thatch roofs are rather too specialised for detailed inclusion in this general discussion. As they are generally confined to very old buildings, an adequate pitch has usually been found in the past through trial and error, even if the original design was inadequate. Thatch roofs give excellent thermal insulation and are also extremely durable if properly maintained. The main necessity is to check the ridge cappings and the eaves for deterioration, perhaps due to water penetration and decay, but generally vermin and bird damage is more severe. In a reed thatch this damage is normally confined to the ridge and eaves and can be readily controlled by regular inspections and closure of access holes. However, the situation is more difficult in a straw thatch and the only reliable protection may be wire netting, although this is not recommended as the thatch cannot be removed easily in a fire emergency. If a thatch is to be replaced it is therefore worth the extra cost of reed.

Wall defects

Water penetration through walls is generally prevented through a number of precautions in design and during construction. Overhanging eaves can considerably reduce rain penetration into wall surfaces except during windy weather and, even where flush eaves are employed, gutters and downpipes are provided to ensure that walls are not exposed to roof water discharge. However, these precautions are not essential as walls should be capable of resisting rain penetration in any case. Two different systems are commonly employed to prevent rain penetration. In solid walls very thick construction is used so that penetrating rainwater can be accommodated without becoming apparent at the internal surface, the accumulating water later dispersing by evaporation to the exterior. The alternative technique is to use cavity walls in which complete penetration of the external skin is expected but precautions are taken to prevent absorption of the penetrating water into the internal skin. In both cases defects can occur and internal damp problems are often encountered.

In solid walls dampness resulting from rainwater penetration can be attributed to insufficient absorptive capacity; thus penetration usually occurs because the wall is too thin, either throughout the building or at local thin points such as window reveals. Alternatively, the penetration results from excessive permeability. In older buildings this permeability may have developed slowly with time, usually due to the action of atmospheric acids on carbonaceous stones; the deterioration of limestones is described more fully in Chaper 4. In older solid walls dampness often follows the pattern of the masonry, perhaps being confined to either the mortar or the stone, whichever is the more permeable. This is seen most clearly if the internal surface is finished in limewash, the custom in old churches in some parts of the British Isles. Construction in impermeable granite presents a particularly interesting example as the mortar courses often suffer from penetrating dampness, even in walls that are particularly thick. The explanation lies in the fact that rainwater flowing down the outside wall cannot be absorbed by the impermeable granite and an unusual amount of water is thus absorbed into the porous mortar. Whilst the mortar might have perfectly adequate capacity if used in combination with stone of average porosity, it has insufficient capacity if required to absorb water shed by the granite. Whilst it is often possible to distinguish between penetration through mortar and that through stone (or brick) if only a limewash finish is used internally, a heavy plaster will conceal this detail and the internal surface may appear to be uniformly damp (Figure 3.3).

Penetrating dampness resulting from inadequate absorptive capacity, that is inadequate thickness or excessive permeability, tends to result in uniform internal dampness. However, sometimes the upper parts of the walls are dry where they are protected from rainfall by the overhanging eaves. If rainwater is absorbed into a wall it will tend, at least in macroporous materials, to drain towards the base, which naturally tends to be wetter. If a damp-proof course is provided it is likely that this draining dampness will accumulate on top of it, giving an appearance very similar to rising dampness, but treatment is entirely different and it is thus essential to ensure that dampness of this type is correctly diagnosed.

If dampness is concentrated at the top of a wall it must be suspected that there is a roof or adjacent gutter defect. This defect may not be immediately obvious. For example, the defect may occur beneath a hip valley gutter, yet a superficial external inspection may fail to detect any explanation for the water penetration. However, a more detailed inspection, either externally or internally, may disclose that a secret gutter is involved, running under the hip valley and concealed by the tiles above. This type of gutter is particularly prone to leaf blockage, especially if the eaves are fitted with sprockets, as the leaves will tend to jam at the bend where the pitch of the slope alters. Dampness at wallheads may be due to poor positioning of the eaves gutters or penetration through the bottom course of tiles or slates because short undertiles have been omitted; it is not sufficient to rely on sarking felt to

Figure 3.3 Dampness in solid walls

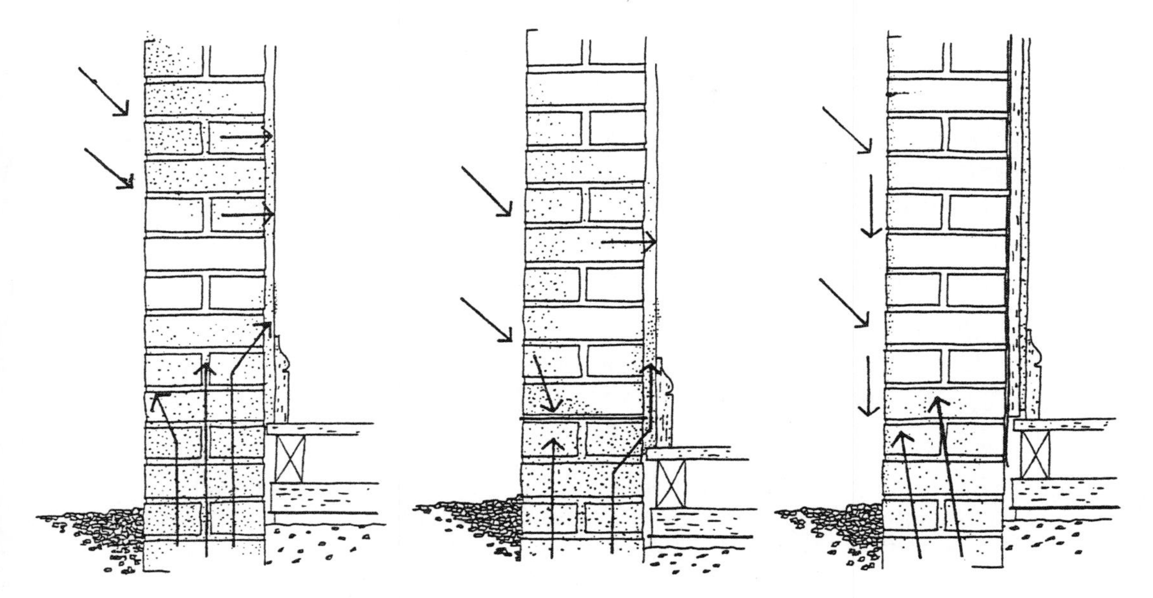

(a) Penetrating and rising dampness in a solid wall.

(b) A damp-proof course in a solid wall; penetrating dampness still present and tending to accumulate on the damp-proof course so that it may cause internal dampness at the base of the wall, whilst rising dampness may occur if damp-proof course is bridged by internal plaster, for example.

(c) Dry lining, an excellent method for controlling dampness in solid walls; plasterboard is fixed to battens which are in turn fixed to the wall over a PVC or polythene membrane. There is no danger of interstitial condensation as the membrane is close to the warm internal surface. However, rising dampness will persist if there is no damp-proof course and a water-repellent treatment should not be used as the rising dampness may accumulate behind it, perhaps exaggerating the danger of spalling due to freezing or salt deposits. If a water-repellent treatment is to be applied it must be accompanied by the insertion of a damp-proof course.

protect the wallhead from this penetration as the felt will deteriorate.

Another design feature that frequently results in dampness at the top of walls is a parapet; rain absorbed into the parapet is able to penetrate downwards if solid walls are involved without damp-proof courses. In some cases the construction has proved entirely adequate in the past and dampness has developed only at a later date, perhaps owing to the development of increased permeability as previously described, but sometimes owing to the development of defects in parapet copings. Alternatively the dampness may be caused by overflowing or leaking parapet gutters, whilst if dampness is apparent at the top of partition walls within a building it is possible that similar defects in valley gutters may be responsible.

These parapet and valley gutters, as well as flat roofs, frequently drain into hoppers which lead to downpipes. In some large buildings the hoppers are fitted with spouts at higher level through which the water can discharge in the event of the hopper becoming blocked. Dampness in many buildings can be attributed to the failure to keep hoppers free from accumulated leaves, falling moss and lichen from roofs, and pieces of stone from deteriorating masonry so that the entire roof discharge eventually occurs through the spouts or if the spouts are absent the system overflows. These blocked hopper defects, gutter leaks and defective downpipes frequently result in vertical patches of dampness in buildings which may eventually cause the development of severe fungal decay in contacting structural woodwork.

Cavity walls

In theory the use of cavity walls should completely avoid penetrating dampness but unfortunately practice is not so perfect as theory! Dampness at the top of walls can arise through roof, parapet and gutter defects, as for solid walls, but even direct penetration can occur, spreading across the cavity through mortar slovens or droppings whch were permitted to accumulate on wall ties during construction (Figure 3.4). At the base of a wall the cavity should preferably extend to well below internal damp-proof course level or, alternatively, if the damp-proof courses are continuous they should be stepped so that the level in the external skin is lower than the level in the internal skin. Occasionally the situation is reversed, through either carelessness or ignorance, and penetrating dampness accumulating on the damp-proof course at the base of the external skin may flow into the internal skin; in one example this defect occurred in most of the houses throughout a large building development!

Figure 3.4 Dampness in cavity walls

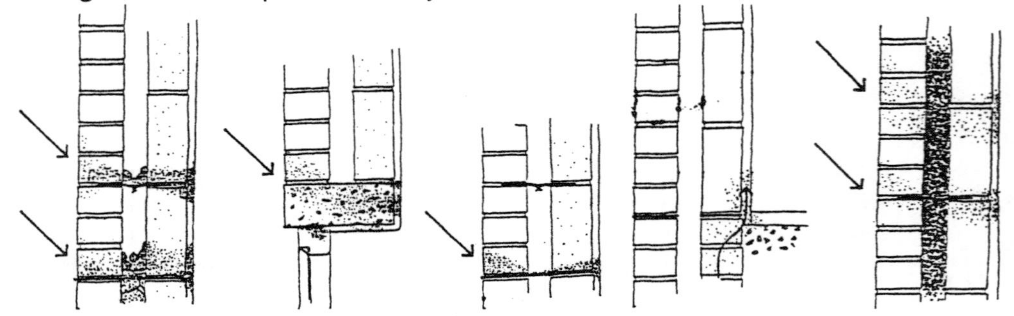

(a) Mortar slovens accumulated at base of cavity or lodged on ties.
(b) Penetration through exposed lintel, caused by failure to provide damp-proof membrane above lintel.
(c) Penetration across damp-proof course; damp-proof course should be stepped up into inner leaf or should not be continuous across the cavity.
(d) Damp-proof course bridged by internal plaster.
(e) Penetration across cavity through unsuitable cavity fill insulation.

Cavity ventilation

It has become the practice in recent years to restrict cavity ventilation to improve thermal insulation. In fact, cavity ventilation has a very important function as vents at the base of the cavity permit humid air to disperse. If these vents are omitted dampness continues to penetrate through the external skin in the normal way and humid air accumulates, tending to descend at the external skin but rise at the warmer internal skin and flow into the roof space. In flat roofs condensation may eventually occur beneath the impermeable roof covering, permitting water to accumulate in the supporting boarding and in severe cases causing dampness staining on ceilings beneath (Figure 3.1).

Cavity fill insulation

Cavity fill insulation is often used today in an attempt to reduce heat losses, but fill materials must be chosen with considerable care; non-wettable pelletised materials such as expanded polystyrene can combine considerable thermal advantages with freedom from disadvantages but some fibrous materials can conduct moisture across cavities. Even foam formed *in situ* can have considerable disadvantages; wettable open-textured foams can conduct moisture across the cavity, whereas closed foams may eventually permit accumulation of interstitial condensation within the inner skin. Some *in situ* foams are water-based formulations which, if incorrectly formulated, can actually cause dampness through the collapse of the foam and the accumulation of moisture at the base of the inner skin.

Rising dampness

Water absorbed by capillarity into the base of a wall from the supporting ground is generally described as rising dampness. Uniform dampness confined to the base of both external and internal partition walls can usually be attributed to the lack of an effective damp-proof course. A damp-proof course may be discontinuous, perhaps round hearths or where old doorways have been blocked, and localised dampness may therefore occur. Rising dampness is not confined to walls but also occurs in solid floors if a proper damp-proof membrane has not been provided. In some cases damp-proof courses are present in walls and damp-proof membranes in solid floors, yet rising dampness still occurs through discontinuity or bridging (Figure 3.5).

For example, damp-proof courses were often positioned above the level of solid floors in old Victorian properties, presumably because it was considered that solid floors were regularly washed and the walls therefore required protection against the washing water as well as dampness rising from the ground. Usually a batten was fixed to the walls immediately above the damp-proof course and provided an edge to the plasterwork above. This batten effectively prevented the plaster from bridging the damp-proof course and also provided a fixing for the skirting which was isolated from

Figure 3.5 Damp-proof course bridging

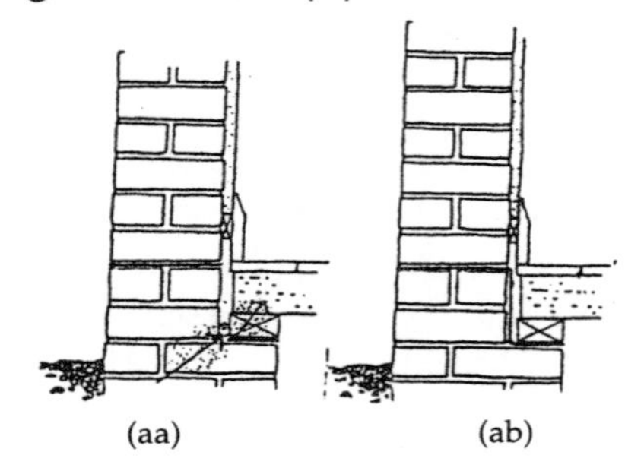

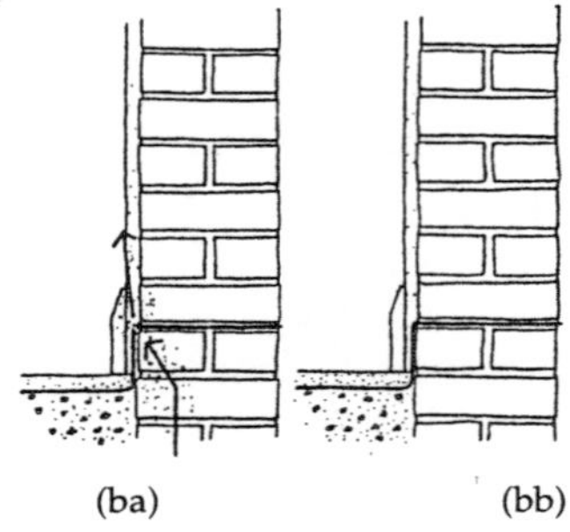

Bridging through slovens accumulating between a wall and an adjacent plate, causing Wet rot typically in the plate. The correct construction in (ab) also ensures that joist ends cannot contact the wall.

Bridging through failure to link the floor damp-proof membrane with the wall damp-proof course. The correct construction is shown in (bb) but it is often easier to fix the skirting to a batten above the damp-proof course, as in (ab).

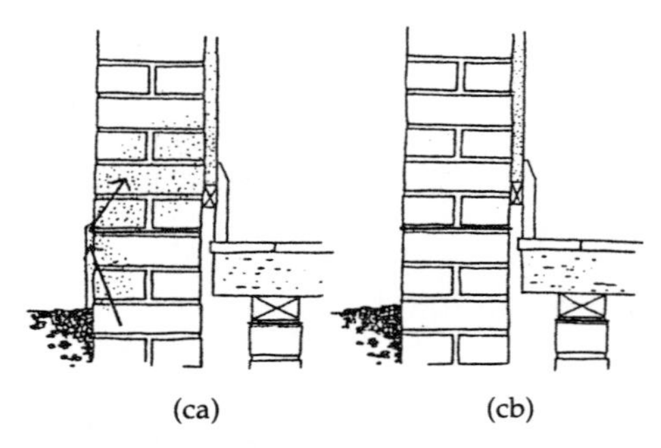

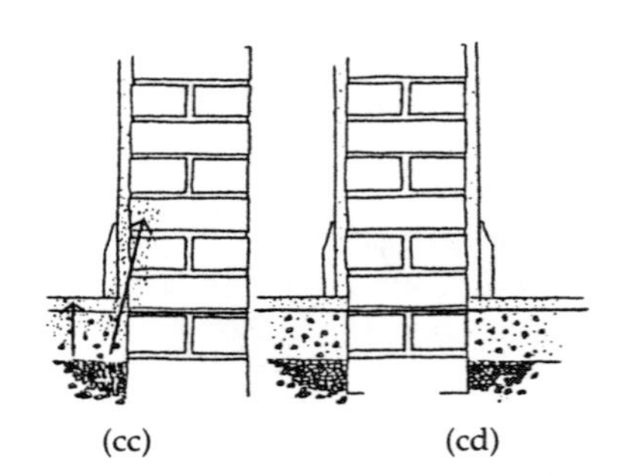

Bridging of a damp-proof course through external rendering or internal plastering, (cb) and (cd) showing the correct construction for suspended and solid floors respectively. In solid floors the damp-proof barrier can be stepped as in (bb).

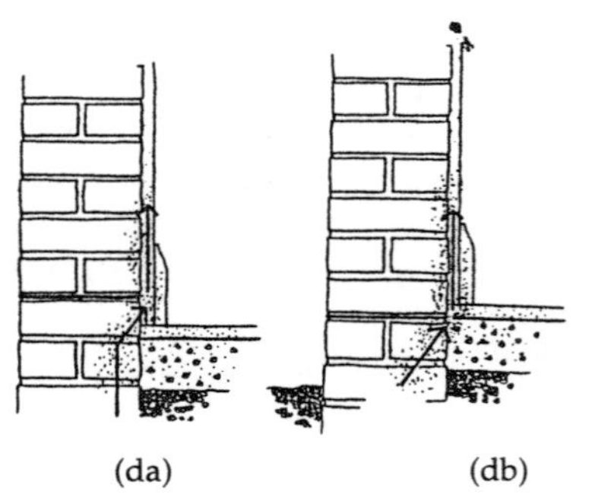

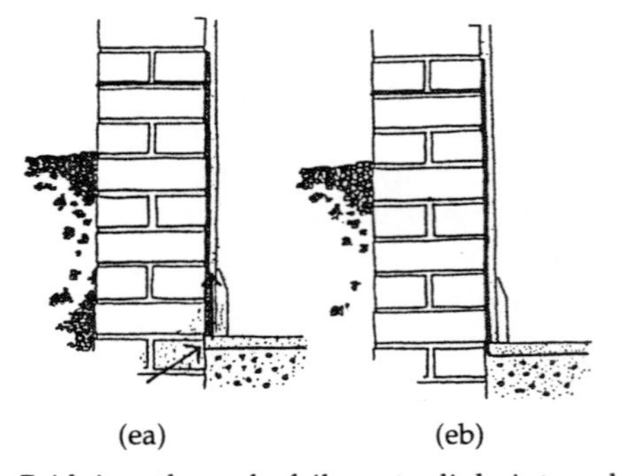

Further examples of bridging caused by failure to link damp-proof membranes in solid floors with damp-proof courses in adjacent wall.

Bridging through failure to link internal tanking with a floor membrane. Note that the damp-proof course in the walls must be above external ground level, whatever the internal floor level.

the wall below the damp-proof course level by an air gap. Whilst this system appears perfectly logical, the door frames extended below the damp-proof course, absorbing water and becoming affected by Wet rot or Dry rot which then extended into the adjacent skirtings. Scarf repairs to the doorposts and skirting replacements were simply followed by further fungal decay and eventually the skirtings were abandoned in many cases and replaced with a dense rendering. The effect of the rendering was simply to bridge the damp-proof course, causing the deterioration in the plaster and decorations above.

In modern construction an exactly analogous situation occurs when a damp-proof course is installed by a remedial treatment company adjacent to a solid floor, except that the main danger in this case is that a new floor screed may remain damp at its edges, despite the presence of a proper damp-proof membrane, through contact with the wall below the damp-proof course level. The correct procedure is to ensure that the damp-proof membrane in the floor is always led up the wall behind the skirtings so that it becomes continuous with the damp-proof course in the wall.

Similar bridging can occur externally if walls are rendered or even if the line of the damp-proof course is pointed for aesthetic reasons. It has already been explained that penetrating water in external solid walls may accumulate on an existing damp-proof course and that this accumulation may be wrongly identified as rising dampness. In many areas the traditional method for dealing with this problem is to render the base of the walls to a limited height, apparently to reduce water penetration resulting from splashing at the base of the wall. Unfortunately this rendering frequently serves only to bridge a perfectly adequate damp-proof course and thus to introduce a new source of rising dampness. Piles of earth against walls can achieve similar bridging. Where internal accommodation is below external ground level, direct water penetration can occur, perhaps under significant pressure in the case of basement accommodation.

Window condensation

Windows usually provide less efficient thermal insulation than the surrounding walls in which they are installed and thus have colder surfaces on which condensation may be observed when the external temperature is significantly below the internal temperature. This condensation process will be explained in more detail, both in this chapter on dampness and in Chapter 5 which is concerned with thermal insulation. Here it is important to appreciate that condensation may occur whenever humid air is cooled, cooling of the air itself resulting in the formation of mist and cooling in contact with a cold surface resulting in the formation of dew. There is also a danger that interstitial condensation may occur if humid air diffuses into solid external walls, or if humid air comes into contact with impervious roof coverings, a defect that has been previously described in connection with wall cavity ventilation. Condensation may also occur on cold water service pipes or on any other cold surface.

Flue condensation

One particular condensation problem justifies special mention. Whilst efficient combustion in boilers is dependent on an adequate air flow which must be vented through the flue, an excessive air flow will result in significant unnecessary heat loss. In slow-combustion coke and anthracite boilers the air flow is restricted to control the rate of combustion and the flow of air in the flue is thus extremely limited. The very humid combustion gases pass slowly up the flue, becoming cooler and frequently condensing, particularly at chimney stack level. This condensation moisture tends to run down the inside of the flue where it is absorbed, causing damage to mortar and susceptible stone through the action of the acid combustion gases. These reactions frequently result in the formation of hygroscopic salts which migrate in the dampness to the plaster and decorated surfaces within the accommodation. Damp and stained patches are frequently observed on chimney breasts containing boiler flues, the dampness being particularly apparent when the boiler is out of use through the absorption of atmospheric moisture by the hygroscopic salts deposited at the evaporation surface.

Modern gas and oil boilers are equally susceptible to condensation. Whilst air flow is never restricted to control the rate of combustion, the boilers are rather more efficient than solid fuel systems and, whilst the flue gas is equally humid, its temperature is rather lower and its water-carrying capacity rather less; this relationship will be explained shortly. In all cases flue condensation can be avoided by increasing the air flow and this is usually achieved by ensuring that gases leaving the boiler pass into a flue which is open to the air at its base, the warm combustion gases inducing convection flow which draws in the additional air that is required to prevent condensation. In many cases boilers are now designed with openings at the bottom of the flue connections. If open flues are not employed the flue must be lined with, for example, stainless steel to prevent absorption of condensation moisture into adjacent brickwork or masonry, and a method must be provided to drain off accumulating condensation water at the base of the flue.

Hygroscopic salts

Flue gases are not, of course, the only sources of hygroscopic salts. Flue gases result in the formation of salts through reaction with the construction materials and exactly similar processes can occur on external wall surfaces in urban areas where combustion gases contribute to air pollution. Salts generated by air pollution seldom caused damage internally, unless they are able to percolate and concentrate in particular areas, but they are a serious cause of external damage, as will be explained in Chapter 4. Hygroscopicity, that is the development of dampness during humid weather through the presence of hygroscopic salts, can take a variety of forms. If the hygroscopic effect is uniform over internal and external wall surfaces it must be suspected that hygroscopic salts were introduced within construction materials, perhaps in bricks or in the sand used for mortar and rendering. If hygroscopicity is localised it can be suspected that hygroscopic salts have been introduced in a flow of dampness which can be attributed to a particular defect, commonly within rising dampness in which the salts may be present at low concentrations, eventually becoming concentrated at the evaporating surfaces until their hygroscopic effect is more serious than the rising dampness itself. Hygroscopic salts also have the effect of restricting evaporation so that dampness rises to a progressively increasing height as salt accumulations occur over the years.

Patches of hygroscopicity can be very difficult to explain and it may be necessary to enquire into the detailed history of the building. The generation of hygroscopic salts within construction materials through flue gases has already been described but it must be appreciated that the salt deposits will persist, even if the flues are removed. In older buildings patches of hygroscopic dampness can sometimes be attributed to combustion gases from old bread ovens or coppers which have since been removed. In other cases hygroscopic salt deposits can be attributed to storage of coke, perhaps on the other side of the wall; coke frequently contains hygroscopic salts and, as it is sometimes wet when delivered, these are able to migrate into the fuel store floor and walls. In other cases hygroscopic salt deposits can be attributed to bacon curing, or storage of salt for use in water softeners or treatment of icy roads and paths. In many cases chemical analysis is an essential preliminary to any attempts to identify the sources of hygroscopic deposits.

Quantifying dampness

Dampness is difficult to define but it is even more difficult to quantify, yet definitions of degrees of dampness are essential before any logical diagnosis of dampness defects can be attempted. Dampness has been previously defined as a moisture content sufficient to cause darkening, staining, mould contamination and cooling of structural surfaces. However, it is very difficult to define the moisture contents at which these defects can develop. It is also essential to appreciate that dampness is concerned not only with the moisture content of the structural material but also with its relationship to the surrounding atmosphere. In many cases the source of dampness can be readily identified but it is essential to be aware of the rather complex relationships that may occur in some circumstances. These complex relationships, the measurement of dampness and the diagnosis of its cause are all discussed in more detail in Appendix 3.

Drying buildings

Dampness in buildings introduced during construction, through roof or plumbing leaks, through flood or through fire-fighting can present a serious problem as it is often difficult to disperse the large quantities of water involved and, as drying is therefore rather slow, there is a danger that wood decay or other spoilage defects may develop. The problems involved in drying buildings are discussed in detail in Appendix 4.

Exposure conditions
There are several aspects of building design, such as tile lap and the thickness of solid walls necessary to achieve thermal insulation or resistance to rain penetration, which depend on the exposure conditions. The assessment of exposure conditions is discussed in Appendix 5.

3.2 Penetrating dampness

Buildings are designed to avoid rainwater penetration. Walls are either sufficiently thick to absorb rainfall and allow it to evaporate later, or constructed to incorporate a cavity which breaks the capillary flow of moisture towards the interior. In addition the amount of rainfall incident on the walls is often reduced by widely overhanging roof eaves, whilst the roof itself is designed to be impervious to rainfall. Despite these various precautions problems often occur and must be remedied.

Roof leaks
Roof leaks are always best remedied using the traditional repair methods of refixing loose slates and tiles and replacing as necessary. Various proprietary remedial processes have been introduced in recent years which are claimed to be far cheaper and more efficient than traditional repairs but, whilst these claims may be valid in some respects, they are usually extremely misleading.

Pitched roofs
Pitched roofs clad in slate or tile are normally extremely durable and reliable. Slate sometimes delaminates but this is very rare and confined to situations in which particular types of slate are used in severely polluted urban or industrial areas. Occasionally tiles may deteriorate by frost spalling, damage that can be attributed to the microporosity of the tiles as explained for natural stone in Chapter 4. In both cases replacement of the original slate or tile is the best solution as it will certainly achieve the most reliable result. However, complete replacement in this way is only rarely required, and refixing and replacing individual loose and damaged slates or tiles is far more usual. Most tiles are fitted with nibs which hook over the battens and ensure that the tiles cannot slip in stormy weather. If slipping occurs the best solution is simply to strip the roof carefully and to refix the original slates or tiles if they are in sound condition. It is often suggested that the opportunity should be taken to replace the original covering with asbestos cement or concrete tiles but these materials are unlikely to be as durable as the original slates or tiles which will probably be sold by the roofing contractor at an enormous profit; second-hand slates are particularly valuable.

It is important to emphasise that an external coating should never be used on slate or tile roofs under any circumstances. It will not itself prevent slipping but will encourage the whole roof to slip rather than individual slates or tiles so that refixing loose slates and tiles will still be necessary and the coating will therefore achieve no functional purpose whatsoever, except the cosmetic effect of changing the appearance of the roof. In addition the coating will seal the roof and may encourage condensation, and even the best coating systems are reliable only for a very short period relative to the normal life of a slate or tile roof. In recent years new systems have been introduced which can be sprayed on the underside of slates and tiles, thus fixing them to the battens However, if slipping is already apparent, the slipped slates or tiles must be refixed in the correct position before spraying and it is really far better to fix these loose items and simply leave the roof as it is rather than to adopt an additional unnecessary system. In any case there must be doubts regarding the long-term reliability of the adhesion to slates or tiles; adhesion will be difficult to establish when tiles are wet or cold and subsequent temperature or moisture content changes will still threaten the reliability of adhesion established in warm dry conditions. Even if reliable adhesion is established the durability of polymer systems cannot match traditional slates or tiles fixed to wood battens with suitable non-ferrous nails. Roof underside treatment systems are usually foams which, in addition to securing loose slates and tiles, are claimed to introduce improved insulation, but there is a danger that they will trap moisture in battens and encourage decay which will not be readily detectable because of the concealing nature of the treatment; if insulation is required it is far more sensible to use loose quilt or particulate systems on the ceilings.

There are obvously occasions when roofs are defective and fundamental repairs or alterations are necessary. If general penetration appears to be present it is frequently suggested that the tiles 'have become porous', but this suggestion should be firmly rejected as such changes do not normally occur in either tiles or slates. A more likely explanation is that the tiles or slates have been laid at a lap that is inadequate for the exposure conditions and, as storms representing the most extreme exposure conditions are very rare, the roof may have been perfectly satisfactory for many years before this defect has appeared. In many cases the defect is confined to the shallower pitch area of tiles over the eaves sprockets, causing local penetration and perhaps the development of fungal decay in the rafter feet and plates. In some cases it is possible to relay just the tiles over the sprockets, increasing the lap by introducing an extra one or two courses.

Gutters and downpipes

Defects in pitched roofs are not usually associated with the slates and tiles themselves and can only rarely be attributed to laying to an inadequate lap. The common danger points are the eaves gutters, hip valleys, parapet gutters and flats, particularly when these are likely to become blocked through leaf accumulations. Annual attention, usually in December, is essential to ensure that blocks cannot occur but problems may still arise through the negligence or ignorance of the person employed to carry out this maintenance cleaning. For example, it may not be appreciated that downpipes are blocked at mid-height, perhaps by a bird's nest washed down from a gutter, and that the blockage can be cleared only by dismantling the pipe. In large properties flat roofs frequently discharge into substantial hoppers which are often fitted with a drain at their base which leads into the downpipe and also with a spout at a higher level which is designed to relieve flooding should the drain become blocked. It has often been found during inspections that staff have carefully kept the spouts clean without appreciating that the actual drains were blocked so that water has discharged down wall surfaces instead of being conducted through proper downpipes. Secret valley gutters concealed by overhanging tiles are a particular nuisance as leaves running down the gutters tend to cause blockages where the gutters change slope onto the sprockets, yet the gutters are concealed and it is difficult to see the leaves and even more difficult to remove them.

Whilst poor design and lack of regular maintenance are the most serious causes of leaks from roofs, inadequate turnings or flashings on soakers are also a frequent cause of problems. Valley gutters and flat roof trays should be at least 150 mm (6 in) deep and the upstand must run under the tiles of adjacent roofs for 150 mm (6 in) or under slates for 225 mm (9 in). Inspection of an Elizabethan manor house in Devon disclosed that gutter boards, rafter feet and plates under the parapet gutters were wet and actively decaying in almost all cases. The lead gutter lining was lifted and it was found that the problem had been present for many years and repairs had consisted in the past of laying new gutter boards on top of the actively decaying boards, simply encouraging further decay. The most recent repair had involved scraping away the decayed wood and casting dense concrete over the remains of the gutter boards. Whilst the form of these repairs showed remarkable ignorance and negligence, perhaps the most striking feature about the whole repair was the way in which the lead had been lifted by cutting along the wall with a bolster rather than picking the turning out of the flashing course in the proper manner. As a result the lead turning was inadequate in height and the entire lead lining needed replacement.

In another case an inaccessible valley gutter was blocked by ivy which then trapped falling leaves from neighbouring trees. The valley gutter overflowed heavily during storms but the occupiers of the house, in their innocence, attributed the overflows to an inadequate rainwater disposal system, completely ignoring the heavy growth, and remedied the defects simply by providing bowls and buckets in the roof space to catch the drips. Eventually the property was vacated and, within a short period, the gutter overflow caused very severe and widespread Dry rot to develop. However, it was interesting that the Dry rot did not develop in the roof space where the water discharge was occurring, apparently because the roof space was well ventilated, but only in timbers concealed within a wall beneath the defective valley.

In yet another case a valley gutter had been repaired on several occasions by taping a crack that had developed, probably because the leadwork had been laid in excessive lengths so that it was unable to tolerate the seasonal thermal expansion and contraction. The development of a severe leak was therefore followed by a further repair of this original defect but, during subsequent storms, the amount of water penetration increased dramatically, causing severe damage to decorations within the accommodation beneath. Careful examination then disclosed that the lead had become perforated where two hip valleys discharged into the end of the gutter, and the leak had become so dramatic because the perforations had rapidly increased in size to a diameter of about 1 cm (1/2 in).

Moss and lichen damage

The development of such perforations is relatively common and usually occurs first where water flow is greatest, such as under hip valley discharges, although the damage is actually caused by acids generated by the lichen and moss growth on the roof tiles.

Metal roof coverings

As all metal coverings to flat roofs and valleys are susceptible to acid attack, it is always advisable that they should be maintained by regular bitumen coating and they are then usually indefinitely durable. However, if metal coverings have become defective due to the development of splits or perforations, it is always advisable that they should be relaid with the original material but with suitable precautions to avoid the redevelopment of the defect. For example, if the roof is susceptible to perforation the metal should always be coated with several layers of tar or bitumen preparation; such coatings are always needed where metal roofs receive the discharge from walls or roofs which are supporting moss and lichen growth. If a metal roof covering is split it must be suspected that it has been laid in panels of excessive length. Metal roofs and valleys must always be provided with movement joints, either by stepping or, in the case of a flat roof, by introducing a movement joint by rolling over a bead.

Coating systems

Coating systems can be used to repair metal roofs but reinforcement must be introduced to span the defect. One method is to coat and felt the roof and another is to use woven glass fabric to reinforce the coating.

Whilst many different coating systems are available, varying widely in price, there is really nothing to beat a simple tar or bitumen coating in terms of cost and reliability; coal tar coatings have better resistance to microbial attack and are generally more durable, although they are not now readily available. The main requirement is to ensure an adequate coating thickness. If reinforcement is to be used, the surface must be coated with the tar or bitumen system, the reinforcement laid on top, the coating stippled through to ensure good adhesion and several further layers of coating applied to build up a full protective layer over the reinforcement. Many roofing contractors use fine chopped-strand glass-fibre reinforcement but this is entirely unsuitable as the high-viscosity bitumen systems are unable to penetrate and establish proper adhesion; only an open-textured coarse woven fabric is suitable for this purpose. Whilst hot tar or bitumen coatings are theoretically suitable, they are unreliable if applied in cold or wet conditions. Emulsions can be used but solutions generally give the best results, although it must be appreciated that an adequate period must be allowed between coats for both emulsion and solvent systems. Laminated felt roofs can be repaired in the same way but in the case of hot bitumen roofs cracks sometimes develop which must be repaired before recoating. Random cracks should be raked out and touched in with the bitumen preparation before general coating but straight cracks usually indicate movement in the structure, perhaps the presence of a steel joist beneath, and serious consideration should then be given to applying reinforcement over the entire roof in order to tolerate this movement. Alternatively a slot can be cut along the line of the crack and filled with mastic as a movement joint.

Condensation in flat roofs

Whilst referring to flat roof problems it is necessary to mention the danger of condensation under an impermeable roof covering, particularly metal. Where flat roof spaces are continuous with wall cavities there is a severe danger that moisture penetrating the outer wall skin may increase the humidity, allowing condensation under the roof covering with a serious danger of fungal decay in the supporting joists and even the development of damp staining on ceilings beneath. These problems are best avoided by sealing the top of the wall cavities to prevent the passage of humid air and also ventilating the roof space and providing insulation on top of the ceiling to prevent unnecessary heat loss.

This condensation under flat roofs can cause damp staining on ceilings, which can be confused with damage from simple leaks through the roof-covering, and care must always be taken to ensure correct diagnosis of the cause, as with all dampness problems.

Parapets

If dampness is confined to the edge of a ceiling under a flat roof it must never be assumed that the roof itself is at fault as there may be a defect associated with the roof parapets. Indeed parapet walls are a particularly frequent cause of damp defects. In solid construction the roof covering should have an upstand of at least 150 mm (6 in) which should be covered by a flashing continuous with a damp-proof course through the parapet. In some cases, particularly in older buildings, a continuous damp-proof course is not used in this way and instead the parapet copings are relied on to prevent descending dampness. Such a system is less reliable and clearly undesirable but it can be effective if the copings are relatively impermeable, carefully jointed and overhanging so that discharging water is thrown well clear of the parapet wall surfaces, although it is always best to lay the copings on a damp-proof course extending to just beyond the width of the parapet wall to avoid any danger of damp penetration through coping joints damaged by thermal movement. Cavity parapets should be treated in the same way but so often damp-proof courses are omitted from the inner skins, despite the fact that this skin is, of course, exposed to the weather when forming the rear of a parapet, and even in modern construction the copings alone are often relied on to prevent rain from descending in the cavity.

If defects in a building suggest that dampness is penetrating down the inner skin from the parapet above, it is always essential to remove the coping to check on the situation within the cavity. If the cavity is continuous the flashing to the flat roof should lead through the inner skin with an upturn within the cavity. In some cases the flashing does not penetrate through in this way or the upturn has been omitted, and usually the simplest remedial action is to reconstruct the inner skin with a proper damp-proof course. However, if it is suspected that dampness is penetrating down the cavities from the coping, it is possible that the coping can simply be laid on top of a damp-proof course spanning both skins, supported on slates to prevent sagging and leakage where lengths of damp-proof course are lapped. Whilst such remedies are obviously most sensible in regularly constructed brickwork, they may not be possible in older buildings with solid parapets constructed in random coursed masonry and it may be more realistic to use water repellents to reduce rain absorption into the exposed parapet surfaces; suitable water repellents are described later in this chapter.

Chimneys and flues

Rain can penetrate down chimney pots but it can also penetrate into the exposed brickwork or masonry of chimney stacks. Whilst a damp-proof course should be present where the stack passes through the roof to control descending moisture, there is obviously a danger that water will flow down the flue, accumulating on the bends and shelves where it will dissolve acids in any soot deposits. If these acid solutions are then absorbed into the adjacent mortar there is a danger that hygroscopic salts will be formed which will eventually migrate to interior decorated surfaces, becoming apparent as damp patches whenever the air is relatively humid. Condensation and hygroscopic salt problems will be considered later in a separate part of this chapter. Clearly measures should be taken in unused chimneys to ventilate flues so that any water deposits are dispersed before they can cause any damage. If flues are on external walls a cowl can be provided as well as a vent to the exterior at the base of the flue. In the case of internal flues it is usual to provide a vent to the sealed fireplace; this vent must be as close to floor level as possible to avoid unnecessary heat loss.

External walls

Walls are traditionally protected from rain penetration by overhanging roof eaves fitted with gutters and downpipes to dispose of the rainwater discharge. However, rain is often blown onto wall surfaces and clearly the danger of penetrating dampness depends on the degree of exposure, that is the total rainfall related to average or peak wind velocity, as described in Appendix 5. On a normal porous wall the surface will be wetted by the rainfall and this water will be absorbed into the wall by capillarity, the force depending on the pore size. If the pores are small the capillary force is very powerful but in all normal porous walls the force is sufficient to ensure that water is absorbed into the wall rather than flowing down the external surface. The total porosity of the wall in relation to its thickness will determine whether the wall can absorb the rainfall before it becomes apparent at the interior surface. In addition it is found that a macroporous (large-pored) wall surface will allow water to evaporate more rapidly to the exterior than a microporous (small-pored) surface, and because the latter is likely to retain moisture for a longer period it is more likely both to cause dampness to the interior and to result in frost damage to the exterior surface. If a wall consists of impermeable blocks such as granite, rain is absorbed only into the mortar courses and, as the mortar represents only a fraction of the total wall volume with a relatively low capacity for water absorption, penetration to the interior is likely. Thus in solid walls the use of a non-porous stone is likely to lead to more severe damp penetration than the use of a very porous stone.

Renderings

One method for reducing water penetration through walls is to increase their absorptive capacity, perhaps by the use of a thick porous rendering, a process that also significantly increases the thermal insulation value of the wall. Unfortunately it is not generally appreciated that the function of the rendering is to be porous and often thin dense or water-proof renderings are employed. These are not usually sufficiently flexible to tolerate the seasonal thermal and moisture content changes in the structure, and eventually fractures develop. Water flowing down the wall can be readily absorbed into the fractures, yet the remaining rendering obstructs evaporation so that the final result tends to be severe dampness internally.

Coatings

Attempts to reduce this dampness often include the use of cement slurry or paint, or water-proof bitumen, oil paint or plastic coatings, but all of these systems should be avoided as they introduce severe condensation dangers; a permeable structure must never be sealed at the cold surface as there is then a danger that interstitial humidity will result in condensation immediately under the water-proof layer, followed by severe frost spalling in cold weather. Interstitial condensation can be avoided by ensuring that the water

Figure 3.6 Water repellency

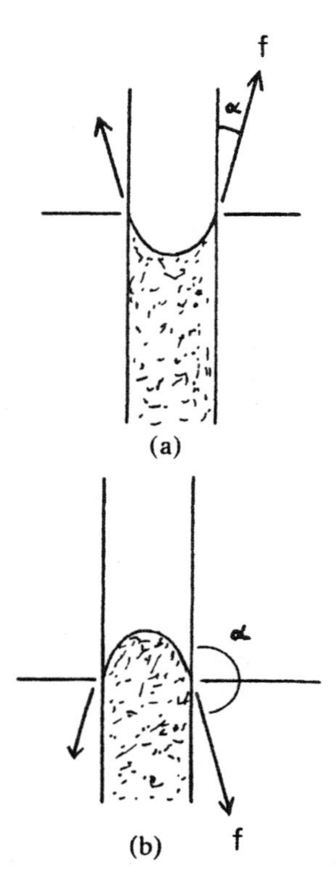

In (a) water is being drawn into a pore by capillarity, the capillary force being the component of the surface tension acting along the pore. If the angle of contact is reversed as in (b) the capillary forces tend to repel the water from the pore. This reversal of the angle of contact also accounts for the globulation of water that occurs on a water-repellent surface.

vapour resistance on the inner side of the main insulation is about five times the resistance for any outer render and decorative system, but with walls constructed from impermeable material such as granite an even greater ratio is necessary as only the limited area of mortar can disperse the condensation; only thin emulsion decoration is suitable and it is preferable to avoid any render or decoration.

Water repellents

Rendering and water-proof coating systems suffer from the severe disadvantage that they completely change the appearance of walls, but water-repellent treatments provide a method for reducing water penetration into porous walls without the condensation and spalling dangers associated with impermeable water-proofing treatments. Water repellents act by lining the pores so that the angle of contact is reversed and the capillarity forces then repel water rather than absorbing it into the pores (Figure 3.6). When first applied water-repellent treatments cause rain to globulate on the surface so that it falls away rather than being absorbed but, with most water repellents, this globulation or 'duck's-back effect' is soon lost and rain is able to wet the surface. However, this wetting is only superficial and, if the water-repellent treatment is applied sufficiently generously, the pores remain water-repellent to an appreciable depth so that water absorption into the walls is prevented. Water-repellent treatment is particularly inexpensive as it involves only a rapid generous spray treatment, but the treated wall has a number of advantages in addition to the simple control of dampness as a dry wall has better thermal insulation properties and remains more uniform in colour during wet weather. It is of course possible for water to accumulate behind a water-repellent treatment, either through limited interstitial condensation or through wind pressure overcoming the water repellency where macroporous building materials are used in extremely exposed conditions, but true water repellents do not seal the pores and any water introduced in these ways is able to disperse by evaporation to the exterior. However, it must be firmly emphasised that only silicone resin water repellents should be used as these are the only compounds that have sufficient durability when applied as the very thin pore linings that are necessary if permeability is to be maintained so that moisture can disperse.

Silicone water repellents for masonry were first classified in British Standard 3826 which defined three principal types of silicone water repellents. Class A was a normal silicone resin treatment but it gave unreliable results on carbonceous substrates such as limestones. Class B is a more expensive silicone resin formulation that will give reliable results on all normal porous substrates. Class C was a siliconate, a water-soluble formulation which could not be generally recommended as it had a comparatively short life. All three British Standard water-repellent systems were designed to ensure sufficient permeability on normal substrates to permit trapped water to disperse, but dampness rising behind a treatment could introduce salt deposits which might cause the treated layer to spall away so that water repellents should never be used in circumstances where walls are subject to rising dampness. This British Standard was replaced in 1984 by BS 6477 which was not restricted to silicone water repellents and defined four classes of treatment based on the properties of the masonry surface. Group 1 is for predominantly siliceous masonry such as clay bricks and sandstone. Group 2 is for predominantly calcareous masonry such as limestone. Group 3 is for alkaline surfaces such as fresh cement rendering, and Group 4 is specific to calcium silicate bricks. The old Class A silicones satisfy Group 1 and Class B silicones satisfy Groups 1 and 2, the only groups that are really significant for remedial works.

Silicone water repellents should never be applied to walls that are covered with moss, lichen or algae, or walls with defective pointing or excessively friable stone; all these problems should be remedied before water-repellent treatment, as described in Chapter 4. However, it is worth describing several special situations in which repellents can achieve particularly dramatic success.

Impermeable stone

Walls constructed from impermeable stone present particular problems, as previously explained, as the entire incident rainfall is concentrated at the mortar joints. In many cases these joints are too fine and are subject to thermal movement, sometimes allowing considerable rainwater absorption and the development of severe dampness at the interior. These problems are not confined to common impermeable stones such as dense quarried granites but apply equally to pebble and knapped flint walls. With these forms of construction pointing is often difficult or even impossible, but spraying the wall generously with water-repellent formulation can often achieve dramatic control of moisture penetration.

The spray must be applied particularly generously so that it flows down the wall surface and is absorbed into all the cracks, crannies and porous areas that will normally absorb rainwater. This method of applying water repellents should be universally adopted. A coarse low-pressure spray is employed, the spray lance being moved horizontally across a convenient width of the stone surface to produce a run-down of perhaps 150 mm (6 in), the next pass being across the run-down to produce a further run-down and so on, so that a continuous curtain of water-repellent fluid flows downwards until that section of the wall surface has been treated. Usually the solvents in the treatment will persist for a sufficiently long time for treated areas to be readily identified, but there are also some products available containing fugitive dyes which usually stain the wall mauve at the time of treatment, the colour then gradually disappearing over a week or so.

Cladding

Water repellents are not, of course, the only remedial treatments that can be used to reduce water penetration into walls. Tile hanging and wood cladding are particularly efficient and have the distinct advantage that they greatly increase the thermal insulation value of the wall by introducing an additional cavity. Indeed, in the case of a 225 mm (9 in) solid wall cladding probably represents the most efficient means for overcoming both dampness and thermal defects, but these systems completely alter the appearance of the property and can only be used subject to planning consent.

Pre-cast concrete

Dampness in concrete structures such as pre-cast buildings is seldom due to the porosity of the materials employed and is more normally a fault arising through failure to pay proper attention to detailing. Joints can cause particular problems and are often sealed with mastic at the internal face so that the open joint can drain and ventilate. However, water-repellent treatments are often applied to concrete, generally to reduce water absorption so that the concrete remains light in colour and attractive, even during rainfall. Reducing the absorption also reduces the danger of permanent staining arising where water flow occurs and the wet surfaces attract airborne dirt and encourage algal growth. Water-repellent treatments are inexpensive but their influence on the appearance of concrete structures such as bridges, multi-storey car parks and cast stone or fair-faced concrete buildings can be dramatic, and it is recommended that water repellents should always be used to maintain the appearance of both concrete and masonry buildings, either when first constructed or following expensive cleaning. Water repellents can also be considered as part of a functional structural design.

New solid walls

If solid walls are used for cheapness for garages, out-buildings and agricultural structures, water-repellent treatments will control rain penetration and improve the thermal insulation value of the structure; the thermal advantages are explained more fully in Appendix 6.

Cavity walls

Water-repellent treatments are occasionally required for cavity walls, usually when damp penetration is occurring through bridging of the cavity. The most common example is bridging through mortar slovens or droppings which accumulate on wall ties during construction, usually leading to patches internally opposite each tie bridge; these patches are particularly prominent if the cavity has been filled to improve thermal insulation as the fill reduces water evaporation from the contaminated ties. Mortar droppings can also accumulate at the bottom of cavities and, if these are poorly designed, bridging can occur which may give the appearance of rising dampness whereas in fact the dampness actually originates through penetration of the external skin and bridging through the mortar droppings. Whilst the most severe accumulations of mortar droppings become more porous with ageing, accumulations can also occur in cavities through progressive mortar deterioration, usually at the internal surface on the external skin. Such accumulations can cause dampness to become apparent at soffits if accumulations on the trays above the lintels are sufficient to cause bridging. In all cases these problems are most easily solved by water-repellent treatment of the external surface, although difficulties may arise in diagnosing the true cause of the dampness. Cavity bridging can also occur with some forms of cavity insulation; in some cases whole groups of houses in which cavity insulation has been installed have required water-repellent treatment to overcome the dampness. This problem is most common in cavities filled with mineral wool; the material is pretreated with a sunflower-oil water repellent to prevent this penetration problem, but the treatment can be destroyed by microbiological action.

Other damp-proofing methods

It has been emphasised that only silicone resin should be used, although several other formulations are available. Wax, resin and stearate formulations tend to cause darkening of the treated surfaces and also tend to cause sealing, thus encouraging frost and salt spalling damage. However, whilst these materials may be unsuitable for use on normal porous wall surfaces, they are often useful for other purposes such as wood treatments and chemical injection damp-proof courses.

Dry lining

Water-repellent treatments cannot always be used or are not always successful in, for example, situations where floor levels are below adjacent ground levels and where rubble masonry walls are subject to severe rising dampness. In such circumstances consideration may sometimes be given to dry lining, but there are a number of precautions that must be observed if such treatment is to be reliable. The most efficient dry lining is relatively simple and consists of battens fixed to the wall on strips of damp-proof course or similar materials with a simple plasterboard finish over the battens. The battens should not be fixed over a continuous damp-proof membrane such as polythene, felt or bitumen coating as this will encourage condensation. This system is remarkably efficient in controlling both damp penetration and condensation, as well as improving the thermal insulation properties of a wall. However, the battens should be preserved, preferably with a water-borne pretreatment such as Tanalith C or Celcure A.

High-capillarity tubes

Finally mention should be made of the high-capillarity tubes which are promoted by several manufacturers as a means for remedying dampness in walls. These tubes were originally developed by the late Lt-Col. B. C. G. Shore, a well-known innovator of conservation processes, particularly for use in conjunction with natural masonry. In theory the tubes are manufactured from a porous material and are then bedded in a porous mortar in the walls affected by dampness, either by running horizontally in the walls or with a slight downward slope towards the outside. Moisture within the walls migrates to the internal surface of the porous tubes where it evaporates, the relatively dense humid air flowing out of the tube and thus inducing a ventilation process which encourages drying. Clearly the tubes must discharge to the exterior as they will otherwise serve only to increase the atmospheric humidity. When properly manufactured and installed, porous tubes can greatly increase the rate of evaporation of moisture from a damp wall. However, the evaporation within the tubes also tends to result in the deposition of salts which eventually reduce the evaporation rate, not particularly by clogging the pores but by their hygroscopic action which actually reduces evaporation. For this reason it is necessary to replace the tubes at regular intervals of about five years to maintain the evaporation rate.

However, it must be emphasised that porous tubes should never be used if an alternative damp-proofing process is realistic; porous tubes simply divert the dampness without correcting the cause, and dampness can be controlled much more reliably by preventing rain penetration, rising dampness or condensation.

3.3 Rising dampness

There are many contractors advertising specialist services to remedy rising dampness by installing damp-proof courses, yet most apparent rising dampness cannot be attributed to the absence or failure of a damp-proof course. If apparent rising dampness is confined to solid external walls it must be suspected that it is penetrating dampness accumulating on an existing damp-proof course, or rising dampness through bridging of the damp-proof course, perhaps due to the presence of a rendered plinth or even a line of pointing bridging the damp-proof course externally. Indeed, bridging of a damp-proof course is perhaps the commonest cause of true rising dampness and is most readily remedied by removing the bridge rather than by attempting to insert a new course which is so often the standard process adopted by so-called specialist contractors.

There is a danger of rising dampness in all parts of a building in capillary contact with the ground, particularly the external and partition walls, the sleeper walls supporting plates and joists, and solid floors. An impermeable damp-proof membrane should be present throughout and, if there is no membrane, the best remedy is to insert one. It is relatively easy to insert membranes under plates and joists but in solid floors it may be necessary to re-lay the floor to incorporate a new membrane or it may be necessary to use a heavy coat of bitumen or resin composition over the surface of the existing floor. Walls are much more difficult and must be considered separately in more detail. All damp-proof membrane systems must be continuous and there must never be a chance of bridging due to a gap between adjacent membranes, such as between wall and floor membranes through bridging within the floor material or wall plaster.

Walls

In walls it is first essential to check for the presence of any bridging, perhaps due to changes in levels between the external ground and floors, or between adjacent floors on either side of a partition wall. If bridging is detected it is necessary to remove the bridge and in addition all affected plaster which will normally contain salts which have accumulated through evaporation of the rising dampness over a prolonged period. If there is no damp-proof course one can be inserted.

Solid damp-proof courses

The traditional method is to rake out mortar courses over a short run and to insert overlapping slates, metal sheets or bituminised felt, as in a new structure, but systems are now available in which saws are used to cut slots in walls and which achieve the same result in a rather more efficient manner. However, in all cases there is a danger that the building will settle and, whilst this settlement may be only perhaps 5 mm ($^1/_4$ in), this can be very significant around doorposts, which must be trimmed if cracking of the structure is to be avoided. Another alternative is to run resin into the slot, a process that can avoid settlement if it is carried out carefully in short runs, leaving intermediate sections which will support the structure whilst the resin cures and which can be separately treated later.

Injected damp-proof courses

These various methods for the introduction of solid membranes are definitely the most reliable systems for installing new damp-proof courses but they are very difficult to use in some circumstances, such as around fireplaces and in rubble masonry walls. Chemical injection damp-proof courses are much easier to use but results can be rather inconsistent. There are two popular methods in use at the present time. The first involves diffusion from drillings. In the original process fairly large-diameter holes were drilled downwards at an angle of 45° and were topped up continuously with water repellent until the desired absorption had been achieved. More recently the frequent topping-up process has been replaced by bottles feeding automatically through tubes. The original processes used solutions of waxes in chlorinated solvents such as trichloroethylene, but in recent years aqueous solutions of siliconates have been preferred. Unfortunately the drillings sometimes connect with fractures or open vertical joints in the walls which allow the treatment to drain away into the footings rather than diffuse into the adjacent walling materials to form a continuous water-repellent barrier. In the Peter Cox Transfusion process bottles are used to feed porous sponges which allow the treatment to flow into contacting walling without flowing to waste into fractures and cavities. In wall materials that are relatively impermeable it is difficult to achieve adequate diffusion to produce a continuous water-repellent zone, whilst in materials that are very permeable the flow of rising dampness may carry the water repellents further up a wall before curing occurs so that the damp-proof course may be discontinuous or formed too high within the wall. In recent years an alternative treatment process has been introduced which involves sealing tubes into holes in a wall and injecting non-aqueous water-repellent solution under pressure until the required zone has been treated. It is again difficult to obtain the necessary distribution without wasting material through fractures and cavities, and most operators prefer to use such processes only in brick walls, injecting the bricks alone and hoping that sufficient diffusion will occur into adjacent mortar.

Plate 3.1 Forming a new chemical damp-proof course using pressure injection (Cementone-Beaver Ltd)

Whilst all these injection damp-proof course processes achieve a reasonable degree of reliability if used and installed conscientiously, they cannot be used as general remedies for rising dampness by inexperienced staff and in some areas their use has fallen into disrepute as a result of the high level of failures. However, a number of precautions can be taken which will ensure the greatest possible degree of success. Aqueous diffusion treatments should be applied to walls that are relatively dry so that curing occurs rapidly and before the treatment has been able to migrate to areas where it is not required and is likely to be ineffective. Pressure processes should be applied using low-viscosity products which will penetrate rapidly over the zone to be treated but which will then cure. In referring to low viscosity it is not sufficient to use a solvent of low viscosity but is equally necessary to use a polymer system with a relatively low molecular weight so that it does not suffer excessive filtration during the injection process. Silicone resins that cure rapidly by reaction with moisture in a wall are most effective but in many relatively porous materials the oxo-aluminium compounds (Manalox) can give very reliable results. Many contractors have an excellent reliability record for chemical injection damp-proof courses and they attribute the relatively low number of failures that they suffer to difficulties that occur in drilling some walls, particularly rubble walls constructed from very hard stone such as granite and flint.

Although pressure injection with organic solvent formulations cannot always be used, it should be preferred as it gives more consistent results than the diffusion methods. Distribution and curing problems with aqueous diffusions systems have already been mentioned briefly, and these aqueous systems are also generally alkalis which can introduce serious salt problems in some circumstances.

Electro-osmotic damp-proofing

Electro-osmotic damp-proofing is widely known as a remedy for rising dampness as a result of extensive advertising but this is rather unfortunate as it is certainly the least reliable of all the methods available; in the author's practice complaints regarding failures of electro-osmotic damp-proofing systems greatly exceed the total complaints against all other systems, and many promoters of electro-osmotic systems are now offering chemical injection damp-proof courses in addition. The use of electro-osmotic systems for damp-proofing is based on the classical observation that water flow can be induced in a capillary when there is an electrical potential difference between each end. In 1807 Professor Reuss in Moscow placed electrodes in wet quartz dust and demonstrated that a direct current would drive the water from the anode to the cathode. It was the two Ernst brothers in Switzerland who adapted this process in 1930 as a remedy for dampness in buildings and applied for patents in a number of countries.

Active and passive systems

Electro-osmotic systems of this type, in which a current is passed through the wall, are known as 'active' systems. It was soon found that these active systems possess the serious disadvantage that the metal anode is destroyed by electrolysis. However, the Ernst brothers observed that electrodes installed in damp walls possessed a potential relative to the earth which, they concluded, might well be inducing the rising dampness. They therefore suggested that the dampness could be prevented by earthing the wall electrodes to remove this potential.

'Passive' electro-osmotic systems of this type have been installed in the British Isles since they were first developed by the Ernst brothers, but more recent publicity has centred around systems which are rather more elaborate and claimed to be more effective. Certainly wall masonry generally possesses a positive potential relative to earth and, if this potential is truly responsible for capillary absorption of dampness, a simple short circuit might induce drying. At the same time the short circuit system might operate in reverse, giving the wall electrode system the same potential as earth and thus enabling the dampness to rise further. Alternatively the flow of water in the wall may be induced by the current flow between the zones of different potential. It can then be suggested that the electrode system provides a low-resistance path which avoids current flow through the water in the wall so that rising dampness no longer occurs, but this theory ignores the fact that a circuit must be complete and that current will flow in the electrode only if a current also flows in the wall, presumably exaggerating the dampness in the process.

Certainly the position with regard to passive electro-osmosis is unconvincing. It is said that the rising dampness is caused by the presence of electric charges and it is suggested that they are associated with the surface tension inducing capillarity. It is also shown that these forces act in a consistent direction, water always moving towards the positive charged area. However, there is no explanation for the fact that the potential is sometimes reversed relative to the direction of water flow and there is also no explanation for the fact that these potentials are absent in laboratory experiments with clean porous materials and deionised water. In these experiments the potentials develop immediately a soluble salt is introduced into the rising dampness and it seems that the potentials are actually caused by an ionic diffusion gradient. Thus sodium sulphate in the wall will dissociate into positive and negative ions within the rising dampness. The negative sulphate ions are rather larger and less mobile than the positive sodium ions and filtration of the rising dampness results in a greater concentration of positive sodium ions at higher levels and a greater concentration of negative sulphate ions at lower levels. Where reversal of potential occurs it is simply an indication that the relative sizes of the positive and negative ions have been reversed. The potential within the wall is not therefore causing the capillary absorption but simply resulting from it, and it seems unlikely that these potentials may be used for the suppression of dampness.

Electro-osmotic damp-poofing systems have been investigated, and there is no evidence that passive systems work or are theoretically capable of working. With active systems the position is less clear. Certainly Professor Reuss established that water flow can be induced between electrodes placed in wet quartz dust. Unfortunately many translators have described the experiment as occurring in sand but simple laboratory investigations are sufficient to demonstrate that electro-osmosis can occur only in a very fine pored substrate. In coarser pored substrates flow appears to be induced through movement of charged ions rather than through electron flow, so that the rate and direction of induced water movement tend to depend on both the nature of the substrate and the nature of any soluble salts that may be present. In experiments in the author's laboratory it has been demonstrated that in relatively coarse pored substrates movement is insignificant with deionised water but that, if soluble salts are introduced, the direction of flow depends on the identity of the salt. The clear conclusion is that passive systems can achieve no control over rising dampness but that active systems may be able to induce reverse capillary flow in one way or another under suitable conditions, but the direction of flow will depend on the identity of the soluble salts that are present. Even if care is taken to ensure that electro-osmosis is used only when water movement can be induced in the required direction, problems can be encountered through destruction of the anodes by electrolysis, unless resistant anodes are used such as carbon or platinised titanium.

These statements form a summary of current information regarding electro-osmotic systems and water flow in walls, but another effect is often observed. When an active system is employed the current progressively falls. Contractors frequently claim that this is due to drying of the wall but most physicists and chemists will add that the major effect results from polarisation of the electrodes. If either explanation is correct, disconnection of the system for hours or days should result, on reconnection, in restoration of the original current, but it is actually found that this is rarely the case and a permanent reduction in current occurs until the flow becomes insignificant after a period of a year or two. This diminution in the current flow is sometimes accompanied by a noticeable reduction in rising dampness, but this is usually observed only in the laboratory or in free-standing walls that are not affected by penetrating rainwater or other sources of dampness. The apparent explanation for this phenomenon is that ions within the wall, particularly calcium and aluminium, have been caused to migrate in the current field and have become redistributed, forming a densified zone between the electrodes which both obstructs further flow and obstructs rising dampness. Whilst this phenomenon has been observed in many walls in which active electro-osmotic systems have been installed, it is not clear that the effect is universal, and it appears to be most apparent in walls constructed from carbonaceous stones or from lime mortars.

A further and most important point is that this effect is observed only for systems in which fairly high field strengths are used. The field strength depends on the applied potential divided by the distance between the electrodes, and it is therefore essential to employ an adequate potential in relation to the distance between the electrodes, which should be kept to a minimum. For this reason this effect is never seen in systems that employ earth electrodes but only in systems where the anodes and cathodes are inserted in the wall, usually with the cathode set in a chase at floor level and the anodes set in a chase in the opposite side of the wall 150–300 mm (6–12 in) higher to induce a field diagonally across the wall. The anode must, of course, be resistant to electrolysis; platinised titanium wire is most suitable but some contractors use carbon fibre or carbon rod which, because of its relatively high resistance, must be fed with current at intervals. Good electrical contact with the wall is essential and great care must be taken to ensure that electrodes are firmly bedded in a suitable mortar; ordinary Portland cement, lime and sand mortars are suitable but with the cement content only sufficient to ensure a reasonably rapid set.

Active electro-osmotic systems can control rising dampness to a certain extent provided that unusual salts are absent from the groundwater and that an adequate field strength is employed using properly designed and installed electrode systems. The current consumption usually falls progressively with time until it becomes insignificant, when it is usually found that, if continuous running electrodes have been employed, an impermeable zone has developed which will permanently resist rising dampness. Investigations of installations by the author's staff have invariably shown that passive systems never work and that, when active systems are employed, any failure to control rising dampness can usually be attributed to the presence of unusual salts, or an inadequate field strength through poor design or installation or through damage to the system as a result of subsequent building works in the property.

Tanking

The use of high-capillarity tubes and dry lining systems has already been mentioned in the section of this chapter devoted to penetrating dampness. However, it is probably appropriate at this point to refer to tanking or a process of dry lining designed to withstand hydrostatic pressure where internal wall and floor surfaces are below external ground level as in basements. Damp-proofing is always most readily achieved by reducing the external ground level and, if the excavation cannot remain, finishing the external surface with a dense water-proof rendering before back filling. A damp-proof course can then be injected at the base of the walls internally and a continuous damp-proof membrane provided in the floors. If the floors are suffering only from dampness rising from capillarity or are only a short distance below external ground level, bitumen or resin compositions used for bedding floor coverings may give sufficient protection but areas further below ground level will require more fundamental remedial measures such as the installation of a damp-proof membrane, preferably below the site concrete but at a depth of at least 100 mm (4 in) below the surface of the floor if it is necessary to resist hydrostatic pressure.

Perhaps the most important point is to ensure a proper seal between this floor membrane and the damp-proofing in the walls. Whilst several wall systems can be employed, the dry lining described in the previous section of this chapter is unsuitable, unless some system of drainage can be provided to dispose of the water accumulating behind the vertical membrane, and the most suitable tanking system probably consists of the use of a water-proof rendering. In this connection it is necessary to warn that some water-proofing agents are, in fact, rapid curing agents such as calcium chloride, and it is actually essential to use proper water-proofers such as silicate or aluminium stearate systems. One difficulty is to ensure reliable adhesion between the rendering and the wall itself to resist any hydrostatic pressure. This is usually best achieved by treating the wall first with a very wet cement and sand slurry which will tend to penetrate into the porous wall and achieve good adhesion. Alternatively, proprietary silicate preparations such as Vandex products can be employed to establish adhesion.

In all cases the greatest difficulties arise in joining the wall water-proofing to the floor membrane; this is probably most easily achieved by water-proofing the walls and floors together. For example, in an older property the walls can be stripped to expose bare brickwork or masonry, the floor concrete slab laid, perhaps on a PVC or concrete sheet laid over the oversite, and then the entire wall and floor surfaces treated with a wet cement and sand slurry or a proprietary silicate product, paying particular attention to the joint between the walls and the floor slab. A water-proof cement and sand mix is then used to render the walls and to form a floor screed, the walls eventually being finished in a lightweight plaster to reduce thermal losses and condensation, with the floor preferably laid on a bitumen or resin composition. Mention has been made of the use of proprietary silicate preparations such as Vandex in place of a wet cement and sand slurry but it is necessary to point out that some proprietary silicate formulations are cheap imitations which are largely ineffective and care must be taken when selecting products of this type.

High-capillarity tubes

High-capillarity tubes are often proposed as a means for reducing dampness in walls affected by rising dampness. Their use has already been described in connection with the treatment of penetrating dampness but it must be

Plate 3.2 High-capillary tubes to encourage water evaporation from walls: a ceramic tube is embedded in special mortar and then fitted with a protective cap (*Doulton Wallguard*)

emphasised that they are particularly unsuitable for drying walls affected by rising dampness as they tend to encourage the flow of moisture and thus encourage the accumulation of any soluble salts that may be present in low concentrations in the rising dampness. These tubes are entirely unsuitable for drying damp basement walls; they serve only to remove moisture from the wall and increase the humidity of the basement atmosphere.

Plastering

Many specialist contractors installing damp-proof courses place considerable emphasis on the need for special replastering systems. In fact, any normal plastering system can be used following treatment against penetrating or rising dampness, provided the damp-proofing treatment is reliable and effective, and the specification of special plastering systems is an admission that the damp-proofing system is not thoroughly reliable. The specification of special replastering systems originated in conjunction with a system of passive electro-osmotic damp-proofing which was completely ineffective. Where such treatments against rising dampness have evidently failed there is a tendency for contractors to claim that the damp-proof course 'is operating in the manner intended' and that the dampness is continuing through failure to observe the replastering specification, ignoring the obvious fact that any normal plaster can be used if a wall is dry.

Suitable replastering can alone achieve control of both rising and penetrating dampness in some circumstances. For example, continuous rising dampness tends to deposit salts in the plaster surface which are often hygroscopic and restrict water evaporation, causing dampness to rise progressively higher in the wall surface. The damp surface of the plaster also has a darkened and stained appearance, perhaps with salt deposits and spalling which emphasise the defect. Simple removal of the plaster will remove the salt accumulations and the dampness will then commence from a low level again where it may well be insignificant until sufficient time has passed for further salt deposits to accumulate in the plaster surface. Thus any form of replastering will tend to remove much of the immediate evidence of dampness. In addition the use of water-proof cement, lime and sand undercoat plaster will hold back considerable penetrating and rising dampness, provided that the background is well wetted before the plaster is applied so that it achieves good adhesion; any normal finishing plaster is used on top of the undercoat. Generally lightweight aggregate plasters are an advantage because of their excellent thermal insulation properties and thus their ability to reduce condensation, but in choosing an aggregate for use in the undercoat great care must be taken to select one which will not interfere with the water-proofing properties. It is actually recommended that proprietary cement-based background plaster should be used such as Tilcon Limelite plasters, but gypsum plasters must never be used on damp backgrounds, even if they are described as renovating plasters, a term that refers to their suitability for thick or uneven plaster coats rather than their resistance to dampness.

In summary, replastering will always reduce the visible signs of dampness and the use of special undercoat plasters can give good control of limited dampness under many circumstances. The use of special plasters in conjunction with other processes to control penetrating and rising dampness will clearly improve the reliability of the damp-proofing system but the use of special plaster systems in conjunction with ineffective damp-proofing can only be described as fraudulent; only the plaster system is then effective in controlling dampness and there can be no justification for the cost involved in the installation of an ineffective or unreliable damp-proofing system.

3.4 Condensation and hygroscopic salts

Condensation

Condensation is inevitable when the temperature of air falls below its dew point which depends on its humidity or moisture content. Condensation can be remedied in two different ways; by reducing the humidity of the atmosphere so that the dew point falls below the ambient temperature of the atmosphere and surrounding surfaces, or by increasing the temperature of the atmosphere and surrounding surfaces. Both these systems are realistic in appropriate circumstances. Condensation occurs particularly in kitchens, bathrooms and laundries which are subject to periods of high temperatures and high humidities. As these periods are often relatively short they do not greatly influence average surface temperatures, so that condensation tends to occur on the walls as well as on the cold water pipes and windows. If humid air is allowed to escape from these rooms into other parts of the accommodation condensation will also occur elsewhere; it is particularly common in relatively cold areas such as toilets, corridors and larders. Where condensation occurs regularly and persistently mould may develop on the affected surfaces and may in turn support small plaster beetles.

Ventilation

In all cases of condensation it is best to rely on ventilation to remove the humid air but this is not really as simple as it sounds. It is essential that air flows from the humid rooms out of the building but this is difficult to ensure if the wind is in the wrong direction and tending to move the humid air into the accommodation. The best solution is to use extractor fans in humid areas, but in practice there is a tendency for these to be either ignored or used for excessive periods if they are manually controlled, and automatic systems are certainly desirable. Sensors are now available which will operate the extractor fan only when there is a danger of condensation, avoiding unnecessary extraction and heat loss. In some more sophisticated systems the entire house is slightly pressurised, usually by installing a blower in the roof space where it draws in air already slightly heated by solar radiation or convection loss from the accommodation.

Adequate ventilation is essential in kitchens fitted with boilers or gas cookers, and this ventilation system frequently results in an unnecessarily cold building, yet the system can also be used to avoid condensation problems. The essential requirement is to provide a well-sealed door between the kitchen and the accommodation, preferably fitted with an automatic closing device and, if possible, two doors with an intermediate vestibule in order to prevent draughts. A boiler in the kitchen, or an extractor fan within a cooker hood (not a recycling odour remover), can often provide sufficient ventilation to prevent excessive condensation by limiting the humidity of the kitchen air without the need for any other precautions, but it must also be appreciated that there is often a danger of condensation occurring within the boiler flue.

Flue condensation

With a slow-burning solid fuel boiler the flue is usually sealed directly into the top of the boiler and there is then a tendency for condensation to occur as the combustion gas is cooled through contact with the flue surface. This condensation is invariably acid owing to the absorption of combustion gases and, if the flue is unlined or parged, these acids are absorbed into the chimney structure, where they attack mortar and masonry, forming soluble salts which often migrate through plaster on chimney breasts to give hygroscopic damp patches. There are two ways in which these condensation problems can be avoided. The first is to line the flue but, if a slow-burning boiler is involved, condensation will still occur and provision must be made to drain this condensation from the bottom of the flue; it is failure to provide this drain that often produces heavy staining of the pipe leading from the boiler to the flue. However, a far better method is to avoid the condensation by introducing sufficient air to the flue to reduce the humidity to a harmless level. In many cases boilers are connected to flues using side entry pipes and air can be easily introduced to the main flue by inserting an air brick to the exterior well below the boiler flue entry point. In modern oil and gas boilers draught compensators or even open flues are employed which draw air from the room in which the boiler is installed, but it is necessary to appreciate that these systems can be efficient only if the room is freely ventilated from the exterior.

Window condensation

Condensaton in living accommodation is usually confined to windows which represent the coolest parts of the structure and are thus most likely to be below the dew point appropriate to the atmospheric humidity. Windows are cold because of their relatively poor thermal insulation properties. Whilst these properties can be improved by double glazing, great care is necessary; in all cases the fitting of secondary frames to existing widows is most thermally efficient because of the wide air gap that is formed, but air leakage through the secondary glazing can result in severe condensation on the original glazing. The replacement of wood-framed windows with aluminium-framed double glazing often increases the danger of condensation on these frames as explained in Chapter 5 and Appendix 6.

Wall condensation

If condensation occurs on wall surfaces this is also due to poor thermal insulation properties. With solid walls the use of an external water repellent will reduce the moisture content and improve the insulation properties most cheaply but it may also be necessary to use lightweight insulating plaster internally. If dampness is particularly severe it must be suspected that the atmospheric humidity is unusually high, perhaps through a flow of humid air from a bathroom or kitchen, and some attempt must be made, as previously described, to control this problem. These comments assume that condensation is occurring uniformly over external walls but cold spots sometimes occur. These can sometimes be attributed to the presence of air bricks in the external skins or lintels causing thermal bridging, but in all cases the same principles apply, generally the use of an insulating plaster to improve the thermal insulating properties.

Hygroscopic salts

Hygroscopic salts are generally introduced into walls in rising dampness and accumulate at the surface where evaporation occurs. Hygroscopic salts are introduced less commonly in penetrating dampness, although occasionally in polluted urban atmospheres penetrating dampness may contain salts which have formed through the atmospheric pollutants reacting with the wall materials. In all cases it is necessary to cure the source of dampness and remove the salt deposits. Some damp-proofing contractors emphasise the need to remove salts from within walls but rising or penetrating dampness usually contains only low concentrations of salts and significant accumulations occur only at evaporation surfaces, where they can be comparatively easily removed by replacing the plaster or by a process of alternating water spray and removal of subsequent salt efflorescence by brushing.

Salts can also be introduced from other sources. Salts on chimney breasts due to flue condensation have already been mentioned; in addition, coke stores, water softening salt and bacon curing are all sources of salt problems in buildings. If plaster is present its removal and replacement will often control the salt problems for a considerable period, but sealing is sometimes necessary. The use of cement and sand slurries or proprietary silicate preparations such as Vandex in conjunction with water-proof lightweight undercoat plasters has already been mentioned in the previous section of this chapter. Bitumen products and foil can also be used, but they are actually much more difficult to plaster. Salt neutralisation treatments are sometimes advertised but salt neutralisation in the chemical sense is actually impossible and these products are invariably water-repellent or sealing treatments, often introducing replastering difficulties.

Interstitial condensation

Condensation is not always visible but may occur within structural components when it is usually known as interstitial condensation. Cooling of humid air below the dew point inevitably results in condensation. In temperate or cooler climates it is normal for the interior temperature to be higher than the exterior and, as normal life processes generate humidity even without the moisture contributed through cooking or washing, there is always a tendency for condensation to occur if warm humid air from the interior diffuses towards the colder exterior. Ventilation with exterior air will always minimise the danger of interstitial condensation as the dew point of this exterior air cannot be higher than the exterior temperature; this precaution is easily adopted for roofs and timber-frame walls. However, it must not be imagined that these interstitial condensation problems are necessarily confined to cooler climates as air conditioning in the tropics can introduce a reverse temperature gradient in which the interior is cool and the exterior warm.

Vapour barriers

Interstitial condensation occurs as humid air passes from a warm towards a cold area and is cooled below its dew point, but in normal porous structures this condensation may be entirely insignificant if moisture accumulating in this way is able to disperse to the exterior by evaporation. Serious problems arise only when unnecessary moisture vapour barriers have been introduced in the cold zone so that the condensing moisture remains trapped. For example, in cool climates a property may be 'weather-proofed' using an impermeable exteror paint, and condensation accumulating beneath this barrier at the cold surface may eventually freeze, perhaps causing severe spalling damage. In a flat roof construction, humidity in the roof space, perhaps diffusing from the accommodation or from the wall cavities, may condense under the impermeable roof covering, perhaps causing severe decay of the supporting boards and joists and even severe dampness and staining of ceilings beneath. In tropical areas condensation may occur beneath impermeable floor coverings or behind wall linings in air-conditioned buildings, perhaps causing severe fungal decay damage. Whilst the damage caused by interstitial condensation is often extremely serious, the problem is quite simply remedied by always ensuring that moisture vapour barriers, if present, are positioned at or close to the warm surface and that substantial insulation is provided between the vapour barrier and the cool surface. With porous solid structures such as solid masonry walls or wood window and door frames, interstitial condensation can be avoided if the water vapour resistance of the internal plaster and decoration is at least five times the value for any external render and paint. If microporous decoration is used externally on wood window frames, an impermeable finish must be used internally.

Painted joinery

There are a number of less obvious situations in which interstitial condensation can cause severe problems. With normal painted exterior wood joinery such as window frames, any moisture entering the wood through minor defects in the paint coating, such as normal condensation running down glazing and being absorbed through a defective coating around the rebate, will be trapped and cannot disperse because of the relatively impermeable paint coating. During cold weather this moisture will be distributed within the wood, tending to concentrate through condensation under the external paint surface where it often causes peeling and blistering of the paint through preferential wetting failure, as explained in more detail in *Wood Preservation* by the same author. It is this effect that leads to the high cost of maintaining exterior painted wood joinery. However, in recent years a number of manufacturers have introduced permeable water-repellent treatments for maintaining the exterior surface so that any trapped moisture can freely disperse, while the interior surfaces are maintained using an impermeable system as previously explained.

Defects arising through condensation or hygroscopic salts have always been the most difficult dampness problems to diagnose reliably. Special care has therefore been taken in Appendix 3 to suggest methods that are most suitable and most reliable. Whilst many damp problems can be diagnosed without the aid of special instruments, reliable measurements are essential to the correct diagnosis of condensation problems.

4 Masonry treatment

4.1 Masonry deterioration

Masonry deterioration can be attributed to a number of chemical, physical and biological factors, many of them closely interrelated. Before considering deterioration processes in detail it is essential to establish the nature of masonry. As a starting point natural stone can be conveniently classified according to its origin into the primary or igneous rocks, the secondary or sedimentary rocks and the metamorphic rocks.

Igneous rocks

The igneous rocks, which are formed by solidification of molten material, include the basic rocks such as basalt and dolerite, often known as whinstone, and the acid rocks such as granite. Whilst whinstone is most commonly crushed for road stone, granite is widely used as a building material but it varies greatly in its nature. It is composed largely of silica deposited as crystals which may be small and open-textured with the appearance of a rather coarse sandstone, or large and compact; the latter is virtually impermeable to water.

Sedimentary rocks

The sedimentary rocks are formed from water or air-borne debris. In sandstones deposits of silica granules, usually from granitic rocks, are consolidated with an amorphous silica or calcium carbonate cementing matrix, often in conjunction with aluminium or iron oxides. The properties of sandstones depend largely on the cementing matrix as the silica granules are virtually inert. The limestones constitute the second major group of sedimentary rocks and are formed either from the accumulations of animal shells in carbonaceous cement or by crystallisation from solution to give a characteristic oolitic structure. Sedimentary stones of intermediate structure also occur and are termed arenaceous limestones or carbonaceous sandstones, depending upon whether silica and carbonate predominates. The limestones used in construction are generally calcium carbonate but magnesian limestones of mixed calcium and magnesium carbonate are extensively used in some areas; York Minster is an example of a magnesian limestone building. Magnesium carbonate is rare in buildings because it is soluble in normal rainfall and very soluble in acid polluted rainfall.

Metamorphic rocks

The metamorphic rocks are developed from all these types under the influence of heat and pressure. Thus the schists and gneisses are formed from the igneous rocks, quartzite from sandstones, marbles from limestones and slates from mud and shale.

Suitability for building depends on a number of factors, the most important being availability and workability. Obviously local stones have been preferred but they have not always been ideal and progressive improvements in communications over the years have caused building stones to be obtained from ever widening areas. For these reasons the masonry conservationist must possess an exceptionally broad knowledge, even if his activities are confined to a particular locality, and in addition to having a knowledge of stone it is essential that he understands the basic principles of construction.

The construction of a natural stone building will depend very largely upon the properties of the material that is used. Most existing stone buildings are in solid masonry constructed of porous sandstones or limestones. In this form of construction resistance to rain penetration depends on the balance between the water absorbed, the capacity of the wall, and the rate of subsequent evaporation, as previously explained in Chapter 3. Heavy showers alternating with bright and windy periods are less likely to lead to major water accumulations than continuous drizzle with exceptionally humid conditions which obstruct evaporation, but almost always dampness will be apparent on areas of thinner wall section, such as window reveals.

Leaching and acid attack

Although buildings may be perfectly dry when constructed in this manner, limestones or carbonaceous sandstones may become progressively more porous as the carbonate content is leached away by rainfall. Simple leaching is usually insignificant, except in the case of the more soluble magnesium carbonate limestones, but rainfall normally consists of carbonic acid formed by the absorption of atmospheric carbon dioxide and this weak acid attacks the carbonates in natural stone to form the more soluble bicarbonates. In polluted urban and industrial areas the much stronger sulphur and nitrogen acids may be formed, greatly increasing the rate of carbonate erosion and even causing the slow deterioration of silica. The rate of deterioration depends on the physical nature of the stone in terms of crystallite form and size, and pore volume and size. The relationship governing the rate of leaching appears to be complex but actually obeys the well-known natural law of mass action; the rate of leaching is greatest for stones with a high total porosity coupled with small crystallite size. Thus an impervious carbonate such as a marble will suffer only surface etching whereas a stone with high total porosity comprising pores of small diameter, a microporous stone, will tend to retain moisture owing to limited ventilation of pores and powerful capillarity. When water retention occurs in this way it permits protracted absorption of atmospheric pollutants and thus prolonged acid attack of the stone components so that deterioration tends to be more severe.

Salt crystallisation

Stone deterioration does not depend on acid leaching alone. Acid leaching results in the formation of soluble salts, sulphates or nitrates if sulphur or nitrogen gases are present as atmospheric pollutants. As these salts accumulate through evaporation at the stone surfaces crystallisation occurs which involves the absorption of water of crystallisation, the number of water molecules absorbed per salt molecule depending on the nature of the salt and thus the nature of the atmospheric acid and the stone with which it reacts. Calcium sulphate has two molecules of water of crystallisation, but calcium nitrate has four, magnesium sulphate has seven and magnesium nitrate has six. The growth of salt crystals depends on the number of water molecules absorbed during crystallisation and, as this crystal growth stresses and damages the stone, it is clear that a magnesium limestone is generally likely to be less resistant to crystallisation damage induced by atmospheric pollution than a calcium limestone.

Durability

The natural durability of a stone depends on a number of factors. One factor is the ability of the stone to resist leaching which will increase its porosity and reduce its cohesiveness. Resistance to leaching depends on the relationship between total porosity and crystallite size, leaching being most severe for microporous stones which tend to hold moisture and thus absorb atmospheric pollutant gases over a longer period. Salts generated by the action of rainfall, or from rainfall that has absorbed pollutant gases, tend to crystallise in the evaporation zone close to the stone surface. Crystal growth invariably occurs owing to the absorption of water of crystallisation, the degree of expansion depending on the nature of the salt involved. The salt depends on the stone and the identity of the pollutant gases, whilst resistance to the crystallisation stress induced by crystal expansion depends upon the porous properties of the stone; whatever the total porosity, a macroporous stone is invariably more resistant to crystallisation stress than a similar microporous stone, apparently because the latter tends to hold water for a protracted period. This relationship also applies to crystallisation stress caused by freezing so that the pore-size distribution within a stone, or the proportions of micro- and macroporosity coupled with total porosity, will generally define stone durability in the most reliable way, although naturally resistance to crystallisation also depends on the cohesiveness of the stone, so that stones of very low porosity tend to be more durable in all cases.

The soluble salts resulting from acid rainfall attack will depend on the nature of the stone so that a stone of given durability must be predominantly macroporous if it is magnesium limestone, perhaps less macroporous in the case of calcium limestone and even less macroporous in the case of a sandstone, these macroporosities being related to the salts that are likely to be formed. However, salts may be derived from other sources, the most obvious being water flow from one stone to another. Thus if a calcium limestone is exposed to washings from a magnesium limestone, it must be particularly macroporous if it is to be durable, just as if it were a magnesium limestone itself. Whilst washings from magnesium limestones onto calcium limestones frequently cause salt crystallisation damage, the best known example of this effect occurs when washings from calcium limestone walls are carried onto sandstone paving when severe deterioration of the latter can occur, even if a stone of normally excellent durability is involved. Salts can also be introduced from other sources, particularly in rising dampness, but here the main problem is to take normal precautions to avoid or remedy the dampness defect.

Typical deterioration

The chemical and physical deterioration that has been described can take various forms. In rural areas leaching may be principally due to carbonic acid formed by the absorption of carbon dioxide into the rainfall. This carbonic acid will have no effect on siliceous stones but a sandstone with a carbonate cementing matrix must be considered as a carbonaceous stone which will be affected in the same way as limestones by leaching in the form of bicarbonate. During dry periods the bicarbonate will decompose to redeposit the carbonate so that leaching tends to involve redistribution of carbonate and its concentration at the evaporation surfaces. As a result the surfaces tend to become progressively denser whilst the stone to the limit of rain absorption will tend to become progressively weaker. Naturally this acid will attack particularly the smaller crystallites and, as these are often the cementing matrix, the stone may become very friable in the area affected by leaching.

Where acid pollutants are present the damage will be caused more rapidly, and although calcium sulphate may contribute to surface densification, the danger will then arise of crystallisation of salts beneath the densified surface, often causing severe spalling. With some predominantly macroporous stones the surface densification is often absent and damage is limited to very slow surface erosion, apparently resulting from stressing of the surface pores during crystallisation of the salts. St Paul's Cathedral in London is constructed from the predominantly macroporous Portland limestone, and on horizontal surfaces catching rainfall the damage tends to be progressive erosion occurring at a rate of less than 10 mm per 100 years. Calcium sulphate does not form a surface skin on horizontal areas exposed to rainfall because it is washed away but, on adjacent vertical areas, a dense surface patina of calcium sulphate can develop, which eventually forms a darkened but apparently durable surface, presumably because it effectively prevents further rainwater absorption. However, where rainwater penetration can occur from another direction, spalling of the vertical surface may occur through crystallisation beneath the densified layer, and very severe spalling of this type is common on stones that are less macroporous and less cohesive such as Bath limestones. This spalling tends to occur to a consistent depth and is thus often described as contour scaling. On stones that are relatively microporous with high total porosity there may be simply a tendency for a very thin veneer to peel away at intervals.

Plate 4.1 Erosion of a Portland limestone statue at St Paul's Cathedral, London: upper surfaces are eroded by rainfall and remain clean, whereas protected surfaces are apparently sound but dirty *(Penarth Research International Ltd)*

Plate 4.2 Contour scaling at Tintern Abbey: this damage is caused by crystallisation of salts at a consistent depth (*Penarth Research International Ltd*)

Reference has been made to the fact that limestones composed of magnesium carbonate tend to be much less durable than those composed of calcium carbonate as magnesium salts are more soluble and their expansion is greater on crystallisation. However, the physical properties of the stone, particularly the total porosity, the degree of macroporosity and the crystallite sizes, may be more important than chemical composition. In particular many traditional building limestones contain magnesium in the form of dolomite, a double carbonate with calcium, and where stones consist entirely of this mineral in its common rhombohedral crystallite form the stone tends to be particularly durable. The Huddlestone used extensively on York Minster is a stone of this type and has proved to be extremely durable. In contrast attempts at conservation during the last 100 years have often involved the replacement of this magnesian limestone with calcareous limestone which has deteriorated extremely rapidly, despite a normal reputation for excellent durability, through absorption of washings from neighbouring magnesian limestone.

Natural variability

Unfortunately this description of stone deterioration has had to be substantially simplified, mention being made only of the principal building stones, and these being assumed to be consistent in composition. There is, in fact, an enormous range of building stones of many different types but, even within a quarry, stones vary from bed to bed and also within a bed. The durability reputation of a particular stone has been established over many years through its reputation in service, and the obvious facts that it is no longer available from a particular quarry and that currently available stone may possess entirely different properties are usually ignored. Thus reputation cannot be considered to be a reliable method for stone selection for either new construction or repair purposes.

Durability assessment

Appearance is of considerable importance but an original quarry may be entirely exhausted and it may be necessary to seek an alternative source of stone, usually from a quarry working the same deposits, as this is the most reliable way to ensure a matching appearance. Samples of the stone must be checked to ensure that they are likely to be reasonably durable, usually by checking the porosity characteristics. Total porosity is of limited significance, except that a high total porosity frequently means a low cohesive strength and comparatively rapid leaching losses. Pore-size distribution is far more important as it enables the water retaining properties of the stone to be assessed, a microporous stone that retains water for an excessive period being particularly susceptible to both leaching and crystallisation damage.

Plate 4.3 Frost spalling in a microporous limestone window-sill: this damage results from expansion of the stone when it freezes in a saturated state, caused in this case by a defective rainwater pipe. (*Penarth Research International Ltd*)

The first method for this type of assessment was known as the determination of the saturation coefficient. This is a relatively simple process in which a stone is permitted to absorb water naturally and the amount of absorption is compared with the porous volume determined by vacuum impregnation. Despite the simplicity of the method the saturation coefficient gives an excellent indication of comparative durability when applied to stones of a single structural type, particularly stones derived from a single bed which require routine checking. Comparative durability assessments on stones of wider types are more difficult and must involve determination of pore-size distribution, that is the proportions of various pore sizes that occur. Usually this is achieved by defining a critical pore size such as 0.005 mm, pores exceeding this size being considered macropores. The normal method involves saturating the stone with water and then applying a pressure or vacuum which will remove water from the macropores alone. In the classical method samples of saturated stone are placed in contact with a suitable capillary bed and a vacuum of 600 cm of water is drawn, representing a pressure sufficient to evacuate the macropores. The preparation of the samples is difficult and the test is consequently expensive. However, a method was developed in the author's laboratory in which the small sample of stone is replaced by a hole drilled in a block of stone. A probe is sealed into the hole and water is injected to saturate the stone, the eventual water flow rate being an indication of total porosity. Air is then introduced at a pressure which will remove water from the macropores alone, the air flow rate indicating the macroporosity. In this way it is possible to determine macroporosity as a proportion of the total porosity, thus indicating the probable durability of the stone.

Pore-size distribution indicates only the probable durability of stones of average porosity as resistance to leaching and crystallisation will naturally depend also on both the total porosity and the cohesiveness of the stone. Thus a stone of low porosity is likely to be much more durable than a similar stone of high porosity and this factor should be taken into account when assessing pore-size distribution results. Whilst saturation coefficient and pore-size distribution tests are usually preferred for routine durability assessments, a laboratory crystallisation test is likely to give more reliable results when stones of different types are being compared. Crystallisation of sodium or magnesium sulphate during a wetting and drying cycle and the consequent weight loss through erosion from the stone surface is the method that is generally used, but freeze–thaw cycles are preferred in some countries. Obviously these methods assess only resistance to crystallisation; acid tests are used to assess resistance to polluted atmospheres and various strength tests are also necessary.

Whilst these various porosity and crystallisation tests have been used for many years for laboratory assessment of limestones, the tests have not been used to any significant extent for sandstones as results tend to be rather inconsistent. A 14% sodium sulphate crystallisation test has been used in recent years as the basis for the classification of limestones into five durability classes A to E which can be related to exposure conditions in service, but research by the author has established that the same system can be used for sandstones to indicate durability in terms of resistance to crystallisation stress through frost or salt contamination. The same research also resulted in a simplification of the durability assessment of porous stones using porosity factors. The most durable stones have a low proportion of microporosity, a low total porosity, and a high cohesive strength which is usually associated with low total porosity. Microporosity is indicated by capillarity, or the amount that is absorbed when a stone sample is stood in contact with water, and total porosity is indicated by the amount of water absorbed by vacuum impregnation; these porosity measurements are expressed as a percentage of the volume of the sample and have no units. A durability factor *D* can be obtained in this way from the square of capillarity divided by porosity; the same factor can be calculated from old test results by the square of saturation coefficient multiplied by porosity. A value of less than 4 indicates a stone that is Class A and which will suffer a crystallisation loss of less than 2% in 15 cycles of the 14% sodium sulphate test; the values for other classes are 4 to 5 for Class B (2% to 5% loss), 5 to 7.5 for Class C (5% to 15% loss) and 7.5 to 12.5 for Class D (15% to 35% loss); values above 12.5 indicate Class E (loss in excess of 35%). These laboratory assessment methods are not limited to natural limestones and sandstones but can be used for all porous materials including clay and concrete bricks and tiles. Samples should be 50 mm square and 25 mm deep; capillarity is determined by standing the block in a tray for 24 hours with the lower face in continuous contact with water so the laboratory equipment incorporates a device to maintain a constant water level.

The method for assessing stone involving the drilling of a hole can, of course, be used on existing buildings to establish the basic porosity characteristics of the stone in order to help with matching of replacement stones. In carrying out such investigations it is important to appreciate that, whilst stone on a building may be showing apparently severe deterioration, this may have occurred over a period of many centuries, indicating a stone that is, in fact, very reliable in service. It is therefore essential to establish the history of a stone building before considering remedial or replacement works.

Atmospheric pollution

Unfortunately stone deterioration is not confined to the relatively simple chemical and physical characteristics of the stone and the surrounding atmosphere that have already been described. For example, the Portland limestone of St Paul's Cathedral in London is exposed to an atmosphere polluted with sulphur dioxide. This generates rainfall containing sulphurous acid, which should react with the calcium carbonate of the stone to form calcium sulphite, but analysis of the stone shows that it contains only calcium sulphate.

Bacteria

Whilst it is possible that this occurs through chemical oxidation it is a strange coincidence that the occurrence of calcium sulphate is always associated with the presence of high concentrations of sulphating bacteria, particularly *Thiobacillus* species. The atmospheric sulphur dioxide originates largely through burning oil; the heavy fuel oil used by power stations and for heating large buildings contains large quantities of sulphur and, whilst the Clean Air Acts have effectively restricted particulate emissions, the present requirements limit only sulphur dioxide concentrations without controlling total emissions. As a result operators blow extra air up flues if permitted sulphur dioxide concentrations are likely to be exceeded so that sulphur dioxide pollution is now a particularly serious problem in urban and industrial areas. In industrial areas nitrous oxide also occurs in the atmosphere, usually through burning coal, but exactly the same oxidation characteristics are observed as with sulphur dioxide; whilst nitrites would be expected to occur on stone, only nitrates are found and nitrating bacteria of the *Nitrobacter* species are invariably present.

In rural areas local concentrations of ammonia often occur, particularly from animal urine in stables and byres. Where the ammonia is absorbed into porous building materials *Nitrosomonas* bacteria are invariably present, which convert the ammonia to nitrites, and *Nitrobacter* species then convert the nitrites to nitrates. This sequence of events frequently causes severe spalling damage to asbestos-cement roofing of stables and byres through nitrate crystallisation that can be attributed ultimately to animal urine through the sequence that has been described.

Sulphate attack

Another interesting deterioration sequence involving bacteria occurs in concrete sewers. Sewage is anaerobic and any sulphates that are present are reduced to hydrogen sulphide by Desulfovibria species. This hydrogen sulphide, or 'bad egg' gas, is released from the sewage and is absorbed into moisture on the exposed concrete above. At this point concentrations of Thiobacillus species occur which cause the formation of sulphuric acid which then reacts with the tricalcium aluminate of the cement in the concrete to form calcium sulphoaluminate, a change that causes considerable expansion. Eventually this calcium sulphoaluminate decomposes to sulphate, a change that results in collapse of the expansion and loss of cohesiveness.

The action of sulphuric acid or sulphates on tricalcium aluminate in cement in this way is known as 'sulphate attack' and can occur under many other circumstances. For example, sulphur pollution acids may react with cement mortar in this way. Bricks sometimes contain significant amounts of sodium sulphate and, if heavy rainfall occurs within a year or two of erection of new brickwork, these sulphates may migrate into the mortar, often causing considerable expansion and severe damage to the structure that may make complete reconstruction necessary. For example, gable walls may expand, lifting the entire roof structure on the purlins. Brickwork may expand round doorways, leaving gaps over the door heads. Sulphate attack confined to one side of a chimney exposed to rainfall may cause considerable torsioning, and distortion caused similarly in other brickwork can lead to very severe damage. In all cases the damage can be attributed to the presence of sulphate coupled with conditions that permit its migration into the mortar, perhaps poor design such as inadequate parapet copings which permit brickwork to become saturated. If there is a danger of sulphate contamination of new construction or repair works, perhaps through the use of bricks with high sulphate contents, it is always advisable to use sulphate-resisting cement rather than ordinary Portland cement.

Sulphate attack is not confined to mortar in brickwork and masonry but can also occur in concrete subjected to rising dampness containing sulphates and if sulphates are incorporated into concrete or mortar mixes. For example, thin stone cladding is frequently positioned during erection using spacer dabs of mortar. If the air gap between the cladding and the background is too large, normal cement or cement and lime mortars will tend to slump and some fixers are tempted to add gypsum plaster to stiffen the mix. This causes slow but progressive sulphate attack and expansion of the dabs which often crack the cladding panels or fracture them away from their fixings. Sulphate attack is one of the most serious problems encountered in building construction and care must be taken to avoid any possibility that soluble sulphates can migrate into any building component containing ordinary Portland cement.

Algae

Bacteria are active only on damp surfaces but the same conditions also support algae. Algae will develop extremely rapidly whenever suitable conditions of dampness, warmth and light occur, usually becoming apparent as a bright green but occasionally dark green, brown or pink coloration, and frequently developing within an hour or two of rainfall. Drying results in death of the algae which then form deposits of dirt or humus on the stone surface. Deposits may arise from other sources, such as wind-blown dirt, bird-droppings and leaves from overhanging trees.

Higher plants

These accumulating deposits permit higher plants to develop, particularly mosses but also liverworts, grasses and eventually trees.

Fungi

At the same time deposits of dead algae and other organic matter permit the development of fungi such as *Cladosporium, Phoma, Alternaria* and *Aureobasidium* species.

Lichens

Lichenised fungi or lichens may also occur. These consist of a fungus, normally an Ascomycete, and an alga in symbiotic relationship, the alga normally being present within the fungus. The fungal hyphae penetrate into the stone by exploring fractures and pores but also by generating organic acids such as oxalic acid. In limestones these acids enable the lichen to dissolve stone material which is usually redeposited as calcium oxalate in or close to the thallus, the part of the lichen on the surface of the stone. In some

Plate 4.4 Typical lichen growth on limestone headstones: general appearance and close-up (*Penarth Research International Ltd*)

cases this accumulating calcium oxalate within the thallus will eventually kill the lichen, although a lichen 'fossil' consisting of the calcium oxalate deposit will remain. In other cases the calcium oxalate is deposited immediately beneath the thallus within the stone so that, in the case of some crustose lichens, the past presence of lichen growth can be detected on stone by examining the surface and detecting the dense calcium oxalate deposits. Acids generated by lichens, and also by mosses, not only damage carbonaceous stones such as limestones or sandstones with a carbonate cementing matrix but also attack silica and cause etching damage on granite and even glass surfaces. Where mosses and lichens occur on roof slopes the acids can cause very severe deterioration of metal gutters, even destroying lead as described in Chapter 3. Infections found on masonry and similar surfaces can be identified using the key in Appendix 2.

Lichens can be classified into three main types depending on the form of their thalli or surface growths. A thallus spreads from the point of germination so the size of a thallus is a useful indication of the age of the infection. The crustose lichens have a thallus that forms a crust in close contact with the stone surface. If the crust is removed by scrubbing or scraping the thallus will regrow over a period of perhaps a year to its original dimensions. In many cases the older or central part of the thallus becomes relatively inactive, perhaps due to accumulations of calcium oxalate, and will fall away to disclose a clean stone surface which will then be recolonised by new growth. In most cases crustose lichen activity is associated with densification of the immediate surface due to deposition of calcium oxalate but severe weakening of the surface may occur at a greater depth and, if crystallisation occurs through the presence of soluble salts or through freezing, the densified surface may spall away. In some polluted atmospheres lichen growth is entirely suppressed but pollution from, for example, fertiliser factories may result in accelerated lichen activity which may in turn cause spalling or contour scaling at frequent intervals, resulting in the progressive loss of stone surface and progressive erosion of any detailing such as lettering. In all polluted atmospheres many sensitive species are controlled but some resistant species may be able to develop such as *Lecanora* and *Candelariella* species, particularly on limestones and other surfaces such as asbestos cement which apparently assist in the neutralisation of the atmospheric pollution. Crustose lichens vary greatly in size, old growths of 300 mm (12 in) across being fairly normal, yet some species are minute, their growth being confined to the interior of the pores and their protruding black fruiting bodies often being mistaken for dirt.

Foliose lichens have thalli in the form of leaves or scales which are attached to the surface by threads, whereas fruticose lichens have branching thalli which are attached to the surface at their base. Whilst all three types of lichen occur on stone, it is the crustose types that are most interesting and also most important in the sense that they cause the surface densification that has been described. Lichen activity increases in humid areas; fruticose species are particularly noticeable as heavy encrustrations on trees close to west-facing coasts. Particular species tend to be associated with particular conditions; species able to tolerate atmospheric pollution have already been mentioned. *Calaplaca* species are usually associated with limestones, whilst *Tecidea* and *Rhizocarpa* species are usually associated with sandstones. Some species tend to be concentrated in nitrogen-rich conditions such as surfaces affected by bird droppings. The distinctive colours of many lichens can enable stone types or conditions to be identified, yet the identification of the lichens themselves is extremely difficult. It is recommended that the reader should not attempt to go beyond the species described by Alvin in *The Observer's Book of Lichens* (Frederick Warne).

Slime fungi

Whilst bacteria, algae, lichen, mosses and higher plants represent the most severe growth problems on masonry surfaces, slime fungi sometimes occur and must therefore be mentioned. These fungi do not possess a distinct structure but simply form a film of slime on the stone surface which frequently incorporates trapped algae which give it a green, dark green or black appearance. The slime fungi can occur wherever continuous dampness exists coupled with a suitable source of nutrient; for example, dark deposits of slime fungi are sometimes seen on interior surfaces in Cornish churches where dampness is derived from penetration of porous granite walls or condensation on granite columns through periodic heating. Slime fungi can also occur in some polluted conditions where they can form a complete dense coating on stone surfaces.

Water and stone deterioration

Stone deterioration, whether it is chemical, physical or biological, is always associated with the presence of water. New buildings must be designed to avoid unnecessary accumulations of water in masonry surfaces or, where water accumulations are unavoidable as in parapet copings, care must be taken to use only stone of adequate durability. Whilst other methods of controlling moisture, such as the use of water repellents, may be useful in some circumstances, it will be appreciated from comments in Chapter 3 that water repellents must be applied with care as they can exaggerate deterio-

Plate 4.5 Typical film of a slime fungus on limestone steps: colour is dark green or black *(Penarth Research International Ltd)*

ration if they encourage crystallisation to occur beneath the treated surface. For this reason water-repellent (or water-proofing) treatments should never be applied to stone in contact with soil moisture such as headstones or masonry. It is also advisable to avoid the use of water repellents as a means to control algal and other growth on stones as the use of suitable biocides will give more reliable results without the dangers associated with water-repellent treatments.

4.2 Biocidal treatment

Biological growth

It must be accepted that moss, lichen and algal growth often gives a building a very attractive mellow appearance but conceals the true appearance of the structure and may cause damage. In fact, damage is unlikely on some relatively inert brick, tile and stone surfaces, but severe structural deterioration can occur if, for example, heavy growth occurs on carbonaceous surfaces or acid washings from heavily contaminated roofs attack lead, copper and zinc gutters and flat roofs. Whether growth needs to be encouraged or controlled depends largely on the circumstances. For example, it might be considered desirable to encourage growth on the surface of a new extension or repair to an old building, or on new cast stone garden ornaments, whereas control of growth may be considered essential in other circumstances, such as when the growth or its acid washings are causing damage as previously described or when the growth conceals important inscriptions on monuments. Lichenologists frequently oppose attempts to control growths on headstones in churchyards, arguing that these growths represent valuable flora which should be preserved. In fact, they are simply saying that it is much more convenient to search for lichens in neglected churchyards than on natural stone outcrops and they completely ignore the obvious fact that headstones and monuments are constructed from stone that should remain durable and possess inscriptions that should remain legible as they otherwise serve no useful purpose.

Deliberate attempts to encourage growth are not usually successful. If surfaces remain wet algal growth will invariably occur but lichen and moss growth will only occur slowly, whatever the circumstances. Some species of lichen can be encouraged by increasing the nitrogen content of the surfaces, perhaps by washing with a 'fertiliser' prepared by stirring manure into water in a bucket or large drum and using the fluid after the solid matter has settled. Alternatively chemical fertilisers can be used but, whatever the system, there will be a tendency to encourage the development of species which develop rapidly on surfaces that are rich in nitrogen, species that are usually associated with bird droppings on buildings, so that the use of a 'fertiliser' of this type will tend to encourage the development of these particular species and not the natural species that account for the normal mellow appearance. Perhaps a more sensible alternative, particularly if an attempt is being made to match new surfaces with old, is to remove the growth from the old surfaces which will then look clean and often far more attractive than the growth-encrusted surfaces. Indeed, in many circumstances growth control is essential, one example being the maintenance of war cemeteries and memorials, as neglect frequently prompts public outrage. The problem is then to decide upon the most economic method for controlling growth.

Controlling dampness

Although lichen growth can survive in very dry conditions, algal growth in particular is entirely dependent on a high moisture content and, as algal growth represents the start of the humus chain that later permits mosses and higher plants to become established, it would seem that the control of moisture content might represent an important means of controlling growth. Whilst this is essentially true it is equally important to appreciate that algal growth is sometimes evidence of dampness defects that require correction. For example, heavy algae may be noticed around a downpipe, indicating that it is leaking, so that repairing the downpipe will both remedy the leak and control the algal growth. Algal growth may occur along the base of a wall, indicating rising dampness, but control of algal growth will simply change the appearance without alleviating the dampness which may result in other deterioration such as crystallisation spalling through the presence of salts or through freezing. Water-repellent treatments can be used to reduce the moisture content of surfaces and can thus achieve excellent control of algal growth, yet there are circumstances in which such treatments can induce severe deterioration such as when a surface is able to absorb water from another source, perhaps from rising dampness; the advantages and disadvantages of water-repellent treatments are described more fully in Chapter 3. Whilst water-repellent treatments may inhibit growth if applied to clean new surfaces, they should never be applied to surfaces on which growth is already present and it is therefore necessary to clean away growth before applying such a treatment or, indeed, a masonry paint, and biocidal treatments to eradicate growth are frequently used as pretreatments.

Scrubbing

Growth of moss, lichen and algae can be removed by dry or wet scrubbing, but this is very tedious and must be considered unrealistic where extensive areas are to be cleaned. The headstones and memorials in Commonwealth War Graves Commission cemeteries were maintained for many years by simple wet scrubbing but it became very difficult to find staff who were willing to carry out this work and, in addition, the water introduced in the cleaning process actually encouraged the redevelopment of algal growth.

Biocidal treatment

In recent years headstones have been treated with biocides and it has been found that no further maintenance is necessary, except in special circumstances such as where local scrubbing of horizontal surfaces is required to remove accumulations of fallen leaves or bird droppings. The biocidal treatment achieves an almost magical result as it is simply sprayed onto the contaminated stone surface, causing the growth to die and become brittle so that it is removed naturally through the action of wind and rain.

This effect is less marked on structures that have been permitted to become encrusted with heavy lichen and moss growth over a protracted period, as only limited penetration of the biocide can usually be achieved from a simple spray. In these circumstances an initial application is usually necessary to cause the death of the growth, with perhaps some brushing after a period of about two months to remove any remaining growth. The clean surface is then treated again with biocide to ensure a uniform and generous inhibitory treatment. If roofs are being treated it is important to appreciate that considerable volumes of growth may become dislodged following treatment and it is necessary to clean gutters at intervals to ensure that they do not become blocked. If a water-repellent treatment is considered desirable for damp-proofing purposes it canbe applied following completion of the biocidal treatment. If suitable persistent treatment is employed it will e found that it will inhibit the growth of lichen for a considerable number of years, but algal growth may redevelo earlier and a relatively superficial algicidal treatment may be considered desirable at fairly frequent intervals, although the complete biocidal treatment may not be necessary.

Plate 4.6 Headstones at a British war cemetery in Belgium showing incomplete biocide treatment, lichens and algae remaining on the dark untreated areas but disappearing on the light treated ares (*Commonwealth War Graves Commission*)

A considerable variety of biocidal treatments is available but many are unsuitable for masonry treatment for a variety of reasons. Whilst the treatments must obviously be toxic to the growth they must also be completely harmless to the operatives and they must cause no damage to the stone, either by direct action on the stone substance or by leaving deposits which, in conjunction with other factors, may result in damage, perhaps by increasing water retention so that frost damage becomes more likely or by chemical reactions which give unsightly discoloration. The activity of many toxic systems, such as calcium and sodium hypochlorite sold as chlorinated lime or bleaching powder and bleach respectively, is only transient. The control of biological growth must involve sterilisation at the time of application, which can be achieved using such systems, but the surface may soon become capable of supporting new growth. In the case of lichens which grow very slowly and mosses which prefer to colonise stone surfaces only after contamination with organic matter has occurred, sterilisation may result in adequate freedom from apparent growth for several years. However, with algae which rapidly colonise unprotected stone whenever the conditions of temperature and moisture content are appropriate, the effects of such treatments may be lost after only a few weeks, and therefore these treatments might be followed by water-repellent treatment to maintain the surface of the stone in a dry condition and thus inhibit further growth.

Other treatments such as sodium pentachlorophenate may leave a deposit within the stone surface but it is slightly volatile and leachable, and whilst this type of system usually possesses activity against a wider variety of organisms, treatments should be considered only as eradicants for use prior to the application of a water repellent or perhaps a special persistent toxic treatment. One of the most efficient eradicant treatments consists of a mixture of sodium methyl siliconate and the sodium salt of a phenol such as pentachlorophenol or orthophenylphenol. The siliconate assists in the toxic action by encouraging the penetration of the phenol into the organism to be controlled and, in addition, the siliconate treatment of the growth tends to prevent susequent water absorption, causing the growth to become dry and brittle so that it is readily removed by light brushing or even becomes detached without further attention during periods of heavy weather. Unfortunately such treatments suffer from the very serious disadvantage that they are likely to cause severe staining on light-coloured surfaces such as Portland limestone headstones, through the development of brown or purple coloration due to reaction of the phenols with iron.

Staining is not the only danger associated with this type of treatment. The introduction of alkali metal salts into stone must always be considered with suspicion as there is a danger of the development of soluble salts which may cause significant crystallisation damage to the stone surface. In addition caustic solutions present a hazard to the operatives, as well as a danger of causing stripping of paint from joinery adjacent to the masonry. Copper, zinc and magnesium salts, as well as fluorides and silicofluorides, are a few of the chemicals that have been used in the past and which must be rejected for certain uses. Although water-repellent components invariably improve the efficacy of biocidal treatments, particularly the rapidity with which eradicant treatments become effective, there are many situations in which water repellents should not be used, as previously explained.

Reliable biocides

Several outstanding treatments have been developed in recent years and are now very widely employed. Amongst the inorganic salts the borates have proved of particular interest. Sodium borates such as borax react with atmospheric carbonic acid to give a deposit of sodium carbonate and boric acid, the latter being almost insoluble in water. There is a danger that the atmospheric sulphur acids will form sulphites and sulphates which can cause crystallisation damage but this is reduced with disodium octaborate (Polybor), which is much more soluble in water than borax but contains less sodium whilst having a higher boric acid equivalent. It is therefore easier to use than borax as well as being more active and less likely to cause damage to stone. Polybor is comparatively inexpensive and is probably the best general purpose toxic treatment available today, but its limitations should be appreciated: whilst its inhibitory action will persist for several years its eradicant action is not very rapid.

For monumental purposes on reasonably porous stone a 4% solution of Polybor will give freedom from algae for about two and a half years and freedom from moss and lichens for a longer period. Concentrations should be varied according to the porosity of the stone, reducing to perhaps 2% on a very porous stone but increasing to 10% on marble. On building structures where moss and lichen are more important than algae the effective life of Polybor treatment will be much greater, perhaps ten years depending upon local conditions. Formulations are now becoming available which form zinc borate on the stone, improving the biological range of the treatment and enhancing both the eradicant action and the long-term effectiveness.

Amongst the organometallic compounds the mercurials are probably most active but they present a serious hazard to operatives and are also unstable under the rather extreme conditions existing on a stone surface, giving only a short effective life. The Group IV metals germanium, tin and lead also give rise to a complex series of organometallic compounds, activity towards fungi and algae being best developed in those which possess the R_3MX structure, where R is an organic group, M is the metal and X is an anion. The biological activity of tin and lead compounds is greater than that of the germanium analogues which are also much more expensive. Tin and lead compounds are active at very low concentrations and ultimate decomposition products are usually white in colour. Of these two there is a tendency to prefer the tin compounds as the lead analogues are less stable and may decompose to give toxic products. Within the preferred group of tin compounds the tri-alkyl compounds are most active against fungi when the alkyl groups contain a total of nine to twelve carbon atoms, but mammalian toxicity decreases as the organic groups increase in size. The optimum trialkyltin compounds contain the tributyltin group, usually as the oxide; this is the tribultytin oxide that has already been described as a component in remedial wood preservation treatments. Indeed, remedial wood preservation involves the sterilisation of brickwork and masonry against the Dry rot fungus.

The sodium phenates that have been previously described are often used for this purpose but have been largely replaced by tributyltin oxide solubilised using a quaternary ammonium compound. Quaternary ammonium compounds (quats) are particular suitable as solubilisers as they possess biological activity of their own, apparently by rupturing the cell walls of algae and lichen as well as by a true toxic action. Generally quats are powerful eradicants but they also possess an affinity for a variety of substrates and are therefore capable of imparting a persistent inhibitory action. Quats have long been used for pharmaceutical purposes against both bacterial and fungal infections. Over the years the costs of the more popular compounds have progressively reduced and they are now used for industrial purposes including biocidal masonry treatments. Good water solubility and powerful biocidal action appear to be associated particularly with compounds containing an alkyl chain of about fourteen carbon atoms together with an aryl group as in benzalkonium chloride. These quats can be used at concentrations as low as 1% (that is 2% of the 50% concentrate usually supplied) to give an effective life against algae of about two and a half years. Mixtures of tributyltin oxide and quat are preferred where lichen and moss are present or where a more persistent inhibitory action is required; this is the type of treatment illustrated in Plate 4.6. However, the use of organotin treatments is now restricted in many countries where quats alone are usually preferred, often at concentrations up to 4% in an attempt to improve their effectiveness; it is probable that all these treatments will be replaced by the zinc borate treatments that have been described earlier. Some zinc borate treatments which were originally developed as wood preservatives contain sodium ions and may damage stone, but treatments are now available which completely avoid these dangers.

In summary it can be said that many of the proprietary formulations marketed for masonry and brickwork treatment cannot be recommended, including bleaching powder and bleach (calcium and sodium hypochlorite), sodium pentachlorophenate and orthophenylphenate, mercurials and generally all metal or alkali metal salts. The best low-cost general treatment is probably provided by the proprietary borate Polybor or the quaternary ammonium compounds which are safe, easy to use and relatively inexpensive. Tributyltin oxide solubilised with a quaternary ammonium compound will give a much more powerful eradicant and inhibitory action and is much more reliable where growth consists of lichen rather than of algae, but the inhibitory action is eventually lost from the immediate surface of the stone, permitting algal growth to develop; this can be readily controlled by applying a simple quaternary ammonium treatment when the growth has developed to a stage when it is considered unsightly. The new zinc borate treatments combine low cost and safety with optimum effectiveness and are likely to be preferred in the future.

Water repellents

Water repellents can be used to inhibit growth on clean surfaces but they should be applied only to new stone or to stone that has already been treated with biocide. Water-repellent treatments should never be applied to surfaces which are liable to internal wetting, perhaps from rising dampness, as there is danger that salt solutions will migrate slowly towards the surface and will be forced to evaporate from beneath the water-repellent treated zone where an accumulation of salt crystals may develop, causing the treated zone to spall. Even if salts are absent from the migrating water there is still a danger that moisture accumulations beneath the treatment may result in damage through freezing. These dangers must not be underestimated and water-repellent treatments should never be applied to headstones, pavings and dwarf retaining walls which lack barriers to rising dampness. In these situations it is best to employ a simple biocidal treatment, preferably one which possesses a degree of persistence which will enable it to prevent re-establishment of biological growth for a protracted period. Finally, growth and dirt may be cleaned from buildings using water, either by high-pressure spray or by scrubbing, and this will encourage the development of algae. In all cases cleaning involving water should be followed by simple algicidal treatment, preferably a quaternary ammonium or zinc borate system to prevent the rapid development of algae after cleaning treatments of this type.

4.3 Cleaning

Need for cleaning

Cleaning is not always considered desirable as some buildings, it is sometimes suggested, were designed to be black. This suggestion is made particularly forcibly with regard to public buildings in Edinburgh and Glasgow as these are often constructed of sandstones which darken very uniformly. However, it must be recognised that many of the important buildings in Scotland were actually constructed before coal came into general use as a fuel and thus before buildings commenced to darken so seriously in this way, and there is really no proper justification for this suggestion. Indeed, all architects wish their buildings to have their most attractive final appearance as soon as construction is complete, and if they are designing a black building it will be constructed from black materials rather than from white or fawn ones which will progressively blacken over a period of a century or more. There can never been any valid aesthetic opposition to cleaning; dirty buildings appear neglected and this is probably the greatest stimulus to cleaning buildings in Britain. A consequence of this is that our towns tend to be haphazard in appearance, a mixture of clean and dirty buildings, whereas in France a law introduced by Napoleon III in 1852 and revived in 1959 requires all buildings to be cleaned every ten years. The result has been a much brighter Paris with a uniform clean appearance. Generally buildings should be clean for aesthetic reasons but it is also necessary to consider whether the cleaning process may reduce or increase the rate of deterioration.

Causes of dirtying

The causes of dirtying are not always clear. Dirt is often attributed to soot deposits but, whilst this was true in the past and carbon particles can often be found trapped in the surface of old buildings, soot has not been a general cause of dirt deposits since the introduction of the Clean Air Act, yet buildings still become dirty. Detailed investigations often show that biological growth is responsible for dirt. For example, even uniform black dirt can be attributed to the fruit bodies of minute crustose lichens and fungi such as *Aureobasidium pullulans*. In many cases simple control of these organisms by the application of a biocidal treatment will result in progressive natural cleaning and it is always worthwhile to treat a small area with a biocide to check whether it may induce the required cleaning without the necessity for any more drastic remedial work. In many cases the biocidal treatment will not be completely effective and the dirt can then be attributed, at least in part, to non-biological factors which will require other remedial measures.

Cleaning limestone

In a rural atmosphere a limestone surface will often change little in appearance with time but it may become slightly dirty and this dirt is often difficult to remove. The calcium carbonate of which the limestone is composed is slightly soluble in the carbonic acid normally formed as rainfall absorbs carbon dioxide from the atmosphere. This slight dissolving of the stone is followed during a dry period by reprecipitation, particularly at the surface of the stone, where there is a danger that the formation of the carbonate precipitate will trap dirt particles which may have been derived from dead algae and other biological growth. Where an area is exposed to direct rainfall the solution action tends to be particularly powerful so that slight erosion of the stone surface is normal and the stone always has a clean appearance. In areas sheltered from direct rainfall, such as under string courses, cornices and sills, this washing action will not occur and dirt will often accumulate.

Cleaning with water

As limestone is slightly soluble in water the easiest method of removing this dirt consists of water soaking followed by light brushing or high-pressure water jet. The soaking period required to achieve the necessary softening will vary from several hours to several days but only a gentle trickle of water is necessary, sufficient to keep the surface damp without resulting in excessive water penetration to the interior or inconvenience to passers-by. If brushes are used for cleaning they should be stiff bristle or nylon; if the deposits are particularly hard, non-ferrous or stainless steel wire brushes can be employed.

In urban atmospheres the condition of limestone will be rather different as atmospheric sulphur oxides dissolved in the rainfall will result in the formation of a patina or surface deposit of calcium sulphate. This deposit can be particularly hard, and it is sometimes said that it protects the stone and that its removal will be harmful. In fact there is a danger when a dense surface of this type occurs that soluble salts will accumulate beneath, eventually crystallising and causing the formation of massive blisters, flakes or spalls. The patina is therefore best removed at intervals before this can occur and the exposed surface of the stone washed to remove salt deposits. This is one of the most significant arguments in favour of regular cleaning of limestone buildings for, if spalling occurs, deep damage may result and stone replacement may become necessary.

Dirt may often become trapped within the calcium sulphate deposits. The easiest method for removing the patina and this dirt is again water spray followed by cleaning with light brushing or high-pressure water jets. In some areas the calcium sulphate encrustations may resist softening for a considerable time and prolonged soaking may be necessary, or even the careful use of mechanical cutting aids. It has often been suggested that high-pressure water jets can be used to reduce the danger of water penetration into the building that is associated with the soaking method, but preliminary soaking is still essential if high-pressure water jets are to remove the surface encrustations without seriously damaging the stone which may be softer than the encrusting deposits. In some situations steam cleaning may be appropriate; although it offers little technical advantage over prolonged water soaking and is much more expensive, it is capable of softening the more persistent encrustations without the use of massive quantities of water. Some years ago steam cleaning was used in conjunction with caustic soda or soda ash but this results in the formation of harmful soluble salts which later crystallise to cause damage to the cleaned surface. In fact, chemical cleaning methods should never be employed in normal circumstances.

Cleaning sandstone

A carbonaceous sandstone can usually be cleaned using water soaking and the techniques normally employed for limestones, but non-carbonaceous sandstones and gritstones will not respond to these methods and it is necessary to employ much more severe cleaning techniques such as dry or wet grit-blasting or the use of hydrofluoric acid.

Hydrofluoric acid

Granite can also be cleaned with dilute hydrofluoric acid but solutions of ammonium bifluoride are preferred; they decompose on the stone to liberate hydrofluoric acid and act in precisely the same way. Great care must be taken in selecting an appropriate cleaning method. For example, hydrofluoric acid must never be used under any circumstances on a limestone or carbonaceous sandstone, and ammonium bifluoride or any other compound likely to form salts within a stone must never be used on any porous stone.

Grit-blasting

Grit-blasting techniques must be carefully controlled to avoid unnecessary damage to relatively soft stones and it is important to remember that dry grit-blasting in particular can generate silica dust, both from the grit in some systems and from silica within the surface, introducing a danger of operative silicosis, so proper protection is essential. The safest technique for cleaning sandstones is to use high-pressure water spray if the desired result can be achieved and wet grit-blasting whenever a more powerful cutting action is required. The use of power tools with wire brushes or grinding discs should always be avoided as they can cause severe damage.

Choice of cleaning method

Concrete, cast stone, rendering and brickwork should be cleaned as if they are natural stones. Thus high-pressure water jets should be used whenever possible. If these do not achieve an adequate result, water soaking should be considered, preferably followed by high-pressure water jet cleaning, but stiff bristle, nylon or non-ferrous wire brushes can be used for localised difficult areas. If these methods are inadequate then carefully controlled wet grit-blasting should be employed, but this should always be entrusted to a specialist company employing operatives who are fully experienced in selecting nozzles, pressures and grits to achieve adequate cleaning without unacceptable damage. This is, in fact, a summary of the approach to cleaning that should be adopted for all surfaces, whether of natural stone or other materials, and all other cleaning methods should be rejected, particularly any methods employing acids, alkalis or salts.

It must be recognised that much of the dirt on buildings is actually biological growth which can perhaps be controlled by simple biocidal treatments. Water is always involved in cleaning techniques and biological growth is therefore encouraged. Indeed, cleaned buildings frequently become bright green through algal growth, to the consternation of the building owner, and biocidal treatments are therefore recommended in all cases following cleaning. Water-repellent treatments can also be considered following cleaning; a water-repellent surface remains lighter and more uniform in colour, and generally free from the patchy darkening and biological growth associated with localised water flows. Water-repellent treatments are thus very efficient in maintaining the appearance of cleaned buildings but they should never be used in circumstances where water is likely to accumulate behind the treatment from a source other than rainfall, such as rising dampness; these dangers are explained more fully in Chapter 3.

4.4 Stone preservatives

> Stone preservatives have been intended to prevent deterioration of the stone without changing its appearance, and there has been, since about 1840, a long succession of proposals to that end. Even so, no stone preservation used as a surface treatment has yet met with any significant measure of success. Some have been followed, sooner or later, by scaling of the treated surfaces and have done more harm than good. The situation now is much the same as it was 100 years ago. This is not for want of apparently promising methods, but because of the fundamental difficulty that no surface treatment penetrates far enough to give the protection looked for.

This comment appeared in Building Research Station Digest (First Series) No. 128 which was first published in November 1959, and surprisingly represents a situation that still prevails today; whilst the need for deep penetration has long been appreciated, it has not yet been reliably achieved through a method that is realistic for application to building masonry.

Water repellents

As water is involved in all masonry deterioration processes, whether they are chemical, physical or biological, it might appear that the most realistic conservation technique would be the use of water repellents to prevent rainwater penetration. However, water can be introduced into masonry from other sources such as rising dampness or interstitial condensation, and deterioration may still occur and may be exaggerated by a water-repellent treatment which may serve only to encourage concentrated crystallisation damage just below the treated zone with a consequent danger of contour scaling.

Need for penetration

These effects have, of course, been discussed previously, both in this and the previous chapter, but a point that has not been emphasised is the way in which this danger of damage decreases sharply if the water repellent penetrates to a greater depth. On a reasonably porous stone it is probable that a normal silicone resin water repellent penetrates to a depth of 5–10 mm ($^{1}/_{4}$–$^{1}/_{2}$ in). However, it appears in theory and has been confirmed in laboratory experiments that an increase in the penetration depth to 25 mm (1 in) or more substantially reduces the danger of damage from crystallisation under the treated zone through salt accumulations or freezing. It is believed that this reduced danger is associated largely with the stabilising influence of the deep treated zone but, whilst a deep treatment may appear to be more reliable in laboratory experiments, it is possible that the rate of deterioration is reduced rather than prevented altogether. Whatever the true explanation, the reliability of masonry preservation processes improves enormously as the depth of penetration increases, and consequently only deeply penetrating treatment should be considered whenever valuable masonry is to be preserved or consolidated; preservation involves treatment, perhaps of new stone, designed to prevent deterioration, whereas consolidation involves an improvement to the cohesiveness of the stone, frequently by the replacement of cementing matrix that has been lost over the years as result of the action of atmospheric pollutants in rainfall.

Crystallite redistribution

In the first part of this chapter it was explained that the rate of deterioration of stone exposed to acid rainfall is dependent on crystallite size and porosity in accordance with the natural law of mass action, deterioration being most rapid with small crystallites which present a very large area to the attacking rainfall relative to their total mass. In this connection it is interesting to consider the work of the American Professor Seymour Z. Lewin. Initially this principle of mass action suggested to him that a stone composed of a small number of large crystallites would dissolve less rapidly than one composed of a large number of small crystallites. He therefore concentrated initially on the development of a process involving prolonged immersion of limestones in a reagent such as potassium chloride which would permit redistribution of carbonate to form a predominantly macrocrystalline structure. However, this process introduced the distinct danger that the microcrystalline cementing matrix might erode, causing disintegration unless a microcrystalline carbonate deposit could be provided within the stone prior to redistribution. Barium carbonate appeared to be more attractive than calcium carbonate as it is attacked more slowly by acid pollutants, and the barium sulphate formed is insoluble so it does not represent the risk of crystallisation damage associated with calcium sulphate formed in polluted atmospheres, or the even greater damage caused by magnesium sulphate on magnesium limestones. A further attraction of barium is the higher solubility of barium hydroxide in comparison with calcium hydroxide. If these hydroxides are applied to stone as baryta or lime water respectively, they will carbonate slowly under the influence of atmospheric carbon dioxide, but Lewin accelerated carbonation by the simultaneous application of urea with the barium hydroxide.

Carbonate formation

Unfortunately the redistribution of the carbonate to form macrocrystallites can occur only if the stone is soaked for a protracted period in the appropriate reagent, a process that is unrealistic in the case of buildings. As a result it now appears that Lewin has since advocated *in situ* formation of barium carbonate as a remedial conservation process, despite the obvious danger that it may cause the stone to become predominantly microporous and thus particularly susceptible to damage through freezing of absorbed water or crystallisation of salt solutions introduced from other sources. Thus Lewin's process is now exactly the same as the baryta water processes proposed about 100 years ago, except for his introduction of urea to accelerate carbonation.

Silicate formation
Baryta and lime water were proposed as stone-consolidant treatments because they ultimately resulted in the deposition of carbonate, a natural stone component. Similarly silicates have been proposed as stone preservatives, acting through the deposition of silica or insoluble silicates, although most silicate and fluosilicate systems have been found to give only limited consolidation. Siliconesters such as ethyl silicate were first suggested in 1861 as a means for introducing silica into stone as a consolidant but they were not used for this purpose until about 60 years later when they were introduced into Britain by G. King and B. C. G. Shore. Although siliconesters enable treatments to be applied in alcohol and other organic solvents, the results are actually no different from those obtained when silica is deposited by other means, the main problem being to ensure that silica is deposited to an adequate depth in a form that is consolidating. In this connection the catalysts used in siliconester formulations are particularly important. For example, acid catalysts have often been used in alcohol preparations, yet they are completely neutralised on carbonaceous stones and thus entirely ineffective; consequently basic catalysts should always be used. In addition only polar solvents such as alcohols should be used as it is found that, when siliconesters are applied from non-polar organic solvents such as white spirit and kerosene, the silica deposit tends to be amorphous and microcrystalline, achieving very poor consolidation and introducing the normal dangers of crystallisation damage associated with microporous stone structures.

Treatment dangers
All these preservative or consolidation systems which involve depositing material within the interstices of the stone suffer from the same disadvantages. Whilst the best treatments are clearly those that result in insoluble deposits without introducing soluble salts, there is still the danger that they will modify the stone structure so that it becomes microporous and more sensitive to crystallisation damage. In addition, all these treatments are capable of only limited penetration and, in reducing the porosity of the surface, they may not significantly reduce water absorption but may seriously obstruct evaporation. Water accumulating behind the treated zone in this way, or diffusing from other sources, may freeze or absorb atmospheric acids with consequent danger of salt crystallisation, the restricted evaporation encouraging slow crystal growth and particularly severe spalling of the treated layer. Certainly one method for reducing these dangers is to always follow a preservation or consolidation treatment with a water repellent to reduce the danger of water accumulation and this is certainly always advisable where there is a danger that a consolidation treatment may lead to restricted porosity of the stone surface. However, from a preservation point of view perhaps a more sensible technique might be to apply a water repellent to a considerable depth, or from the consolidation point of view to apply a water repellent which also possess consolidating properties.

Polymers and monomers

It has frequently been suggeted that organic polymers should be employed for the preservation and consolidation of masonry but it must be appreciated that, if they are applied at sufficient loadings to achieve adequate consolidation, they are likely to completely seal the stone surface with a consequent danger of damage through interstitial condensation behind the treated zone. If organic polymers are applied from dilute solutions to ensure that the porosity of the stone is preserved they tend to lose much of their consolidating power and they also become particularly susceptible to loss by volatilisation and oxidation. At the present time only the silicone resins can avoid these difficulties, but the normal water-repellent silicone resins consist of relatively large polymers which cannot penetrate deeply into stone, particularly the microporous stones that are generally those that require urgent treatment, even if their carrier solvents can penetrate.

Silanes

Saturated polymer systems are essential if adequate life is to be obtained from treatments applied as dilute solutions in order to ensure that permeability of the surface is maintained, and silicones give much longer life than normal organic systems. As polymer solutions will not achieve adequate penetration the logical solution is to apply monomers and form polymers *in situ*. At the same time it must be appreciated that the purpose of the treatment, at least in most instances, is to achieve consolidation to an appreciable depth using a polymer system which also possesses water-repellent properties. These requirements limit the choice of monomer systems that are available but considerable success has been achieved in laboratory tests using trimethoxymethylsilane. This is the basic monomer used in the Brethane process developed by the Building Research Establishment. When this monomer is applied to stone, hydrolysis of the methoxy groups occurs to form methanol, which is then lost by volatilisation, leaving hydroxy groups which condense with the loss of water to form a polymeric system. This process is encouraged by the incorporation of suitable catalysts, but there are a number of problems that require further consideration. For example, the high proportion of methoxy groups in the monomer limits the effective silane concentration, and the large quantities of methanol that are produced may represent a potential danger to the sight of operatives applying the process so that it would be safer if the triethoxy monomer was employed instead. In addition, the monomers are very volatile and substantial losses can occur before polymerisation is complete, further limiting the effective silane concentration. In the circumstances it is more sensible to employ short-chain polymers perhaps formed from between four and eight monomers. This idea is very attractive for several reasons. The elimination of a large proportion of the alkoxy (i.e. methoxy or ethoxy) groups enables the effective concentration of the system to be greatly increased, and triethoxymethylsilane polymerises in exactly the same way as tetraethoxysilane which is, in fact, another name for the siliconester ethyl silicate which is readily obtainable in partially polymerised form. The ethyl silicate product does not possess any water-repellent methyl groups but these can be added by blending with a proportion of a methyl or dimethyl silane monomer.

Silane systems are being more widely adopted as masonry preservative and consolidant systems now that account has been taken of these problems, although clearly further development work is essential before their effectiveness and reliability can be fully established. There are a variety of proprietary formulations, many giving excellent performance in laboratory tests, but it remains necessary to evaluate them under realistic service conditions before their reliability can be established. These comments may appear unhelpful but it is obviously foolish to submit buildings to treatments which are intended to be preservatives or consolidants but which are actually ineffective and perhaps damaging. This description of preservation and consolidation processes has been included to emphasise the difficulties that have been encountered in the past and to indicate the way in which developments are likely to move in the future.

5 Thermal insulation

5.1 Heat losses

Remedial methods to improve thermal insulation have been introduced only comparatively recently, stimulated by rising fuel costs and government encouragement to conserve dwindling energy resources. Unfortunately many advertisements for insulation processes grossly exaggerate the savings that can be achieved, probably through ignorance in most cases, but it must also be suspected that the current interest in energy conservation has attracted some attempts at deception. Improved thermal insulation is generally an attempt to conserve heating energy in temperate or cold climates, but insulation may be just as essential in desert areas, for example, to avoid rapid increases in interior temperatures when buildings are subjected to strong sunlight. In both cases there is a need to improve the resistance to heat energy transfer between the interior and the exterior of the building.

Conduction, convection, radiation

This energy transfer can occur by conduction, convection or radiation, but usually in combination. For example, in cavity walls convection heating at the warm inner skin causes circulation of the air and transmission of the energy to the cool external skin, whereas conduction is involved in the transfer through each skin. Heat loss can be reduced by obstructing these convection currents by the use of cavity fill, or by reducing the thermal conductivity of the skins; conductivity is generally related to density, so that the incorporation of air in lightweight concrete reduces density but also reduces conductivity and thus improves insulation. Radiation is usually the least significant factor in heat loss but is still important in some situations and the insertion of a reflecting surface, such as aluminum foil backing on plasterboard ceilings, can significantly reduce heat loss.

Heat losses from buildings

The purpose of introducing additional thermal insulation is to minimise heat losses, but a building must be considered as a whole rather than as individual parts. For example, there is no point in installing very expensive double glazing if much cheaper ceiling insulation will actually achieve a much greater reduction in heat loss. Indeed there is no sense in a householder installing any insulation unless the cost can be recovered through fuel savings. In this connection there are two very important points to consider. Firstly, if thermal insulation is very good already, additional insulation will achieve only a marginal improvement. Secondly, improved thermal insulation may result in the householder maintaining higher temperature levels without increasing fuel costs rather than maintaining the same temperature levels and reducing fuel costs. It can be considered that improved thermal insulation results on average in only half the anticipated fuel savings, the other half being devoted to improvement in comfort. Indeed, the greatest financial savings occur only to the highest energy users, those with the lowest fuel consumptions usually taking the entire benefits in the form of increased living temperatures and thus improved comfort.

Thermal transmittance

Thermal insulation methods are usually marketed by giving examples of the savings that can be achieved, often under rather extreme conditions, and there is rarely any attempt to consider the way in which the installation may affect a particular building. Clearly a responsible building defect surveyor must approach the situation in a rather different way, assessing the present situation and proposing remedial measures that are likely to result in improvements which will justify the cost involved. Calculations involved in assessing the thermal properties of buildings are considered in detail in Appendix 6. It is necessary to appreciate that heat loss through a building component depends on the thermal transmittance or *U* value (units W/m^2 °C) for the component coupled with the difference in temperature between the external and internal air. The rate of energy loss per degree of temperature difference through a component such as a wall depends on the area multiplied by the *U* value, but in addition considerable amounts of heat are

Figure 5.1 Heat losses from a typical house

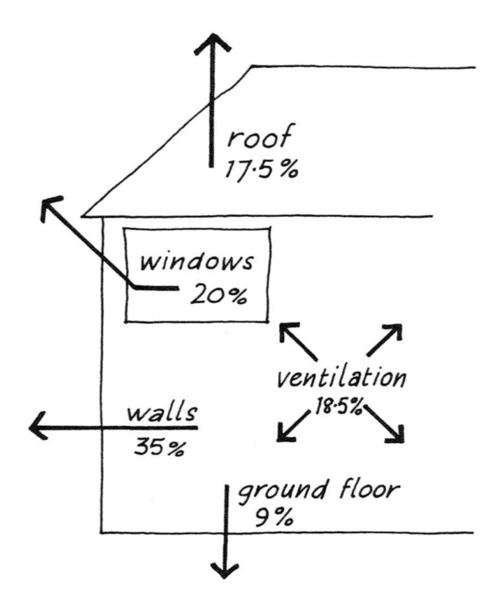

lost through ventilation. Table 5.1 and Figure 5.1 show typical heat losses from a two-storey semi-detached house of about 100 m^2 gross floor area, but the form of construction and other factors will also considerably influence the situation, as will be appreciated from Appendix 6. For example, if the house is in a particularly exposed situation the *U* value will be influenced to a far greater extent by the construction details. In the case of the roof a*U* value of 1.50 has been assumed, a typical value for a pitch roof of tiles on battens with felt over the supporting rafters and an aluminium foil-backed 10 mm plasterboard ceiling. If felt is omitted the *U* value increases to 2.1 so that the omission of felt alone would increase heat loss through this structure by 40%! If a 50 mm depth of glass-fibre insulation quilt is installed between the ceiling joists the*U* value is reduced to 0.50, indicating that this very simple insulation has reduced heat losses through the roof structure by about 66%.

Table 5.1 Heat losses from a two-storey semi-detached house of about 100 m^2 floor area

Fabric heat loss		
Roof, 50 m^2, *U* value 150	75	(17.5%)
Ground floor, 50 m^2, *U* value 0.76	38	(9.0%)
Walls, 100 m^2, *U* 1.50	150	(35.0%)
Windows, 17 m^2, *U* value 5.00	85	(20.0%)
Total fabric heat loss (W/°C)	348	(81.5%)
Ventilation heat loss		
1.0 air change/hour (240 m^3 x 1/3 W m^3 °C)	80	(18.5%)
Total heat loss (W/°C)	428	(100.0%)

In the case of the ground floor a *U* value of 0.76 has been assumed in Table 5.1, a typical value for a solid floor with linoleum or plastic tiles. However, if this floor is covered with a carpet and underlay about 10 mm thick the *U* value is reduced to 0.66, representing a reduction in heat loss of about 13%. If a suspended timber floor is involved it will have a typical *U* value of 0.68, reducing to 0.65 with cork tiles or thin carpet but to 0.60 with the heavy 10 mm carpet previously described. The *U* value of 1.50 assumed for the walls is typical for a wall comprising outer and inner brick skins of about 112 mm ($4^1/_2$ in) and a 50 mm (2 in) cavity, finished on the inner surface with 10 mm ($^1/_2$ in) of gypsum plaster. However, the current Building Regulations for England and Wales which apply to new dwellings specify a maximum wall *U* value of 0.6, usually achieved by replacing the brick inner skin with aerated concrete blocks, using cavity insulation and perhaps using lightweight plaster to achieve further insulation. Clearly the replacement of the inner skin with lightweight concrete blocks is impracticable in existing buildings but considerable reduction in the *U* value of a cavity wall can be achieved by introducing cavity fill to restrict convection losses, as previously described. Unfortunately cavity fill insulation is not always as effective as anticipated; foam materials sometimes collapse after installation through the use of incorrect mixes, and some foams encourage water penetration through cavity bridging.

A *U* value of 5.00 has been assumed in Table 5.1 for windows. In fact windows consist of both the glazing material and a frame which occupies 20–23% of the total window area. In single glazing the glass thickness is largely insignificant in comparison with the insulation arising through the internal and external surface resistances; these expressions are explained in more detail in Appendix 6. With double glazing the *U* value decreases and thus the insulation value improves as the air gap increases, although there is no significant gain in thermal insulation with an air space beyond 20 mm, but larger gaps may be necessary if sound insulation is also required; these problems are discussed in *Defects and Deterioration in Buildings* by the same author. With the most efficient thermal insulation of windows, such as the use of secondary glazing fixed to existing wood-framed windows, the *U* value may be reduced to 2.5, halving the heat loss through the windows.

This may appear to represent an enormous improvement in insulation but it must be placed in proper perspective; windows account for only about 20% of the fabric heat loss from a typical building so that even this dramatic improvement can reduce fabric heat loss only by about 10%. In addition, this saving relates only to the fabric heat loss or the loss that occurs through the building components and it does not take into account the very considerable heat loss that occurs through normal ventilation in a building. In the example given in Table 5.1 the ventilation heat loss represents about 20% of the total heat loss but this assumes a ventilation rate of 1 ach (air change/hour). In fact, ventilation can be as much as 2.5 ach if ventilation is encouraged by flues and ill-fitting windows and doors, representing 50% of the total heat loss from the building in Table 5.1. On the other hand, careful draught control can reduce ventilation to only about 0.5 ach, reducing heat loss caused by ventilation to only 10% of the total in a typical building, so that draught control actually represents the cheapest and most effective method of improving thermal insulation. Indeed, there are many simple ways in which thermal insulation may be similarly improved at comparatively low expense. Of these the use of carpeting has already been mentioned. It is not generally appreciated that curtaining also can be extremely efficient as it introduces an air space over windows, achieving the same effect as double glazing. For example, a normal wood-framed single-glazed window will have a *U* value of about 4.3, but well-fitting closed curtains will reduce this to between 2.9 and 2.4, representing a reduction in heat loss through the windows of 43%! Even net curtains can have a very significant influence, particularly if they are closely fitting against the window frame. The problem with curtains is that humid air from the accommodation can diffuse into the air space and cause condensation on the windows, a problem that also applies to inadequately fitted secondary glazing.

Figure 5.2 Internal surface temperatures in relation to temperature differences between interior and exterior air.

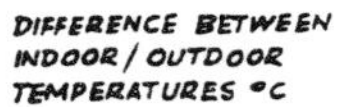

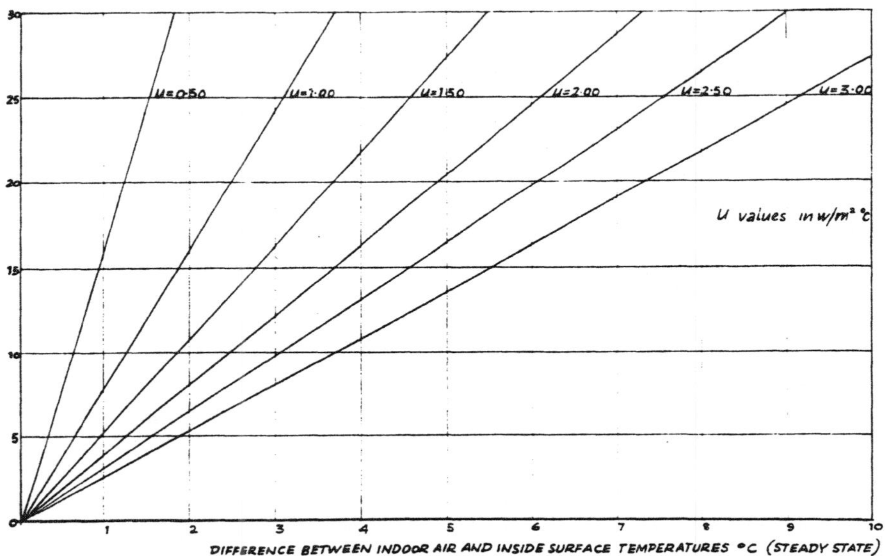

This diagram can be used to estimate internal surface temperatures for various interior and exterior air temperatures. Alternatively temperature measurements can be used to determine the R (or U) value of an external wall, door or window component. It is assumed that the internal surface has high emissivity (it is not polished or reflective) with a surface resistance of 0.123, and the external surface resistance is 0.055 (high emissivity, normal exposure). $R = 0.178$ ($U = 5.6$) therefore represents minimum thermal resistance, appropriate to single glazing and painted or anodised metal window frames

Measuring heat loss

This brief description will be sufficient to emphasise the care that must be taken in assessing heat losses from buildings when considering possible improvements in thermal insulation. Clearly heat losses can be estimated by calculation using the methods described in Appendix 6 but in many cases it is more convenient to determine the U value of structural components by measurement during inspection. These measurements are possible only in winter months when the interior temperature is relatively stable and high relative to the exterior temperature. The measurements involve the use of a special thermometer to determine the temperature of the interior surface of the component under investigation. The difference between this surface temperature and the interior air temperature can then be related to the difference between the interior and exterior temperatures in order to determine the *U* value using the graph in Figure 5.2.

Measuring surface temperature is not easy and involves the use of a flat-end probe; a suitable probe is manufactured by Protimeter Limited for use with various moisture meters. This probe will take a minute or two to stabilise before the temperature can be determined and during this waiting period it is advisable to move the probe around on the surface to avoid the danger that the probe may itself influence the surface temperature. The probe should then be used to measure the air temperature, again allowing sufficient time for the probe to stabilise; some manufacturers suggest that a different thermometer should be used for this purpose, but this introduces an unnecessary error which is completely eliminated if only one temperature measuring instrument is involved.

This method for determining the *U* value is particularly useful for windows as it enables the *U* value of the frames and glazing to be determined independently, often dramatically illustrating the relative heat losses through these components. For example, the internal surface temperature tends to be high on wood frames, clearly indicating good insulation value. On metal frames the temperatures are usually much lower, unless they incorporate a thermal break, but they are never as efficient as wood frames. The advantage of secondary double glazing with a single-glazed frame can be clearly demonstrated, as can the comparison between the heat losses through windows and the surrounding walls. The same instrument can, of course, be used for checking surface temperatures in connection with condensation investigations as discussed more fully in Chapter 3 and Appendix 3, and is particularly useful in detecting cold spots on external walls through the presence of thermal bridges in the structure such as lintels or local cold spots arising opposite air vents in external skins to cavity walls.

5.2 Improving insulation

There is really no point in spending money on insulation unless the cost can be justified in terms of reduced fuel costs. Thermal insulation will be most efficient in financial terms if it is used where the greatest heat losses occur or where the least expensive insulation methods can be used. The proportional heat losses through the various parts of an average house have been detailed in Table 5.1. Clearly the greatest heat loss occurs through the walls, but it must be appreciated that, if a modern house is constructed with lightweight concrete block inner skins to the external walls, the *U* value (transmittance value or rate of heat loss) is already reduced by a third or more. Thus cavity fill insulation has the greatest influence on heat losses when applied in an older property with a brick or 'breeze' (clinker block) inner skin rather than lightweight block.

Cavity fill

Foams formed in *in situ* are most efficient as they give greatest control over convection currents in the cavity. Generally polyurethane foam is much more reliable than the cheaper urea formaldehyde foam; in fact, all foams give unpredictable results unless applied by very experienced operators always using the correct mixes and ensuring proper distribution within the cavities. Granulated expanded polystyrene is slightly less effective but gives much more reliable results; dry granules are introduced on a current of air and the dangers of incorrect mixing are thus avoided. Pelletised glass-fibre and mineral wool are also usually applied in air streams, but their insulation value is slightly less and they are also slightly more difficult to install.

When a typical polyurethane foam cavity fill is installed in a modern house with lightweight block inner skins to the external walls, it will take about 25 years to recover the cost through fuel savings, whereas in an old house with a brickwork inner skin the cost can be recovered in only $12^1/_2$ years. However, interest is ignored in these calculations and there is really no chance that any wall insulation can be economically effective if interest is taken into account, unless inflation and energy shortages lead to cost escalations which may falsely justify the installation. In these comments interest is assumed to be at a building society mortgage rate or, at most, at a clearing bank overdraft rate, and wall insulation can never be justified under any

Plate 5.1 Installing granulated expanded polystyrene cavity fill insulation (*Isodan (UK) Ltd*).

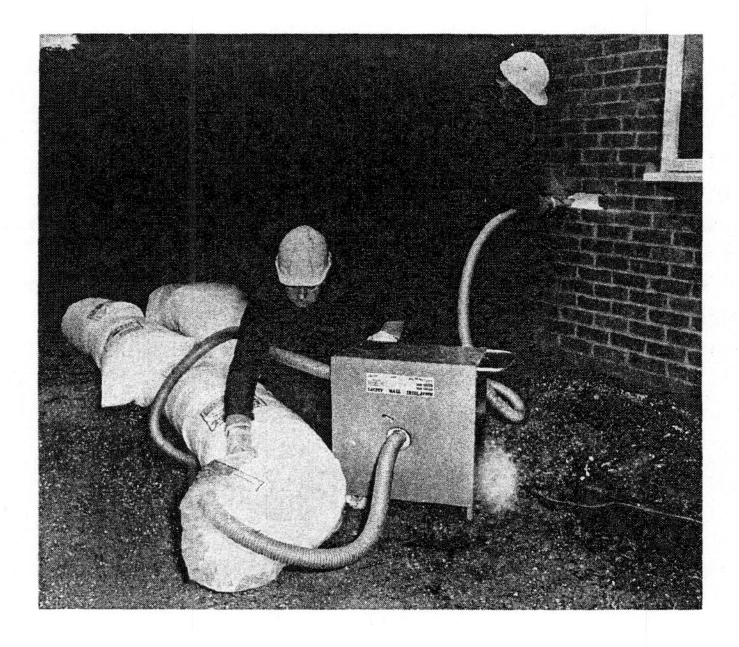

circumstances if higher rates of interest are involved, such as those offered by many installers through finance houses.

These doubtful economics must be carefully considered if cavity fill insulation is contemplated. It must also be recognised that some *in situ* foam systems may involve dangers of damp penetration through bridging if unsuitable foams are used and perhaps a short effective life if incorrect mixes are used. The greatest reliability can probably be achieved by using systems that are approved by the British Board of Agrément (BBA) and installed by approved contractors, but even approved cavity insulation systems may cause bridging. If ties are contaminated by mortar slovens or droppings, a cavity fill will reduce ventilation and dampness may pass across the ties to a sufficient extent to cause isolated damp patches on decoration. Some mineral-fibre cavity insulations are treated with a water-repellent oil to prevent cavity bridging but the humid conditions in the cavity encourage bacterial activity which can destroy the water-repellent effect. However, there are other methods of cavity wall insulation that can be considered. If the wall cavity vents into the roof space, sealing the top of the cavity will reduce heat losses through the wall by about 8%. If the wall is being replastered, the use of lightweight plaster in place of ordinary gypsum plaster will reduce heat losses through the wall by about a further 8%.

Double glazing

Windows account for perhaps 20% of the heat loss from a typical house. However, this comment is appropriate only if the windows are virtually draught-free. Badly fitting frames may result in draughts which may double or treble the ventilation rate and thus the ventilation heat losses, and it is therefore most important that windows should be draught-proof. The frame of a window represents about 20% of the window area for a metal frame but 30% for a wood frame, a very significant factor. In the typical house illustrated in Table 5.1 it is assumed that the windows have a U value of 5.0 but metal-framed windows may have an even higher U value of perhaps 5.6 whereas wood-framed windows may have a U value of only 4.3. Thus, whilst the heat loss through the glazing is clearly very important, the loss through the frame also has considerable significance. The U value for single glazing is about 5.6 but for double glazing this reduces to 4.0 for an air space of 3 mm, 3.0 for an air space of 12 mm and 2.9 for an air space of 20 mm or more, but the glazing represents only 70–80% of the total window area. A metal frame is normally considered to have a U value similar to that for single glazing but a wood frame has a very low U value.

Replacement windows

Many double-glazing 'specialists' offer complete replacement windows which consist of 'factory sealed' double glazing in well-engineered and often very attractive aluminium frames. Unfortunately, the double glazing often has only a small air gap and, whilst it is more efficient than single glazing, it is far less efficient than double glazing with a wider air gap. In addition, the heat loss through an aluminium frame will remain the same whether the frame contains single or double glazing, unless the frame incorporates a thermal break. In fact, typical replacement windows consisting of 'factory sealed' double glazing in aluminium frames may achieve no significant improvement in insulation compared with single-glazed wood frames. Obviously replacement windows with wider air gaps and some attempt at a thermal break in the frame may be more efficient, and uPVC frames, now often termed PVCu frames, have better insulation properties than aluminium frames, although they may be much less efficient than advertised because of the thermal bridging effect of the metal reinforcements that are necessary within the uPVC sections.

Secondary glazing

It is invariably most efficient to double glaze a building by the installation of secondary glazing on the interior of existing windows, particularly if this secondary glazing is installed in conjunction with existing wood-framed windows. In addition, even the most sophisticated secondary glazing is always cheaper than the cheapest replacement windows, and secondary glazing can be very cheap indeed. In its simplest form it can consist of a polythene sheet attached to the inside of an existing window, but glass cut to size, finished with a plastic edging strip and mounted on the windows using simple clips can achieve the best thermal insulation, and it is easily removed when it is not required in the summer months, an important advantage as windows can then be readily opened.

Draughts

Double glazing is economically effective only if it is simple do-it-yourself secondary glazing, but it must be added that, if existing windows are draughty, secondary glazing may also be the best method to reduce draughts and ventilation heat losses.

Curtains

Conversely, if windows are tight and relatively free from draughts, heat losses may be substantially reduced without the use of any double-glazing system, perhaps by the use of closely fitted curtains which remain open in the daytime when some solar energy gain may occur but are always closed at night, producing an insulating cavity very similar to a double-glazed system.

Ventilation control

Ventilation is a very significant heat loss factor in houses and it is always worth attending to doors and windows. Secondary glazing may provide the best method for draught control of poorly fitted windows, as previously explained. Flues, however, always present a problem. Open fires are most difficult. In older houses grates are often fitted with hoods which are provided with a damper at the back, and these should always be closed when a grate is not in use so that the damper reduces flue ventilation. Boilers should preferably be installed outside the living accommodation, perhaps in a separate boiler house or in the garage, laundry, cellar or kitchen. Describng the kitchen as being outside the accommodation will perhaps cause surprise but the kitchen should never be connected to the accommodation in order to avoid unnecessary humidity, ventilation and odours. The kitchen is, in fact, an ideal situation for a boiler which needs a continuous supply of air. This implies continuous ventilation, which is useful in a kitchen as it will help to control humidity and odour during cooking and washing. The boiler heat also helps to keep the kitchen warm, but a radiant heater is perhaps desirable to provide additional heating when the kitchen is actually in use. These comments apply only to normal flued boilers; sealed-flue boilers installed on external walls have no connection with the accommodation air and cannot assist with kitchen ventilation.

Roof insulation

The roof structure in a typical house accounts for about 17.5% of the heat loss, according to Table 5.1, but the heat loss will be much higher if there is no felt or sarking under tiles. However, close-fitting slates may improve thermal efficiency, provided there is no deliberate or accidental ventilation into the roof space. The thermal insulation of a roof can be easily improved with ceiling insulation, which can be a do-it-yourself task at comparatively low cost. In some cases insulation quilt is laid over the entire ceiling area and covering the joists, but it is far better for insulation to be laid between the joists. The trimming of glass-fibre or mineral-wool quilt to fit between joists is unrealistic, and if material of appropriate width is not available it is far easier to use a pelletised material such as vermiculite or preferably more efficient granular expanded polystyrene which is available as ceiling insulation or cavity wall fill under the trade name Isofil. It was considered for many years that 50 mm (2 in) of glass-fibre quilt was adequate, equivalent to about 57 mm ($2^1/_4$ in) of granular expanded polystyrene, 68 mm ($2^3/_8$ in) of mineral-wool quilt or 98 mm ($3^3/_4$ in) of Vermiculite, giving a typical U value of about 0.5. It is now often recommended that the ceiling insulation should be increased to 100 mm (4 in) or even 150 mm (6 in) of glass-fibre quilt but this only reduces the U value to about 0.3 or 0.2 respectively, and although heat loss through the roof is reduced in this way by about 40–60%, the total heat loss from the house is reduced only by about 5–7%, perhaps an insignificant gain relative to the increased cost. The greatest gain is possible from ceiling insulation if it is installed in an old house with no felt or sarking as it can then reduce total heat loss from the house by as much as 20%, and it is certainly one of the most worthwhile imrovements that can be achieved.

Floor insulation

It is largely impracticable to reduce losses through existing floors except by carpeting, and here, as with any insulation technique, the insulation value increases with thickness so that very heavy underlay is always advisable. Carpets can also be valuable by reducing draughts through ill-fitting floorboards or under skirtings and in this way carpeting can often provide the most realistic method for reducing heat losses in some houses.

Heating systems

Whilst this chapter is really concerned with methods for reducing heat losses in buildings, the householder's main purpose is to reduce heating costs and a few comments on heating systems may be appropriate. Electric heating has the distinct advantage that it does not involve a flue which may introduce unnecessary heat losses by greatly increasing ventilation rates, yet some ventilation is always essential for comfort and to control humidity. The problem is to balance the ventilation requirements for adequate combustion and flue flow with the need to restrict ventilation to avoid unnecessary heat losses. Thus under-floor ventilation should generally be used for open grates whenever possible whilst boilers should be situated away from the normal living accommodation, perhaps in a kitchen where gas and oil boilers with good ventilation can assist in reducing humidity and consequent condensation problems.

The efficiency of boilers varies considerably. Whilst slow-burning solid-fuel boilers may appear to be very economic, the boiler performance is controlled by restricting the air flow and there is thus a tendency for only partial combustion to occur so the performance is relatively inefficient. In contrast, oil and particularly gas boilers achieve very efficient combustion and in modern designs good heat exchange is also achieved so that the minimum amount of heat is wasted in the flue. Whilst these basic combustion problems are important, much greater fuel savings can be achieved in existing installations by altering the control systems. For example, many central heating and hot-water boilers are operated by thermostats installed in the boiler or pipes, the boiler firing whenever the pipes reach a certain minimum temperature. The boiler therefore fires at intervals to keep the pipes warm and this heat energy is then wasted if the pipes are not performing a central heating or water heating function. Much better control is achieved if a hot-water boiler is controlled by a thermostat installed in the water storage cylinder so that if water is not being used the boiler will not fire until the cylinder has significantly cooled, usually a very slow process if the cylinder is properly insulated. Similarly a central heating boiler should be controlled by a space thermostat which should be installed in an area which is relatively free from draughts and not affected by any additional heating, and this thermostat should isolate the supply to both the hot water or warm air circulator as well as the boiler firing unit; in many systems the space thermostat controls only the circulator pump, the boiler being controlled by a thermostat installed in the pipework, so that in the summer months there is a danger that the boiler will fire regularly to keep the pipes warm.

There is considerable interest at the present time in the use of solar heating as an alternative to both central heating and hot-water boilers but it must be emphasised that solar heating is unrealistic for the very simple reason that it is most efficient in the summer months but is required in the winter months, and there is currently no effective method for storing heat for this length of time. At present solar heating is realistic only for situations where heat energy is required when the sun is actually shining so that it is particularly suitable, for example, for the heating of swimming pools. It is true that solar heating can be used to reduce hot-water heating costs but there are two fundamental problems that must be recognised; the savings will be insufficient to justify the installation costs and, as savings will be restricted to the summer months, solar heating can have no effect on peak energy demands, a particularly important point in terms of national electricity and gas production capacity.

Economics of insulation

In summary, thermal insulation should be adopted only if the cost can be justified in terms of fuel savings. In a draughty house heat losses can be significantly reduced by draught control. Whilst particular attention should be given to doors and windows, flue problems must not be overlooked. Secondary glazing may be the most efficient method for reducing draughts through ill-fitting windows, so draught control may be the greatest value of secondary glazing. Ceiling insulation is always advantageous as the cost is relatively low but it is most effective in houses which lack felt or sarking under the tiles. Double glazing can achieve only limited reduction in the total heat losses from a house and can be justified only if the cost is relatively low, as with do-it-yourself secondary glazing systems of a relatively simple type; however, as previously explained, these may be more effective in reducing heat losses through draughts than radiation losses through windows. Replacement window frames can never be economically justified and should be installed only for aesthetic reasons; they have a certain snob value at the present time, largely because of their fantastic cost! Close-fitting curtaining can be more efficient than many double-glazing replacement frames and is certainly much cheaper. Cavity wall insulation can never be economically justified; it is never worth considering under any circumstances in a modern building with lightweight concrete block inner skin, but it may be argued that it is valuable from a comfort point of view in an older property with a 110 mm ($4^1/_2$ in) brickwork or breeze block inner skin. Methods for improving the thermal insulation of solid walls are discussed in Appendix 6 and will not be considered here as they are of limited general importance. The cost of modern carpeting is often little different from that of linoleum or tile coverings and carpets are therefore preferred because of their much greater insulation value. Thick carpets with thick underlays are particularly valuable from the insulation point of view, but carpets may reduce heat losses more significantly by restricting draughts between floor-boards and under skirtings.

Any person considering thermal insulation, or any contractor considering a licence for a thermal insulation process, must beware of misleading claims, particularly those of manufacturers who state that their products will reduce heat losses through a building component such as a window by a very great percentage, completely ignoring the fact that the windows account for only a relatively small proportion of the total heat loss from a house. In all cases insulation should be considered only if the cost can be justified in terms of the savings that can be achieved.

6 *Health and safety problems*

6.1 Introduction

There are many health and safety problems associated with buildings. Indeed, most accidents and illnesses occur in the home, although this is not an indication of the extreme danger associated with a house and its contents but simply a reflection of the amount of time that we spend in our homes. It is intended not that this chapter should be a comprehensive review of health and safety problems associated with buildings but that it should mention some of the particular problems that may be encountered when inspecting buildings for remedial treatments or when carrying out those treatments.

6.2 Ventilation

Inadequate ventilation is one of the most serious problems with modern buildings, arising mainly through fundamental changes in living habits over recent years but exaggerated by an unreasonable preoccupation with thermal insulation as a means to achieve energy conservation and improve comfort.

Ventilation, which is normally measured in air changes per hour, is necessary to ensure an adequate supply of oxygen for breathing and combustion, but also to remove the products of our living processes, that is carbon dioxide and moisture from breathing as well as additional moisture from perspiration, cooking, bathing and laundering. Ventilation problems were unknown before World War I as it was normal for all living spaces, including bedrooms, to be provided with fireplaces and flues, the flues inducing adequate ventilation even without a fire. Indeed, a flue would normally induce excessive ventilation and fireplaces were fitted with dampers to seal the throat when the fire was not in use. The most common type of damper comprised a tilting flap at the top of the fireback or a damper shelf projecting from the back of a hood which could be pushed in to close the flue. There was sufficient air flow around a closed damper to induce adequate ventilation in a room through the window frames or door. Following World War I the style of fireplaces changed progressively and dampers were abandoned, leaving open throats which induced excessive draught, and this change and the need for reduced costs resulted in the construction of houses without fireplaces and flues in the bedrooms. The limited ventilation encouraged condensation problems and induced a feeling of stuffiness, and some local authorities introduced building by-law requirements for wall ventilators designed so that they could not be closed for bedrooms without flues.

The progressive introduction of full central heating after World War II further reduced the dependence on open fires with the introduction of automatic oil- and particularly gas-fired systems which could be relied on to maintain adequate temperatures, and eventually open fires were completely omitted from many new houses. Central heating was only realistic because of the relative low cost of oil and gas fuel, but the rapidly escalating demands on fuel strained resources and threatened long-term reserves, creating the petroleum crisis and an increasing awareness of the need for energy conservation whilst maintaining the improved comfort to which people had become accustomed. Reductions in structural heat loss were achieved relatively easily but the need for reductions in ventilation heat loss has not been approached in such a realistic way, any savings resulting from the relatively unscientific wish to reduce draughts.

As a result modern houses are carefully sealed and there is virtually no ventilation. A few years ago useful continuous ventilation was provided by a boiler installed in the kitchen but it became fashionable to have a boiler in a laundry, garage or even a separate boiler room or compartment, and most modern boilers have air inlets and flues ducted directly to the exterior with no connection with the interior air, all these changes reducing ventilation. As a result the air in modern homes in the winter suffers from a low oxygen content, high carbon dioxide content and high humidity which are apparent as a feeling of stuffiness, with excessive condensation even in well-insulated structures. Although open fireplaces in living rooms are again becoming popular, difficulties are often encountered in maintaining adequate draught because of the sealing of the rest of the accommodation and the lack of adequate combustion air. Down-draughts are sometimes encountered with severe smoke problems in the accommodation but these are an indication that the air pressure in the accommodation is lower than the exterior air pressure. The usual explanation is the presence within the accommodation of a powerful extractor fan, usually in the bathroom or kitchen, often the cooker hood. An alternative cause of low internal pressure is a boiler which is installed in the accommodation, perhaps in the kitchen or utility room, but which has not been provided with an air inlet vent from the exterior. All boilers must have an air inlet in the form of a vent through the exterior wall in the room in which the boiler is installed; the vent should be positioned as close as possible to the boiler to avoid unnecessary draughts. If the vent is omitted the boiler air flow is obstructed when the accommodation doors are closed and this can lead to inadequate oxidation and the generation of carbon monoxide which may affect the accommodation through the inadequate air flow in the flue, and if the accommodation doors are open or leak the boiler air flow will cause down-draughts in any flues and prevent the proper use of fireplaces.

All accommodation requires adequate but not excessive air change. One method which is particularly suitable for a flueless house is a small fan feeding air from the roof space into the accommodation through the ceiling of the landing or, in a bungalow, the hall, an arrangement that slightly pressurises the internal accommodation and achieves ventilation without draughts and without unnecessary heat loss. Another alternative is a passive system in which ventilators are provided which are fitted with a special membrane constructed from a material which will allow air diffusion as well as condensation dispersal; the membrane is cold in relation to the interior air so that condensation is encouraged, but when condensation occurs on the membrane it is absorbed and then dispersed to the exterior by evaporation as the membrane is warm in relation to the exterior air.

6.3 Accidental fire dangers

Accidental fire in buildings is almost always initiated in the contents rather than in the structure, the most common cause being ignition of furnishings through an electrical fault, lighted fuel falling from a fireplace or a dropped cigarette. Structural fire precautions are necessary to protect exit routes and to reduce fire spread; masonry materials, concrete and gypsum plaster are particularly useful as they are non-flammable, although wood is an excellent fire barrier and solid wood will perform particularly well in fire doors because of its low thermal conductivity and slow fire penetration properties. Fire precautions will not be considered in detail, although it is necessary to draw attention to certain problems that may be encountered in connection with remedial treatments, not only during inspection but during treatment works.

It is particularly important to prevent a fire originating in the day accommodation from spreading to the bedrooms, and fire in one house from spreading to another in semi-detached and terraced properties. In conventional accommodation these requirements are usually achieved by the use of plasterboard ceilings and fire gables in roof spaces constructed of brickwork, blockwork or timber frame covered with plasterboard. It is not generally appreciated that it is essential for these fire barriers to be absolutely intact without any gaps which may allow fire penetration. It is not sufficient for the fire barrier to be fitted tightly under the roof sarking felt, and a strip of mortar should be laid between the battens on top of the felt to continue the barrier up to the roof tiles. If a plasterboard barrier is used the joints must be covered with strips of plasterboard to prevent penetration. If inadequate or damaged fire barriers are observed during remedial treatment inspections, they should be reported. In addition, care must be taken during remedial treatments to avoid any damage to fire barriers.

Remedial wood preservation and damp-proofing treatments of buildings often involve the use of flammable solvent formulations. There is obviously a risk of accidental fire whilst applying these formulations, particularly with wood preservation treatment, and subsequently whilst the solvent evaporates. This risk can be reduced substantially by ventilation but ignition danger should also be minimised, by prohibiting smoking and by isolating wiring in the treated area, but also by using high-flash-point solvents with flame-proof lamps and equipment. These standard precautions are observed by all prudent and competent remedial wood preservation and damp-proofing contractors; they are, in fact, part of the Code of Practice of the Remedial Treatment Section of the British Wood Preserving and Damp-proofing Association. Unfortunately accidental fires are still caused by remedial treatment, even when these precautions are observed. One possible cause of fire is an electrostatic spark, caused by the flow of solvent in the hose and its discharge through the spray nozzle. This cause of ignition is very rare; ignition through cold spray contacting hot light bulbs is by far the most common cause of fire through operatives using installed lighting or unsuitable lead lamps. Another frequent cause of fires is auto-ignition. If a fibrous material such as glass-fibre insulation quilt is saturated with solvent, the enormous surface area of solvent in contact with oxygen in the air will result in slow oxidation which will be sufficient, in these circumstances, to progressively increase the temperature of the interior of the quilt until ignition eventually occurs. Obviously insulation materials must always be removed before applying organic solvent preservatives, partly to avoid this danger but also to expose the timbers so that they can be properly treated.

Methane gas, the 'fire-damp' that is so feared as a cause of explosions in coal mines, is emitted naturally from all soils containing organic material. Methane emissions from rock fissures above oil deposits were the cause of the eternal fires of Baku, and today similarly formed methane is collected and utilised throughout the world as 'natural gas'. Methane, or 'marsh gas' as it is sometimes called, is generated by the anaerobic (oxygen-free) bacterial decomposition of organic matter and can be seen sometimes as bubbles rising to the surface of stagnant ponds. The methane often contains traces of phosphine and diphosphine which are formed by similar bacterial action; the diphosphine ignites spontaneously as it mixes with oxygen in the air, in turn igniting the methane to cause small explosions or popping noises as gas bubbles to the surface of stagnant ponds, but also causing the flickering lights over marshy areas that are sometimes known as 'will-o'-the-wisp'. This spontaneous ignition is probably the cause of the extensive slow fires that sometimes destroy large areas of peat bog and similar land when weather conditions have made it unusually dry.

Methane from these natural sources is not usually a danger in dwellings, probably because they are not constructed on soils with high methane emissions, but dangers can arise if buildings act as collectors of methane from a wide area. Incidents involving dwellings are usually associated with construction on landfill where methane is generated, sometimes on a massive scale, through bacterial decomposition of buried rubbish, this decomposition also causing serious subsidence damage. Although ignition of methane accumulations in fill is possible, and today fill at risk is usually carefully vented and monitored, the fill is open to the air and danger only arises through accumulations of explosive mixtures of methane and air in buildings where there are many sources of accidental ignition. In 1986 a house at Loscoe in Derbyshire was completely destroyed in this way, despite the fact that it was not constructed immediately on top of the rubbish tip that was the source of the methane that caused the explosion, clearly demonstrating the danger of methane tracking through piped trenches and inadequately sealed entries into buildings.

6.4 Radon dangers

The dangers to health arising from exposure to ionising radiation from radioactive materials have only been recognised relatively recently, particularly as a result of injuries caused by the use of early nuclear weapons and emphasised since then by incidents such as the Chernobyl accident. Such incidents involve radioactive materials originating from mineral sources, and it is not therefore surprising that we are continuously exposed to radiation from natural minerals in addition to radiation from space and from man-made sources such as X-ray equipment. Natural mineral radiation affects us mainly in the form of radon gas diffusing from the soil beneath buildings and to a lesser extent from building materials.

Radon gas, or more correctly radon-222 to identify the isotope involved, is colourless, odourless and tasteless. The gas 'decays' or changes to another element through radiation to produce a series of decay products which are often commonly described as radon 'daughters'. Although radon-222 is a gas, these decay products are all solids which become attached to particles and droplets in air as they are formed, settling in buildings and accumulating within the lungs. The radon-222 gas concentration in the air in a building can be minimised by ventilation but, whilst this ventilation reduces the concentration of the parent from which these decay products or daughters are derived, it does not remove any decay products that have already been deposited within the building or the body. Radon-222 is therefore important as the source of these decay products, but radon-222 is itself derived from uranium-226 and ultimately from radium-238. Radon decay products therefore occur at highest concentrations in association with rocks containing unusually high concentrations of uranium, particularly in certain igneous rocks in, for example, south-west England and north-east Scotland. Mining represents the greatest hazard but a significant risk can arise in buildings constructed on such rocks or, to a lesser extent, constructed from them.

The significance of radiation to health depends on the intensity of the radiation and the period of exposure, producing both 'acute' and 'chronic' symptoms. Radiation intensity is not significant for the low levels of radiation that may arise in buildings from natural mineral sources but radiation has an accumulative effect and it is the total dose, or the product of the intensity of the radiation and the period of exposure, that is important in relation to health. The levels of radiation at which action is necessary are recommended from time to time by the International Commission on Radiological Protection and in the United Kingdom by the National Radiological Protection Board. These recommendations are based mainly on assessments of the risk of fatal lung cancer due to this radiation in relation to other threats to health arising in dwellings. In some areas the levels are high enough to justify precautions in new buildings, and in a few areas the levels are exceptionally high and justify remedial works to existing buildings.

The radon-222 decay products which represent the main radiation threat to health in dwellings are all formed from radon-222 gas. The main sources of this gas in buildings are the soil on which the building is constructed, the building materials, the water supply and the gas supply. Generally the soil in areas of uranium-rich minerals represents the main source. Building materials are much less significant, and the amounts of radon-222 likely to be introduced through water and gas supplies are less than the contribution from the normal ventilation air, except when the water supply is derived from uranium-rich sources or natural gas is being used close to its extraction point.

The main precaution is therefore to reduce the diffusion of radon-222 gas from the soil into the building accommodation. This requirement can be achieved most efficiently by ensuring that there is a continuous vapour barrier to isolate the accommodation from the soil, coupled with adequate ventilation of the accommodation. In practical terms normal damp-proof courses in walls and damp-proof membranes in solid floors will act as adequate vapour barriers, and ventilation must be provided in any case for normal breathing, combustion and avoidance of condensation. However, damp-proof membranes must be continuous to provide reliable protection against radon-222 gas diffusion from the soil; movement joints and service entries must be sealed, and particular care taken to link membranes in solid floors with damp-proof courses in walls. With suspended floors it is theoretically possible to provide similar oversite protection but discontinuities are likely to occur at walls and the sub-floor space should be well ventilated to disperse any leakage. For this reason it is probably best to avoid the use of suspended floors for new houses in high-risk areas. In existing buildings a concrete slab will provide a good barrier to diffusion but cracks, movement joints, service entries and other openings to the soil must be carefully sealed to reduce radon-222 gas diffusion if tests show that the radiation exceeds the action level for remedial works. Accommodation ventilation may vary widely and in older buildings it may be as high as 4 ach (air changes per hour) even with the doors and windows closed. In modern buildings without flues and with efficient doors and window seals, rates as low as 0.1 ach may occur. Bedroom ventilation rates lower than 0.5 ach and living area rates lower than 1.0 ach usually lead to stuffiness and discomfort, and much higher ventilation rates are essential for gas cooking and flued combustion systems. It is only in extreme cases that ventilation rates may be too low for adequate radon-222 gas control in conjunction with a vapour barrier; ventilation rates are significant only if a reliable vapour barrier cannot be provided, particularly in remedial works in existing buildings where electric extractors are sometimes necessary.

Radon risks are discussed more fully by the author in the book *Defects and Deterioration in Buildings*. Advice on whether remedial works are necessary can be obtained usually from the local Environmental Health Department or direct from the National Radiological Protection Board.

6.5 Microbiological dangers

The need to eliminate pathogenic micro-organisms from water supplies has generally been recognised since 1854 when John Snow established that infectious disease can be transmitted through water contaminated by sewage, the death of Prince Albert in 1861 from typhoid fever prompting the rapid introduction of water purification schemes. Water is normally treated by filtration and chlorination, preventing any risk from drinking mains water, although water-borne infections still occur and are increasing due to changes in the use of water.

Although drinking water is normally supplied direct from the high-pressure mains supply, hot water is generally supplied through a low-pressure system involving a cistern or header tank, the cold supply to baths, basins and showers usually being taken from the same low-pressure system to ensure reliable temperature control with tap and shower mixers. This system does not involve any risk with reasonable use as chlorine in the water supply is sufficient to keep the header tank free from harmful organisms. However, if this system is not used for a period, such as in a hotel which is closed during the winter, organisms may develop in the header tanks, as well as sometimes in leaking taps and shower heads, which may be harmful, particularly if the water is drunk during teeth cleaning or showering.

This problem attracted attention in July 1976 when delegates attending an American Legion convention in Philadelphia were affected by a mystery disease, subsequently traced to a bacterial infection of the water supply caused by a previously unknown bacterium which was later named *Legionella pneumophila*. Legionnaires' disease now accounts for about 2% of pneumonia in Britain with about 200 cases diagnosed annually and a fatality rate of about 10%, men being three times more likely to develop the disease and those particularly at risk being between 40 and 60 years old. However, legionnaires' disease is not commonly caused in Britain by infected low-pressure water supplies but is mainly due to open cooling towers which are widely used in air-conditioning plants for large buildings. In these cooling towers the hot water percolates over a fill and is cooled by a current of air induced by a fan. Droplets of water are often blown into the atmosphere, spreading infection over a wide area if the tower becomes infected. The risk is now very high because of the extensive use of these cooling towers; there are more than 200 installed in the Westminster area of London alone, and there have already been a series of major infections in London which have involved fatalities. Open cooling towers of this type are safe if the water is regularly treated with bactericides to prevent infection, although the danger could be eliminated completely by the adoption of alternative cooling systems in which the water is not exposed to the air.

Allergic reactions can be very complex and can be extremely dangerous if a person develops sensitisation to a stimulant, that is when initial exposure sensitises the individual so that subsequent exposure generates a massive reaction. Allergic reactions generally involve irritation of the skin, respiratory tract or eyes, usually through chemical stimulation but sometimes as a reaction to particles of a particular size or shape. Most people react to excessive exposure to dust but persons are only considered to be allergic if they react unusually severely to particular stimulants. Many people are very sensitive to the musty smell of a damp building, usually because they react to certain fungal spores. The Dry rot fungus *Serpula lacrimans* is particularly troublesome in this respect and is recognised as a major cause of asthma.

6.6 Asbestos dangers

Asbestos is the general name for a group of silicate minerals which have been used extensively as fibrous reinforcement in the manufacture of high-density sheet materials such as building boards and artificial slates, and with cement and other binders as low-density fibrous insulation manufactured as sheets or formed by spray-gun application to pipes, tanks, boilers, undersides of roofs, structural steelwork and wherever thermal insulation or fire protection is required. The asbestos used in the United Kingdom is mined mainly in Canada and South Africa, and in recent years it has been realised that it contains fibres which, because of their needle form and size, can cause irreversible lung damage and cancer or asbestosis. Whilst asbestos fibres always represent a risk to health when present as dust in air, the most severe risks arise through exposure to fibres of crocidolite and amosite, commonly known as blue and brown asbestos respectively. In health hazard terms, chrysotile or white asbestos is considered to represent a much lower risk and can be tolerated at two and a half times the air concentrations of blue or brown asbestos.

The most serious risks are associated with working with asbestos products. Generally a building contractor must take appropriate precautions when working with asbestos, and these precautions are most stringent when persons are exposed to blue or brown asbestos with air concentrations above critical action levels. A contractor has the option to prove by testing that persons will not be exposed to such levels of blue or brown asbestos, or alternatively to adopt precautions as if these materials will be present.

Occupants of a building containing undisturbed asbestos are not normally at risk, even if the asbestos is in the most troublesome low-density insulation form with a blue or brown asbestos content, as the critical fibres must be suspended as dust in the air to present a hazard to health. Asbestos fibres can usually only occur in accommodation air through asbestos work dust that has not been removed properly or from air passing through or over insulation which is shedding fibres through breakdown of the binder. If it is suspected that asbestos fibres may be present in air, perhaps through breakdown of a sprayed asbestos roof insulation, samples of the insulation should be examined microscopically to check that it actually contains asbestos and to see whether blue or brown asbestos fibres are present. If these checks suggest that there may be a risk, air sampling should then be used to check whether fibres are present in the air at levels that represent a significant health hazard. Whilst publicity in recent years has rightly emphasised the very serious health hazards that arise through exposure to asbestos, it has not been emphasised that it is dust in air that represents the main hazard and that occupants in buildings are not normally at risk from undisturbed asbestos.

6.7 Wood preservative dangers

In recent years the health and environmental dangers associated with wood preservation have attracted particular attention. Restrictions on the use of existing preservatives and the requirements for approval of new preservatives have become increasingly stringent and are now causing serious difficulty to the industry. These changes have not necessarily resulted in reduced risks to health and the environment as the development of safer preservative systems are now discouraged by the costs involved in submitting new preservatives for approval, and it has been necessary, for economic reasons, to extend the life of established preservative sytems which would not be acceptable if they were submitted for safety approval today.

All wood preservatives contain toxic components but there is no justification for their prohibition. Regulations should specify the precautions that are necessary to ensure their safe use in terms of the hazards to operatives during formulation and use, to the users of treated wood, and to the environment. In some cases these precautions may mean that it is uneconomic to use a particular product and realistic control is therefore achieved.

Remedial wood preservation contractors have a duty to ensure the good health and safety of their own operatives, as well as the occupiers of treated buildings. Proper protective clothing must be provided for operatives, although it is difficult to ensure that it is worn at all times. For example, roof spaces may become rather warm and operatives may be tempted to remove clothing, perhaps during spraying. Operatives may then leave the roof and sit in the sun without a shirt or vest and there is a danger of mild sunburn, accompanied by considerable irritation if the skin is affected by organic solvents. Nose bleeding can also occur when some preservative vapours are encountered in high-temperature conditions, but it must be appreciated that some operatives are more sensitive than others, and particularly sensitive persons should never be employed for this type of work. Masks and barrier creams are often recommended but neither is really advisable. Simple gauze masks tend to absorb treatment fluids, perhaps exposing the user to abnormal concentrations of toxicant vapours; without a mask the operative would probably take more care to avoid unnecessary breathing of spray or vapour. Barrier creams can also give operatives unjustified confidence and it is far better to train them to take necessary precautions. Obviously all operative gangs must be aware of the health and fire dangers and must be aware of the action that should be taken in an emergency.

There are several important points that should be borne in mind when applying organic solvent remedial treatment preservatives in buildings. Low-pressure sprays should be used with coarse jets to ensure that the maximum volume can be applied to the timber surface without the excessive volatilisation of solvents that occurs if high pressures and fine jets or air-entrained paint sprays are used. Whilst it is essential to achieve the maximum loading of preservative on the wood to ensure maximum penetration, dripping of excess fluid must be avoided and care must be taken to ensure that electrical cables are not treated unnecessarily and preservative does not enter junction boxes or other electrical fittings. Treated areas must be freely ventilated to disperse solvent vapour which is a fire hazard and which may affect electrical cables and cause staining around ceiling roses and wall switches. Electrical installations in the treated areas should be disconnected during treatment and even for several days afterwards as there is a danger that sparks may ignite solvent vapour. Smoking, naked lights and plumbing activities must be prohibited in the area for seven to fourteen days depending on the nature of the solvents involved, and notices to this effect should be posted at the entrances of the property and at the entrances to roof spaces and other treated areas. Insulation materials must always be lifted before treatment and replaced later after the solvent has completely dispersed, certainly not less than seven days after treatment in any circumstances, and insulation must never be sprayed with preservative as there is then a real danger of spontaneous combustion.

Some of the phenolic preservatives, particularly the chlorophenols, can cause treatment operatives severe respiratory and dermal irritation, particularly if excessive spray pressures are used which result in preservative atomisation and spray drift. Dermal irritation problems are often due to the solvents alone and enquiries usually disclose that the individuals are sensitive to similar solvents such as gasoline, kerosene, white spirit and turpentine; such problems are usually associated with fair skin and are aggravated by exposure to sunlight. Dermal irritation is also aggravated by some preservative biocides, particularly chlorophenols such as pentachlorophenol (PCP) and organotin compounds such as tri-*n*-butyltin oxide (TBTO), although sensitivity to these biocides varies enormously and, if normal precautions are observed in use, problems are only encountered in particularly sensitive individuals. In extreme cases respiratory irritation can occur and cause coughing and bleeding from the nose but such reactions are usually related to extreme exposure such as spraying preservatives in roof spaces during very hot weather; these problems can be reduced by using coarse low-pressure spray application to flood the surface of the wood with preservative which will then be absorbed by capillarity, avoiding high-pressure sprays which cause atomisation and rapid volatilisation, but improved ventilation may also be necessary.

If sufficient ventilation is provided following treatment, the solvents will rapidly disperse, leaving only the preservative components of low volatility which do not normally cause persistent odour. There have been various suggestions in recent years that these treatments are dangerous to health but it is clear from the excellent record of health in the industry that they are not generally harmful. Mandatory controls have only been introduced comparatively recently but all major manufacturers had accepted voluntary controls for many years under the Pesticides Safety Precautions Scheme, manufacturing and labelling preservatives only in accordance with the guidelines of the Health and Safety Executive. The health risks associated with current preservatives have therefore been carefully assessed and there is no reason to suppose that they present significant risks, either to treatment operatives or to persons resident in treated buildings. Obviously treatment operatives are severely exposed and would be expected to suffer most seriously from any health hazards but, although there are perhaps 5000 to 10000 operatives employed in the remedial wood preservation industry in the British Isles, reports of problems are very few indeed, despite the fact that most operatives work within the industry for many years; on the contrary, it seems that operatives suffer less from some common illnesses such as colds and influenza!

There were proprietary remedial treatment preservatives some years ago which were based on *o*-dichlorobenzene or on mono- or dichloronaphthalene. These biocides are oils which can be readily absorbed through the skin and they certainly present a danger of liver damage to treatment operatives. The dangers were much less with the solid polychloronaphthalene waxes which could not be absorbed in this way and no illnesses were reported despite the extensive use of the waxes at very high concentrations over many years. Pentachlorophenol attracted attention in the past because of its pungent and irritating odour when applied, but in recent years attention has concentrated on the dioxin impurities that may be present in chlorophenols although there is no evidence that these present a problem in relation to remedial wood preservation treatments; it is the chlorophenol itself that presents the most serious risk, particularly if it is absorbed excessively through the skin in hot conditions, such as during spraying or contact with treated wood during hot weather, or through bathing in water contaminated with preservative.

Reactions to the volatile components in solvent wood preservatives are sometimes reported following remedial treatments. Some complaints certainly have a psychosomatic origin, essentially fear generated simply by the odour of the preservative involved, but some complaints are justified. The usual cause is excessive application of preservative, perhaps deliberately in the belief that a more effective treatment is being achieved, although usually carelessness is the true explanation. A typical problem is preservative accumulated on oversites beneath suspended floors through excessive treatment, by which fumes are released over a very protracted period instead of the volatile components being lost rapidly as intended; this problem can also occur through careless installation of chemical injection damp-proof courses. Extreme temperatures through spraying central-heating pipes or treating roof spaces during strong sunshine can cause exceptional volatilisation of components such as organotin and chlorophenol compounds which are only slightly volatile in normal conditions, but most problems are certainly associated with sensitive individuals. Reactions to organotin compounds are usually suffered by fair-skinned persons who also react strongly to any solvent, but some persons seem to be sensitive to the musty odour produced by naphthenates and particularly the slightly sickly odour of the acypetacs compounds which have now largely replaced these naphthenates, with other persons being particularly sensitive to contact insecticides such as Lindane. When problems occur through excessive application of preservative, the building can often only be occupied if materials treated with the preservative are removed; in the case of a floor treatment it is usually sufficient to replace the boarding alone to reduce the odour to a tolerable level, even though the joists and oversites beneath may also have received treatment.

6.8 Sick building syndrome

Sick building syndrome is a condition in which the occupants of particular buildings suffer abnormal levels of sickness. Some of these symptoms are associated with 'wet' air-conditioning systems and can be related to allergic reactions to spores or toxic reactions to bacteria, as previously described in Section 6.5. Mites associated with carpets and soft furnishings may also cause allergic reactions. However, there are other symptoms that cannot be explained in this way. Headache and lethargy are sometimes reported in air-conditioned buildings and can be related to very low carbon dioxide levels. Air-conditioning equipment sometimes includes recirculation through scrubbers which are designed to reduce carbon dioxide levels, but any 'wet' recirculation system will function in this way. If carbon dioxide levels are too low, office workers and other sedentary persons will suffer from inadequate stimulus of respiration and low blood oxygen levels will then cause the observed symptoms.

In recent years a further series of symptoms have been recognised in sick building syndrome involving runny noses which become blocked, dry throats, thirst, tightness of the chest and difficulty in breathing, dry itchy eyes, perhaps with swelling and dry skin. These symptoms are actually caused by abnormally 'dry' air, that is air with a very low relative humidity. Modern air conditioning usually omits humidifiers to avoid 'wet' problems, particularly legionnaires' disease and some of the other microbiological problems previously described in Section 6.5. In cool winter weather the external air has a high relative humidity only due to the low temperature and it has a very low humidity or moisture content as explained in more detail in Appendix 3. When this air enters the building it is warmed, reducing the relative humidity and, if the humidity is already low through low exterior temperatures, the result is excessively low relative humidity in the building.

It can be seen from Figure A3.1 in Appendix 3 that, at a night temperature of 5°C, air has a maximum moisture content of 5.4 g/kg (100% relative humidity), but at 20°C in a building this represents a relative humidity of only about 37%, rather lower than the normal comfort level of 45% to 60%. However, many large buildings have mechanical ventilation which is excessive; a fresh air ventilation rate of 1 l/s per person is sufficient to ensure adequate oxygen supply and carbon dioxide clearance, but a minimum of 5 l/s is required by some codes with 8 l/s recommended, although 4 l/s is adequate with properly distributed ventilation to remove even body odours. In many modern buildings a fresh air ventilation rate of 12 l/s per person or more is used, a level appropriate to a conference or public room with heavy smoking, but many systems add the fresh air to recycled air, and total ventilation rates of 25 l/s are not unusual. These excessive ventilation rates result in 'wind chill' and temperatures must be increased in compensation to maintain comfort, typically to 23 or 25°C. The effect of increasing the temperature from 20 to 25°C in the previous example is to reduce the relative humidity in cold winter weather to about 27%, far below comfort level and the cause of the 'dry' sick building symptoms which are now such a problem in some buildings.

These problems are easily remedied by using only fresh air for ventilation so that the ventilation rate can be substantially reduced, perhaps to levels of only about 6 l/s per person in normal office accommodation. This change will avoid the wind chill and permit air temperatures to be reduced to about 20°C, increasing relative humidity and avoiding 'dry' sick building symptoms except in very frosty weather. These changes also reduce energy consumption for heating and air circulation.

Appendix 1 *Identification of wood-borers and other insects infesting buildings*

Most householders are familiar with common household pests such as silver-fish, earwigs, house-flies and wasps, perhaps even cockroaches, house crickets and carpet beetles. Similarly most remedial wood preservation surveyors are familiar with common wood-borers such as Common Furniture beetle and perhaps House Longhorn beetle, Death Watch beetle and even Powder Post beetle. The problems arise when insects or damage are discovered which cannot be readily identified or explained.

The first part of this appendix consists of a key which will enable most insects found in buildings in north-west Europe to be readily identified. Wood-borers of structural importance are described in detail but there are only brief notes on less important species. Further information on pests of stored food products can be obtained, for example, from the appropriate publications of the British Museum (Natural History).

The identification key is particularly simple to use. In the case of important species a reasonably comprehensive account is given, including details such as size and colour which are features which will confirm the identification. Size can be checked with a transparent scale but it is simpler to examine specimens on a piece of millimetre graph paper. Some characteristics, such as antenna details, can be seen only under magnification, particularly those of very small insects. In the laboratory a low-power binocular microscope can be used but on site a x10 or x15 hand lens is usually adequate. Good light is essential and an illuminated magnifier is an advantage if its magnification is sufficient.

One unusual and particularly convenient feature of this key is the fact that most critical features can be checked from a simple dorsal (top) view of the specimen which then only needs to be turned on its side or back to view confirmatory features.

The second part of the appendix consists of a key to wood-borer damage based upon flight holes, galleries, bore dust and, when available, larvae. It has become apparent from recent checks on commercial remedial treatments that identifications of damage are frequently incorrect. It is also unusual for any serious attempt to be made to confirm that an infestation remains active and, as a result, many current treatments are probably entirely unnecessary.

Identification key and general descriptions

It must be appreciated that this key includes only insects likely to be found in buildings in north-west Europe. Insects have six legs, so spiders, mites and woodlice are excluded. Marine borers, which are molluscs or crustaceans, are also excluded, but the second part includes a description of damaged wood as salvaged pieces are occasionally found in buildings. The general appearance of an insect must be compared first with the illustrations to establish its order. Within each order there is either a general description or, as in the very important Coleoptera (beetles), a key to permit identification as well as details of the most important species. When a key has been used it is important to confirm the identification by reference to size, colour and habitat as given in the description. The visible body of an insect consists of the head, prothorax and abdomen. The wings, if present, are usually folded over the abdomen. The antennae project from the head and there may be appendages projecting from the tail or end of the abdomen.

Order Thysanura (bristle-tails)

Primitive wingless insects, tail with long 'bristles', antennae long and multi-segment, abdomen eleven segments, no metamorphosis (young resemble adults).

Figure A1.1 Silver-fish, *Lepisma saccharina*

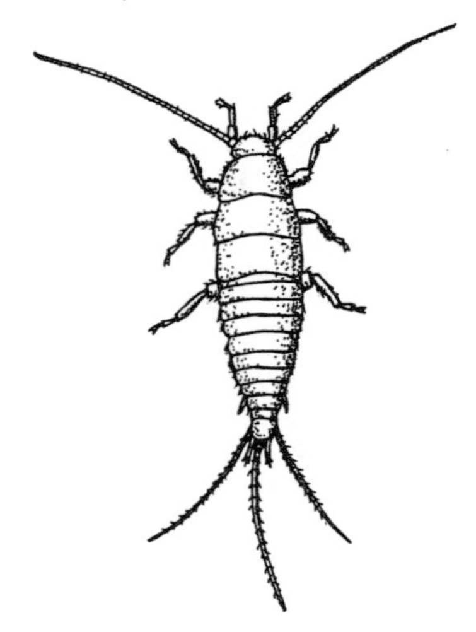

There are only two important species. The silver-fish or silver-moth, *Lepisma saccharina* (Figure A1.1), is shining grey, about 12 mm long, nocturnal and very active. It is a general scavenger but favours starchy materials such as old book-bindings and damp materials such as paper and cellulosic fabrics. The firebrat, *Thermobia domestica,* is similar in habits and general appearance, but it is distinctly mottled and has much longer antennae.

Order Collembola (springtails)

Small primitive wingless insects, stout forked tail folded under abdomen enabling them to jump, antennae four segments, abdomen six segments, no metamorphosis (young resemble adults), gregarious.

Always found in damp situations and jump when disturbed.

Order Dictyoptera (cockroaches and mantids) (previously included in the Orthoptera)

Antennae long and multi-segment, tegmina (forewings) hardened but the left tegmina overlaps the right along the mid-line when at rest (cf. Orthoptera, Coleoptera), hindwings membranous and folded under tegmina, wings sometimes reduced or absent, metamorphosis slight or absent (young resemble adults).

Figure A1.2 Oriental cockroach, *Blatta orientalis*

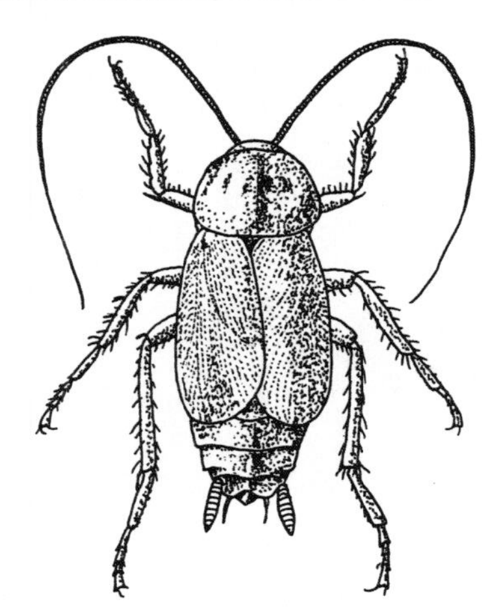

The only family of importance is the Blattidae (cockroaches), which are commonly encountered attacking all forms of food as well as leather and animal skin goods. There are four common species in Europe, which can be identified using the following key:

1 (a) Dark brown, almost black, prothorax uniform colour, tegmina not as long as abdomen (male) or very small (female), 25 mm long (Figure A1.2) — Oriental cockroach, *Blatta orientalis*

(b) Red-brown or dark yellow, prothorax bicoloured, tegmina as long or longer than abdomen — 2

2 (a) Dark yellow, prothorax with two dark narrow longitudinal stripes, 12–14 mm German cockroach, *Blatella germanica*
(b) Red or dark brown, prothorax without longitudinal stripes, adults 23–35 mm 3
3 (a) Prothorax with yellow circular band enclosing distinct bilobed black spot, tegmina with narrow yellow stripe at base, 23–25 mm Australian cockroach, *Periplaneta australasiae*
(b) Prothorax red-brown with central darker area, tegmina without base stripe, 29–35 mm long American cockroach, *Periplaneta americana*

Figure A1.3 House cricket, *Gryllulus domesticus*

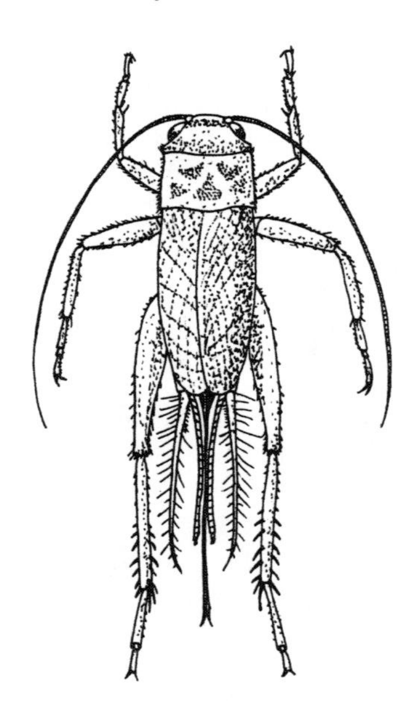

Order Orthoptera (crickets, grasshoppers, locusts)

Although grasshoppers are found occasionally in buildings during the summer months when doors and windows are open, only the house cricket, *Gryllulus domesticus* (Figure A1.3), is important in this order as it attacks food and refuse. It is very active normally, and 12–20 mm long when fully grown. The right tegmina overlaps the left (cf. Dictyoptera, Coleoptera).

Order Dermaptera (earwigs)

The earwigs can be easily recognised by their overall shape and the large claw-like appendages at the end of the abdomen or tail. Only the common earwig, *Forficula auricularia* (Figure A1.4), is important. It is dark brown, 15–20 mm when fully grown, and its tegmina are very short and meet at the mid-dorsal line. It attacks plants as well as other insects, living or dead. Tropical species may be introduced in food products.

Figure A1.4 Common earwig, ***Forficula auricularia***

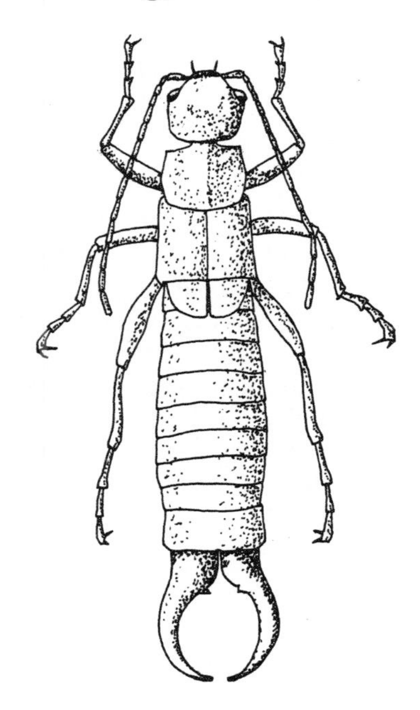

Order Isoptera (termites)

The termites or white ants are the most serious wood-borer pests in all tropical and sub-tropical zones. They are all social insects living in large communities, with both winged and non-winged forms or 'castes'. Although they are known as white ants they are not related to the true ants, which are in the order Hymenoptera. Generally the workers are soft bodied and wingless, confined to the ground or wood, where they devote their energy to feeding, foraging and building. Soldiers, which are also wingless, serve a defensive role alone and are equipped with large heads and jaws. Winged reproductive castes are produced at times and disperse to found new colonies.

About 2000 species of termites have been identified of which more than 150 are known to damage wood in buildings and other structures, but only three species of the genus *Reticulitermes* are found in Europe north of the Mediterranean area. *R. lucifugus* is found approximately south of the

Figure A1.5 The subterranean termite, *Reticulitermes santonensis:* (a) worker and (b) soldier, the most common castes

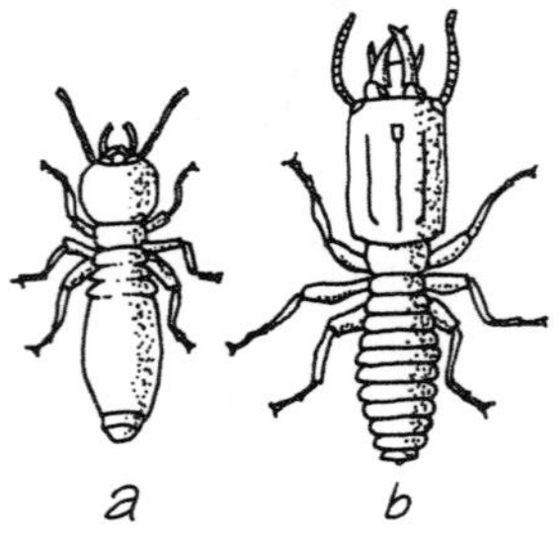

Gironde river in France; it is a subterranean termite sometimes causing extensive damage to interior and exterior woodwork, certainly the most serious wood-destroying pest in southern Europe. *R. santonensis* (Figure A1.5) was originally classified as a variety of *R. lucifugus* but it is now considered to be a true species, distinctly more active and more resistant to adverse dry and cold conditions. It is found attacking wood in buildings in the western coastal area of France between the Garonne and Loire rivers, and has now spread along the connecting railway to Paris, where it is becoming an increasingly serious problem. *R. flavipes*, a North American species, has also been introduced into the Hamburg area. Termites have not been identified elsewhere in north-west Europe. A more detailed account of wood-destroying termites and the countries in which they are found is given in *Wood Preservation* by the same author.

Order Psocoptera (booklice and dustlice)

These are very small and very active soft-bodied white, pale grey, brown or

Figure A1.6 Booklouse, *Trogium pulsatorium*

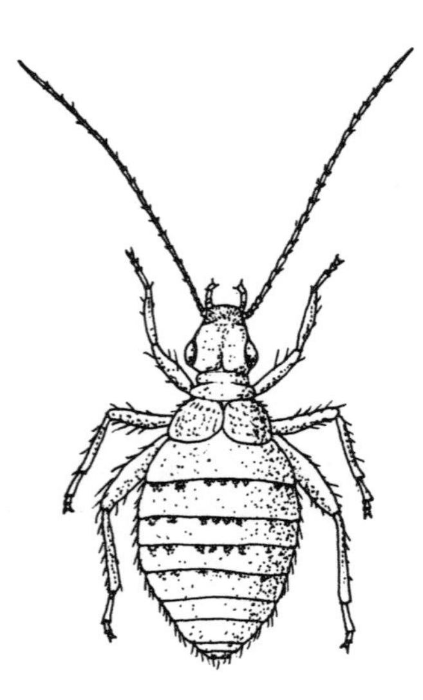

black insects. All species live on fungi growing in humid conditions on paper, plaster, leather or wood; they are common on books where natural glues and sizes have been used in production. Their presence is an indication of dampness. The species most commonly found in buildings can be identified using the following key:

1 (a) White, 1 mm fully grown — 2
(b) Light brown to black, 1 mm fully grown — 3
2 (a) Small wing pads, distinct spots on abdomen (Figure A1.6), can produce a regular tapping noise like the tick of a watch (cf. *Xestobium rufovillosum*, Death Watch beetle, order Coleoptera) — *Trogium pulsatorium*
(b) Small wing pads, no spots on abdomen — *Nymphopsocus destructor*
3 (a) No wing pads, light brown — *Liposcelis divinatorius*
(b) Small wing pads, dark brown or black — *Lepinotus inquilinus, L. patruelis*

Order Anoplura (biting-lice and sucking-lice)

The Mallophaga or biting-lice are mainly associated with birds, living on fragments of feathers or skin. They are usually found in buildings in association with birds' nests, pet birds or domestic fowl. They are active, small, 0.5–6 mm with flat bodies. *Menopon pallidium* and *Lipeuris caponis* are usually associated with domestic fowl (Figure A1.7), whilst the slender bodied *L. baculus* is almost always found on pigeons. The related genus *Trichodectes* is found on domestic mammals: *T. canis* on dogs, *T. subrostratus* on cats, and other species on cattle and horses.

Figure A1.7 Biting-lice associated with domestic fowl: (a) *Lipeurus caponis* **and (b)** *Menopon pallidum*

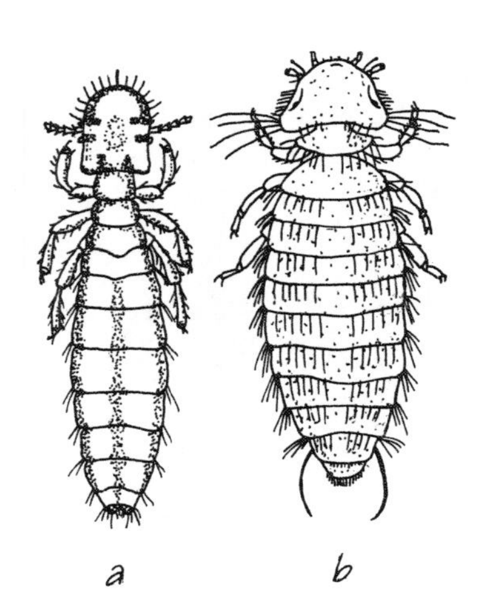

Figure A1.8 Common louse, ***Pediculus humanus***

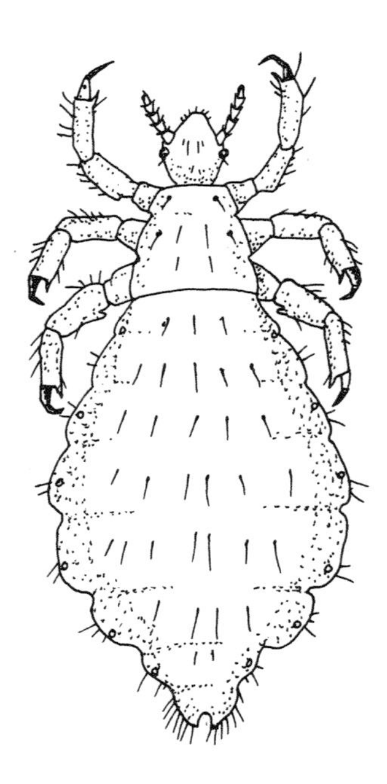

The Siphunculata or sucking-lice are blood-suckers of mammals, the best known species being found on humans and therefore also in buildings. The common louse, *Pediculus humanus* (Figure A1.8), occurs as two races which were previously classified as separate species, but it is now known that the race is decided over several generations by the conditions in which the insect exists. *P. humanus capitis* is the head louse, a small grey insect which infests the body hair. It is smaller and darker than the body louse, *P. humanus corporis,* which infests unhygienic clothing. The crab-louse, *Phthirus pubis,* is confined to the pubic hair and is a minute grey rounded insect.

Order Hemiptera (bedbugs, white-flies, scale-insects and plant-lice)

The Hemiptera are most easily recognised by their mouth parts, which include a very long pointed segmented rostrum or beak for penetrating skin or plant tissue and which is folded beneath the body when not in use. This is a large order of considerable economic importance in agriculture and horticulture, but in buildings only a few species are likely to be encountered.

Two species, *Lyctocoris campestris,* 3.5–4 mm, dark brown with yellow-brown legs and forewings, and *Xylocoris (Arrostelus, Piezostethus) flavipes,*

Figure A1.9 *Reduvius personatus*, **a predator of the bedbug and other insects**

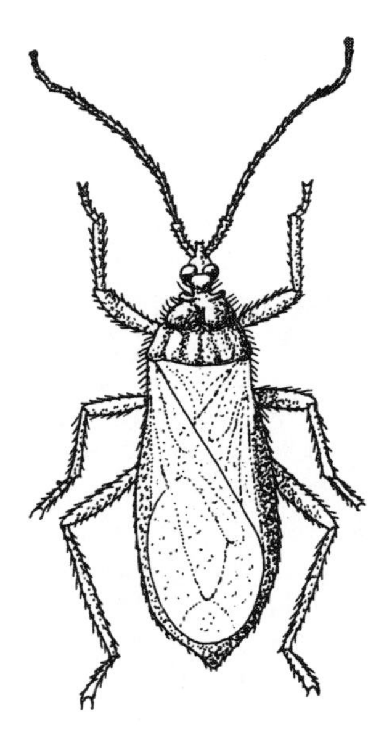

2 mm, red-brown, are found in grain and sometimes in other situations; they are predators of mites and the larvae of beetles and moths. The much larger species *Reduvius personatus*, 16–17 mm (Figure A1.9), is sometimes found in buildings preying on *Cimex* (bedbug) and other insects as well as occasionally attacking humans, when it causes severe pain. The related *Peregrinator biannulipes*, 6–7 mm, is also sometimes found.

Figure A1.10 **Bedbug,** *Cimex lectularius*

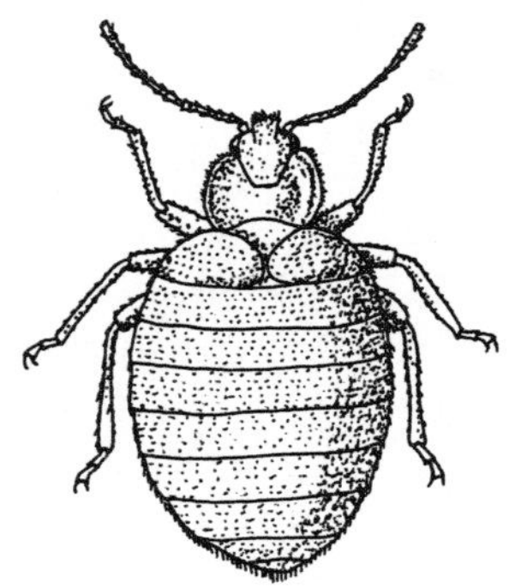

The bedbug, *Cimex lectularius*, is, of course, very important when it is found in buildings. It is about 4 mm, flat, round and dark brown (Figure A1.10), an appearance which accounts for one of its common names of mahogany flats.

Bedbugs are nocturnal and only found on premises where hygiene standards are poor, when they can attack humans, causing swellings and irritation. They possess stink glands which operate when they are disturbed. Related species may be found in birds' nests, for example Oeciacus hiradinis which lives in martins' nests.

Order Lepidoptera (butterflies and moths)

The butterflies and moths are well known and easily recognised. However, the species causing damage in buildings are not easily identified. A number of small moths may be found infesting stored food products and several species of moth may be found attacking fabrics, but the use of synthetic fibres and moth-proofing treatments has severely reduced infestations in recent years. The best known species are the brown house-moth *Hofmannophila pseudospretella*, sometimes found in kitchens, and the common clothes-moth, *Tineola bisselliela*, which attacks untreated wool fabric. When larvae are found they are sometimes thought to be woodworm, although moth larvae are typical straight-bodied caterpillars with prolegs on the abdomen in addition to normal legs.

There are two families of the Lepidoptera with wood-boring larvae. The Cossidae (goat-moths and carpenter-moths) are nocturnal fliers which lay eggs on bark or within flight holes from which moths have emerged. *Cossus cossus (lignaperda)* may be found at the base of oak, elm, willow and poplar trunks. In Europe and North America the most important species is probably the Wood Leopard moth, *Zeuzera pyrina (coesculi)*, which bores in branches, usually in fruit trees. It has sometimes been listed as a cause of damage in structural wood, but if it is found in buildings it is most likely to have been introduced in fire logs. The Sesiidae (clear wings) are not readily identified as moths as on superficial inspection their slender bodies and transparent wings suggest a close relationship with wasps. However they can be recognised by a band of scales around the edges of the wings. The Pyralidae (pyralid moths) can also be mentioned for, although they are not wood-borers, their webs and pupal chambers are often found in deep open joints and splits in old roof structures, where they are sometimes confused with fungal growth by inexperienced surveyors. The two most important species are the bee moth. *Aphomia sociella*, and the honeycomb moth, *Galleria mellonella*, so named because they also infest the combs in beehives and wasps' nests.

Order Coleoptera (beetles)

The beetles are probably the most important insects found in buildings. They always possess two pairs of wings and the forewings are hardened to form elytra or wing cases which meet along the dorsal mid-line (cf. Dictyoptera, Orthoptera) and usually completely conceal the folded hind-wings and the abdomen beneath. Damage caused by wood-boring beetles can be identified using the key in the second part of this appendix. Adult beetles can be identified using the following key, but it must be appreciated that in view of the complexity and importance of this order the key is rather complex with secondary family keys, and some of the descriptions are rather long.

1 (a) Head forming narrow snout in front of eyes — Curculionidae 2
(b) Head without snout — 3

2 **Curculionidae**

2.1 (a) Elytra short with last abdominal segment visible, antenna eight segments — Grain weevil 2.2
(b) Elytra covering last dorsal segment, antenna seven segments — Wood weevil 2.3

2.2 (a) Prothorax punctures distinctly oblong, hindwings absent, 3–4 mm — Granary or Grain weevil, *Calandra granaria (Sitophilus granarius)*
(b) Prothorax punctures dense and round, elytra usually with four reddish spots, hindwings present, 2.3–4.5 mm — Rice weevil, *Calandra (Sitophilus) oryzae*

Figure A1.11 Wood weevils: ***Euophryum confine*** **and head of** ***Pentarthrum huttoni***

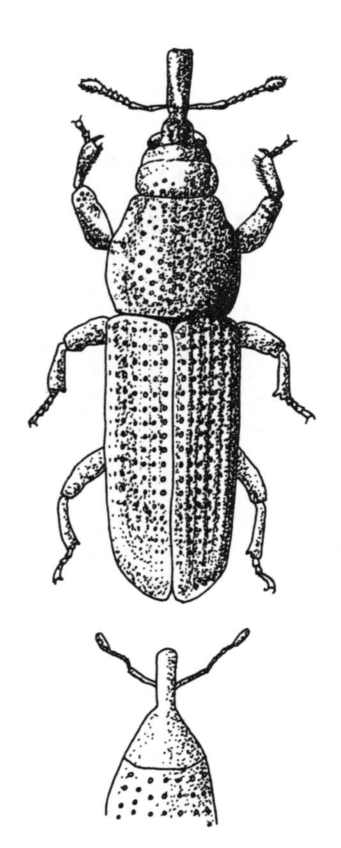

2.3 (a) Elytra with edges towards rear distinctly flexed upwards, noticeable constrictions of head behind eyes and between head and prothorax, 2.8–4.8 mm (Figure A1.11) *Euophryum confine*

(b) Elytra without edges flexed upwards, no noticable constrictions of head, 3–5 mm, (Figure A1.11) *Pentarthrum huttoni*

Wood weevils are only found in wood that is damp or decayed. Unlike most wood-boring beetles the adults tunnel actively. *Pentarthrum huttoni*, a small black shining weevil, is a native of north-west Europe usually found in buildings attacking decayed floorboards and panelling, as well as in old casein-glued plywood. *Euophryum confine* was apparently introduced to Britain from New Zealand in about 1935. It is not so dark in colour but has spread widely, apparently because it is able to infest wood which is not signfiicantly decayed and which may have a moisture content as low as 20%. Another native species, *Caulotrupis aeneopiceus*, is occasionally found in buildings in association with very decayed wet wood in cellars and sub-floor spaces. Infestations of *Cossonus ferrugineus* and *Rhyncolus lignarius* have been reported, but only rarely.

Figure A1.12 *Anthicus floralis*

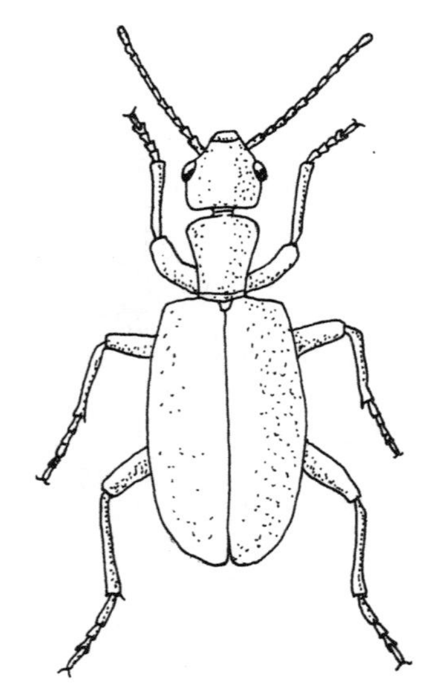

3 (a) Distinct narrow neck between head and prothorax; Anthicidae *Anthicus floralis*, 3.0–3.5 mm (Figure A1.12) is found in decaying organic matter.

(b) Distinct broad neck between head and prothorax Cerambycidae 4

(c) No apparent neck between head and prothorax, which are therefore closely fitted 5

4 **Cerambycidae**

4.1 (a) Antennae shorter than body, elytra with distinct light patches (Figure A1.13), prothorax with two black nodules on either side in female (white marks in male), 10–20 mm, dark brown or black *Hylotrupes bajulus*

Hylotrupes bajulus (Callidium bajulum), the House Longhorn beetle, is a serious pest in dry softwood in mainland Europe as far north as south Sweden, but it is confined to the south-east in England.

(b) Antennae as long as the body 4.2

Figure A1.13 House Longhorn beetle, *Hylotrupes bajulus*

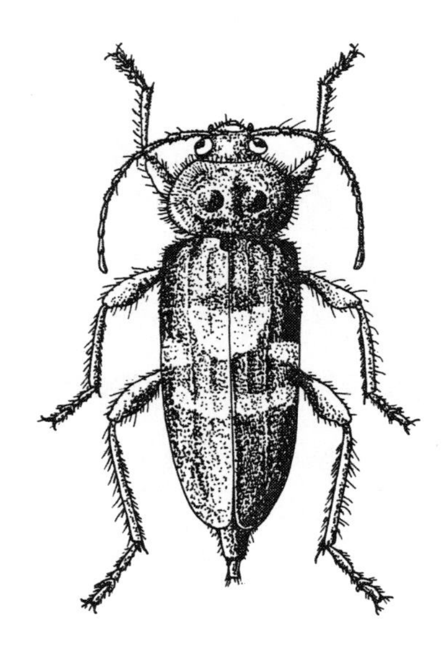

4.2 (a) Length about about 18 mm including antenna, brown *Phymatodes testaceous*

Phymatodes testaceus, the Oak Longhorn beetle (Figure A1.14), is able to attack European hardwood, particularly oak, when dry but only if the bark remains.

(b) Length 3-6 mm including antennae 4.3

4.3 (a) Elytra long, hindwings concealed *Gracilia minut (minuta)*

(b) Elytra half length of abdomen, hindwings exposed *Leptideela (Leptidea) brevipennis*

Gracilia minut and *Leptideela brevipennis* can also attack dry wood but they are confined largely to wickerwork.

Occasionally active infestations of *Eburia quadrigeminata*, yellow or pale brown, 18–24 mm, may be found in North American oak and *Ergates spiculatus* in Douglas fir many years after conversion, but these are examples of exceptional survival by forest species.

Figure A1.14 Oak Longhorn beetle, *Phymatodes testaceus*

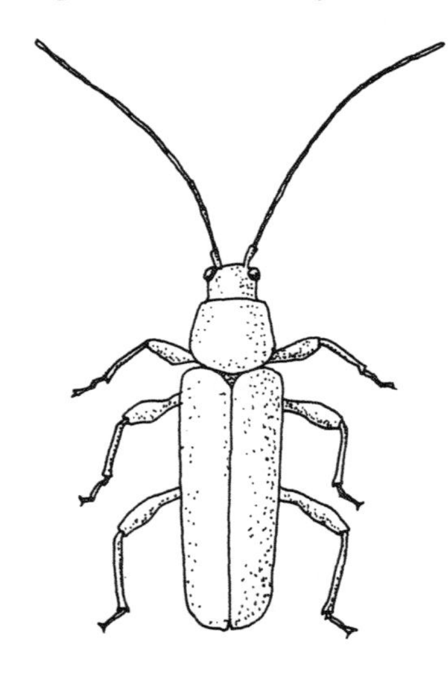

5 (a) Elytra leaving at least one abdominal segment exposed 6
(b) Elytra covering all abdominal segments 17

6 (a) Elytra very short, abdomen flexible, at least six dorsal segments exposed Staphylinidae 7
(b) Elytra moderately long, abdomen inflexible, not more than three segments exposed 8

7 **Staphylinidae**

The rove-beetles, black or brown, usually long and narrow (Figure A1.15), 0.5–15 mm, very active, found in decaying organic matter. Similar in appearance to earwigs (Dermaptera) but without the large claw-like appendages at the end of the abdomen or tail. Very flexible abdomen, sometimes folded forward over the back in a threatening attitude.

Figure A1.15 Devil's coach-horse, *Staphylinus olens*, **a rove-beetle**

8 (a) Antenna with distinct angle between large first segment and rest of shaft — Histeridae 9
(b) Antenna not angled — 10

9 **Histeridae**
These beetles have black or dark-brown shining oval bodies, 1.5–4 mm (Figure A1.16), but are occasionally more brightly coloured and up to 9 mm. They are found in decaying organic matter, sometimes in association with stored food, sometimes beneath bark.

Figure A1.16 A Histerid beetle, *Carcinops quattuordecimstriata*

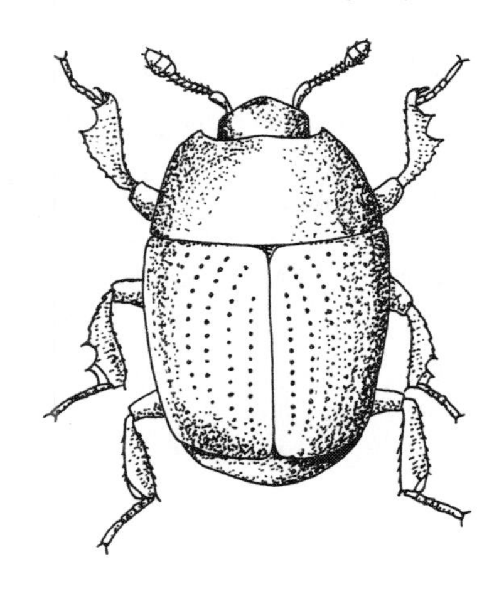

10 (a) Antenna ten segmented with large single-segment club, small narrow insects — Monotomidae 11
(b) Antenna eleven segmented, distinct club absent or multi-segment, longer and broader insects — 12

11 **Monotomidae**
The *Monotoma* species are 1.8–2.5 mm long and found in decaying vegetable matter, sometimes beneath bark. Their tarsi have three segments.

12 (a) Antenna with large three-segment club, abdomen with two or three dorsal segments exposed, elytra non-striate — Nitidulidae 13
(b) Antenna club absent or small, not three segment, abdomen with one dorsal segment exposed, which may be vertical and invisible, elytra distinctly striate — 14

13 **Nitidulidae**
These Sap-feeding beetles feed on tree sap or fruit juice, particularly when partly fermented.

14 (a) Antenna last three segments longer than other segments Anthribidae 15
(b) Antenna last three segments not distinctly longer than other segments Bruchidae 16

15 **Anthribidae**
These beetles are found in decaying vegetable matter, such as decaying wood. *Araercerus fasciculatus* is 3.0–4.5 mm long and known as the Coffee-bea weevil, although not a true weevil.

16 **Bruchidae**
These beetles usually infest legumes, and species less than 5 mm are commonly known as pea or bean weevils, although not true weevils.

17 (a) Elytra at least half blue or blue-green, metallic lustre, antenna distinct club, less than 7 mm Cleridae 18
(b) Brown or black, perhaps with metallic blue lustre, more than 12 mm 19
(c) Elytra not blue or blue-green, less than 12 mm 24

18 **Cleridae**
The 'chequered' beetles are of moderate size and generally brightly coloured and pubescent (furry). They are predators and their identification may confirm the presence of a host insect. Some species, such as *Necrobia rufipes*, have been reported feeding on cured meats such as ham and mature cheeses.

Figure A1.17 *Paratillus carus*, **a predator on Lyctid Powder Post beetles**

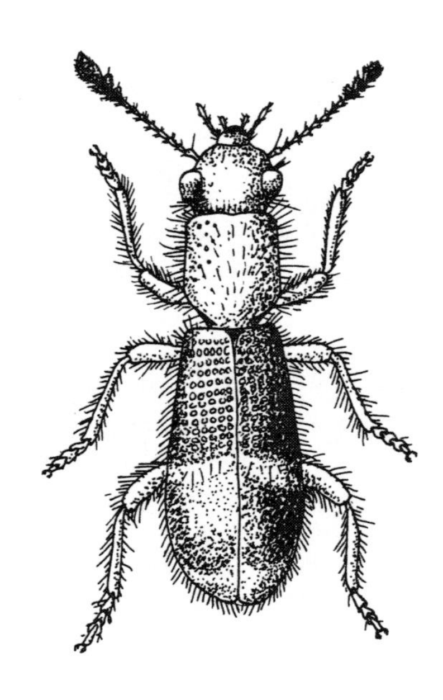

18.1 (a) dorsal and ventral surfaces as well as legs dark brown, 5–6.5 mm *Thanoclerus buqueti*

This species is a predator on the Anobid beetle *Lasioderma serricornia.*

(b) Elytra entirely or partly blue or blue-green, perhaps with transverse light band 18.2

18.2 (a) Elytra with transverse white or yellowish band and sometimes reddish on ends or edges, head red, 5–7 mm (Figure A1.17) *Paratillus carus*

This species is a predator on the Lyctid Powder Post beetles.

(b) Elytra with two white transverse spots near ends rather than distinct band, 5–7 mm *Tarsostenus univittatus*

This species is a predator on the Lyctid Powder Post beetles.

(c) Elytra without transverse band 18.3

18.3 (a) Prothorax and adjacent part of elytra red-brown 4–6 mm *Necrobia ruficollis*

(b) Dorsal surface entirely blue or blue-green 18.4

18.4 (a) Legs and antennae black or dark blue 18.5

(b) Legs and antenna base segments yellow or red 18.6

18.5 (a) 4–4.5 mm *Necrobia violacea*

Figure A1.18 *Korynetes coeruleus,* **a predator on the Death Watch beetle,** *Xestobium rufovillosum*

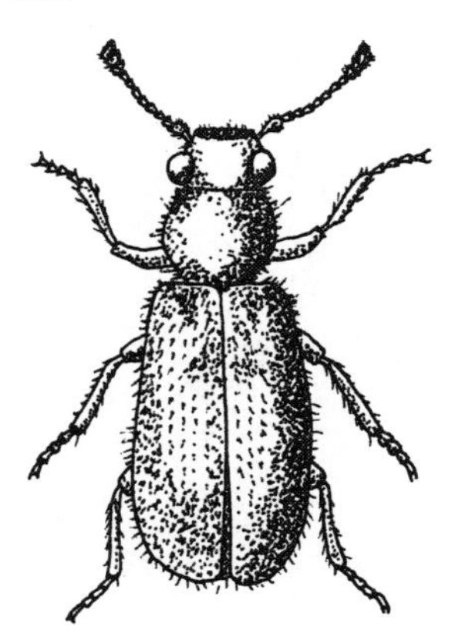

(b) 7–8 mm, prothorax with distinct sharp 'corners' adjacent to head and elytra *Korynetes coeruleus*

This species (Figure A1.18) is a predator on the Anobid beetle *Xestobium rufovillosum*, the Death Watch beetle.

18.6 (a) Antenna club segments as broad as long and dark brown or black, eyes separated by much more than width of eye when viewed dorsally, legs red, ventral (underside) abdomen dark blue, 4–5 mm *Necrobia rufipes*

Figure A1.19 Copra beetle, or Red-legged Ham beetle, ***Necrobia rufipes***

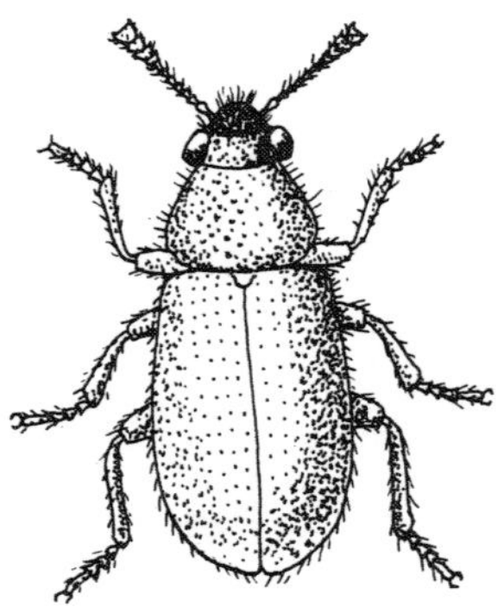

This species (Figure A1.19) is known as the Copra beetle, or Red-legged Ham beetle as it is sometimes found on cured meat.

(b) Antenna club segments three times as long as broad and yellow-brown, eyes separated by less than width of eye when viewed dorsally, ventral abdomen pale red or yellow, 4–5 mm *Korynetes analis*

One other Clerid beetle is occasionally reported. *Opilio mollis*, 7-8 mm, is a predator on the Anobid Common Furniture beetle, *Anobium punctatum*.

Figure A1.20 Rear tarsi of (a) a Carabid (five segments) and (b) an Oedermerid (four segments)

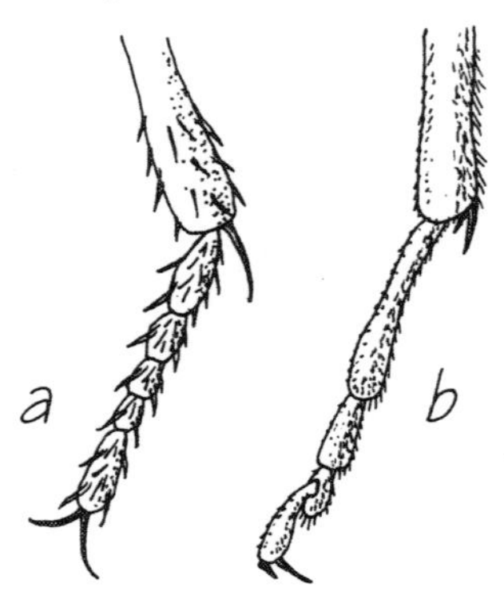

19 (a) Antenna filiform and eleven segments without club 20
(b) Antennae clubbed, rear tarsi four segments (Figure A1.20), 14–25 mm Tenebrionidae 21

20 (a) All tarsi five segments (Figure A1.20) Carabidae 22
(b) Rear tarsi four segments (Figure A1.20) Oedermeridae 23

21 **Tenebrionidae**

Most species infest decaying organic matter, including rotted wood, but a few are predacious and some are the meal-worms found in flour and other stored cereal products.

21.1 (a) Black 20–24 mm, antenna third segment more than twice length of fourth, elytra not striate *Blaps mucronata*

This species is known as the Churchyard beetle. It is often found in cellars and outbuildings in damp conditions.

(b) Brown or black with metallic blue lustre, elytra distinctly striate, usually 20–25 mm, though 12–20 mm has been reported, perhaps a separate species *Helops coeruleus*

This species sometimes occurs in wood in buildings, usually in hardwood such as oak already damaged by the Anobid *Xestobium rufovillosum*, the Death Watch beetle, but it confines its activity to very damp wood distinctly softened by brown rot.

(c) Brown or red-brown, sometimes black, less than 18 mm, antenna with third segment only slightly longer than fourth, elytra striate 21.2

21.2 (a) Elytra red-brown with three black bands at base and middle and apex, 2.2–2.5 mm *Alphitophagus bifasciatus*

This very small species is known as the Two-banded Fungus beetle.

(b) Elytra single coloured 21.3

21.3 (a) 14–18 mm *Tenebrio* spp. 21.4

(b) Less than 7 mm 21.5

Figure A1.21 Confused Flour-beetle, *Tribolium confusum*

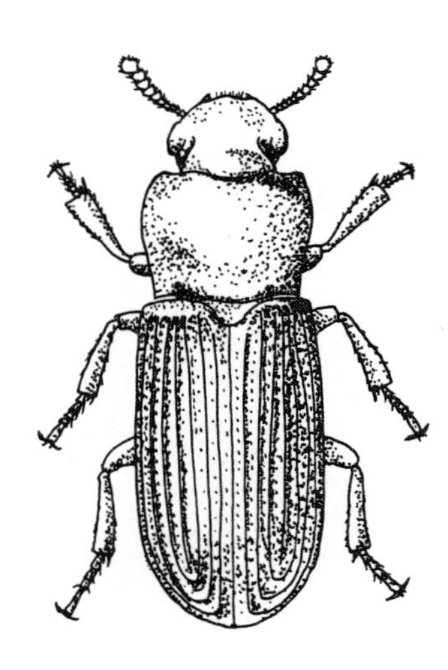

21.4 (a) Dorsal surface shining, antenna last segment as broad as long and third segment slightly longer than fourth, 15 mm *Tenebrio molitor*
This species is known as the Yellow Meal-worm.
(b) Dorsal surface dull, antenna last segment broader than long and third segment twice as long as fourth, 14–18 mm *Tenebrio obscurus*
This species is known as the Dark Meal-worm.

21.5 (a) Body narrow (Figure A1.21), 3–6 mm, elytra ridged longitudinally *Tribolium* spp. 21.6
(b) Body narrow (Figure A1.23), 2.5–4.5 mm, elytra not ridged 21.9
(c) Body distinctly broader, 4.5–7 mm, dark-brown or black *Alphitobius* spp. 21.8

21.6 (a) Antenna with distinct three-segment club, 3–4 mm *Tribolium castaneum (ferrugineum, navale)*
(b) Antenna with segments broadening to form indistinct five- or six-segment club 21.7

21.7 (a) 4–4.5 mm, red-brown *Tribolium confusum*
This species (Figure A1.21) is known as the Confused Flour-beetle.
(b) 5–6 mm, dark-brown or black *Tribolium destructor*

Figure A1.22 Side view of eye of (a) *Alphitobius laevigatus* **and (b)** *A. diaperinus*

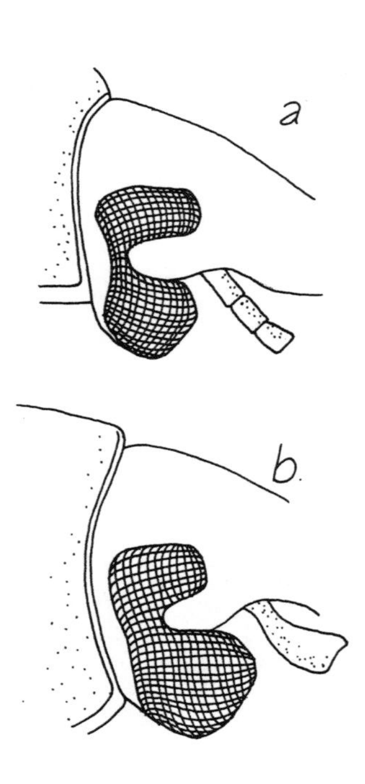

21.8 (a) Eye viewed from side (Figure A1.22) completely or almost completely divided horizontally, antenna fifth segment normal, rear tibia apex slightly hardened, 4.5–6 mm
Alphitobius laevigatus (piceus)

(b) Eye only partly divided by indentation at front, antenna fifth segment with projection, rear tibia apex distinctly broadened, 5.5–7 mm *Alphitobius diaperinus (piceus, ovatus)*

21.9 (a) Eyes more or less round 21.10

(b) Eyes with vertical diameter much greater than horizontal and partly divided by indentation at front 21.11

21.10 (a) Eyes small and round, 2.5–3 mm
Small-eyed Flour-beetle, *Palorus (Caenocorse) ratzeburgi*

(b) Eyes large and vertical diameter slightly greater than horizontal, ridge on front of head flexed upwards, ridge concealing front part of eyes, 2.5 mm
Depressed Flour-beetle, *Palorus (Caencorse) subdepressus*

Figure A1.23 Long-headed Flour-beetle, *Latheticus oryzae*

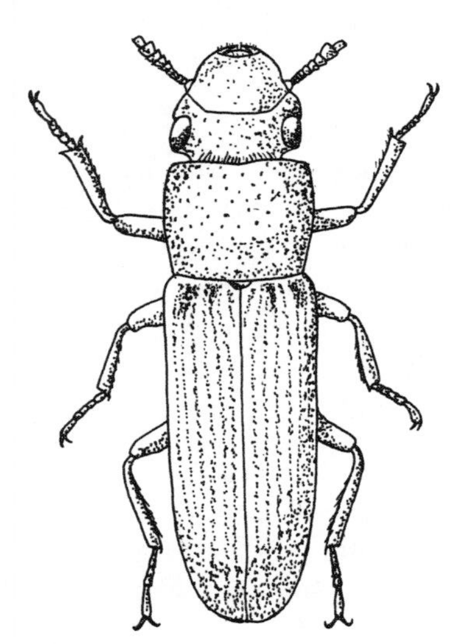

21.11 (a) Head longer than antenna, antenna with distinct five-segment club, hind tarsi with first segment not as long as length of both second and third segments, 2.5–3 mm (Figure A1.23)
Long-headed Flour-beetle, *Latheticus oryzae*

(b) Head shorter than antenna, antenna with loose four-segment club, hind tarsi with first segment as long as length of both second and third segments 3–4.5 mm *Gnathocerus* spp.
This genus includes the Broad-horned and Slender-horned Flour-beetles.

Figure A1.24 *Harpalus rufipes*, **a Carabid sometimes found in warehouses**

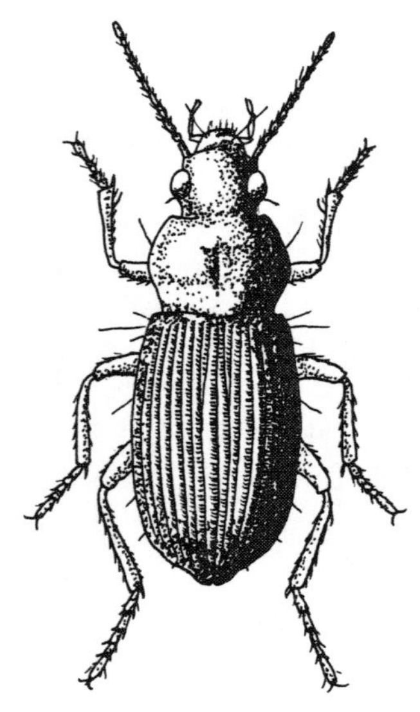

22 **Carabidae**

A family of ground beetles (Figure A1.24), usually occurring in soil, under stones, in rotting wood and under bark. The elytra are fixed together along the mid-line and the hindwings are atrophied in some species. They are similar to the Tenebrionidae in general appearance but may be distinguished by the tarsal segments (see 19 and Figure A1.20). They are principally carnivorous but some species have been recorded as attacking grain and seeds.

23 **Oedermeridae**

Although the adults of this family are often found on flowers or other plants the larval stages usually occur in rotted wood. The Wharf borer, *Nacerdes melanura* (Figure A1.25), is the only species of importance as it is able to attack wood which is reasonably sound and suffering only incipient decay. This species favours wood wetted by sea water and is sometimes found in great numbers in streets and in buildings in areas close to docks or sea defences. It is 6–12 mm, elongate, red-brown with distinct black tips on the elytra, and has long antennae. It is sometimes confused with the carnivorous soldier beetle, *Rhagonycha fulva*, which occurs commonly on summer flowers but the latter is redder with a wider prothorax which lacks the flanges found on the Wharf borer.

24 (a) Elytra oval, prothorax with distinct waist close to junction with elytra, legs long with tarsi five segment, antennae long and eleven segmented without club, 1.8–4.5 mm Ptinidae 25

Figure A1.25 Wharf borer, *Nacerdes melanura*

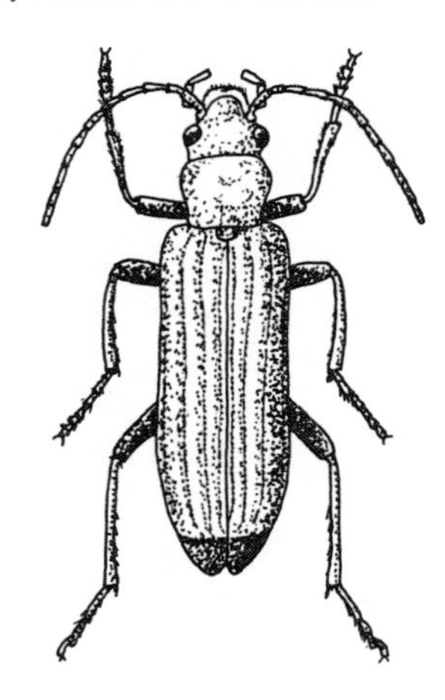

(b) Body (elytra and prothorax together) distinctly round or oval and distinctly convex 26
(c) Elytra with parallel sides 32
(d) Elytra or elytra and prothorax without distinct oval or parallel shape, prothorax without distinct waist 41

25 **Ptinidae**
These are known as the spider beetles because of their relatively long legs. The antennae are situated on the front of the head between the

Figure A1.26 *Eurostus hilleri,* **a spider beetle**

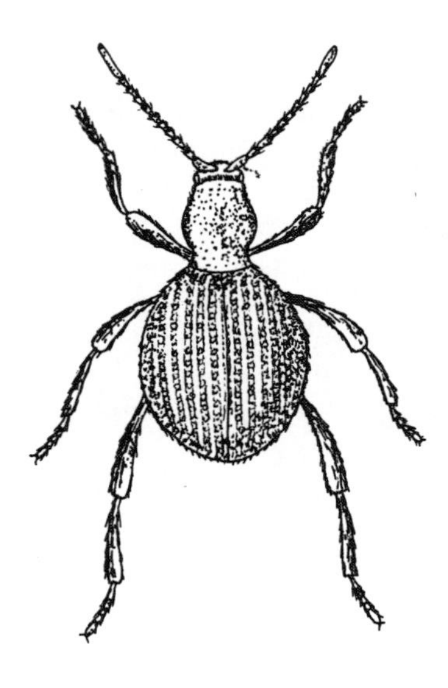

eyes and are generally very close together. The bodies are generally hairy. They cause damage to stored food products, books, furs, woollen goods, natural drugs, etc.

25.1 (a) Space between antennae bases very small 25.2
(b) Space between antennae bases greater than half length of first antennal segment 25.6

25.2 (a) Scutellum (on mid-line at base of elytra) small and indistinct, 1.9–2.8 mm (Figure A1.26) *Eurostus hilleri*
(b) Scutellum large and distinct 25.3

Figure A1.27 Australian spider beetle, *Ptinus tectus*

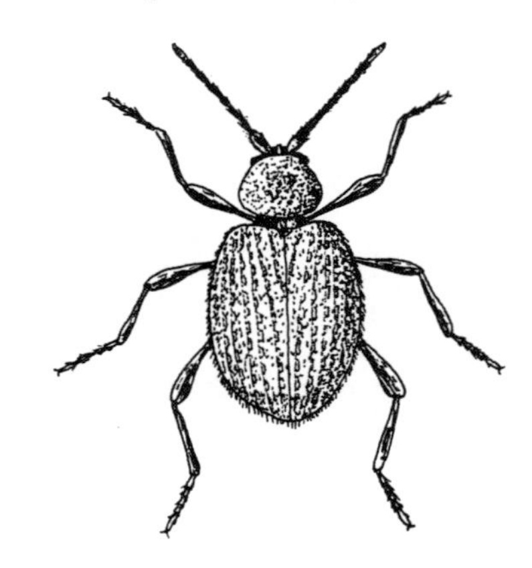

25.3 (a) Dense brown or gold-brown hairs concealing lines of punctures on elytra, 2.5–4.0 mm (Figure A1.27) Australian spider beetle, *Ptinus tectus*
(b) Hairs not dense enough to conceal puncture 25.4

25.4 (a) Prothorax with distinct patches of paler hair on either side, elytra with white scales, 2.0–4.3 mm White-marked spider beetle, *Ptinus fur*
(b) Prothorax evenly covered with uniform coloured hair 25.2

25.5 (a) Elytra without white hairs or scales, 2.3–3.2 mm (Figure A1.28) Brown spider beetle, *Ptinus hirtellus (brunneus)*
(b) Elytra with a few white hairs or scales, 1.8–3.0 mm (Figure A1.29 *Ptinus pusillus*

25.6 (a) Elytra moderately hairy with distinct lines of deep punctures, 1.8–3.0 mm (Figure A1.30) *Tripnus unicolor*
(b) Elytra densely hairy with punctures concealed 25.7

25.7 (a) Elytra grey- or yellow-brown with patches of dark hairs, prothorax with central longitudinal depression, 2.0–4.0 mm *Trigonogenius globulus*

Figure A1.28 Brown spider beetle, *Ptinus hirtellus*

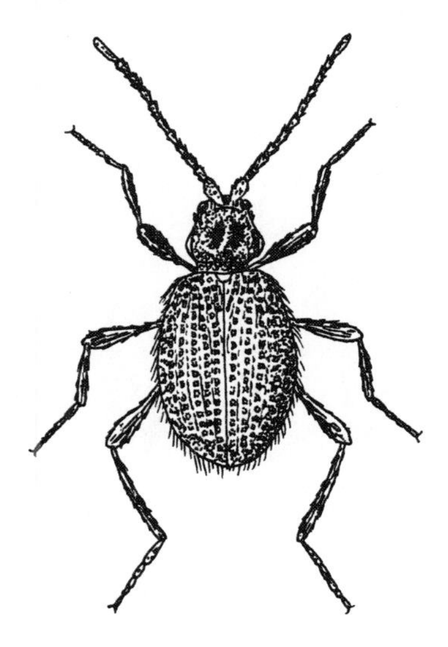

Figure A1.29 *Ptinus pisillus,* **a spider beetle**

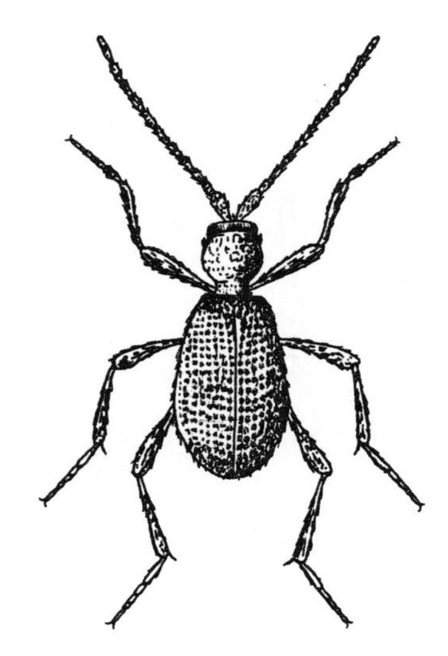

Figure A1.30 *Tripnus unicolor*, **a spider beetle**

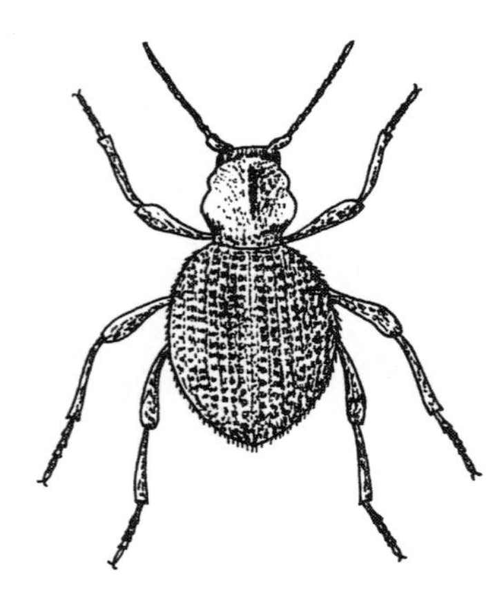

Figure A1.31 Golden spider beetle *Niptus hololeucus*

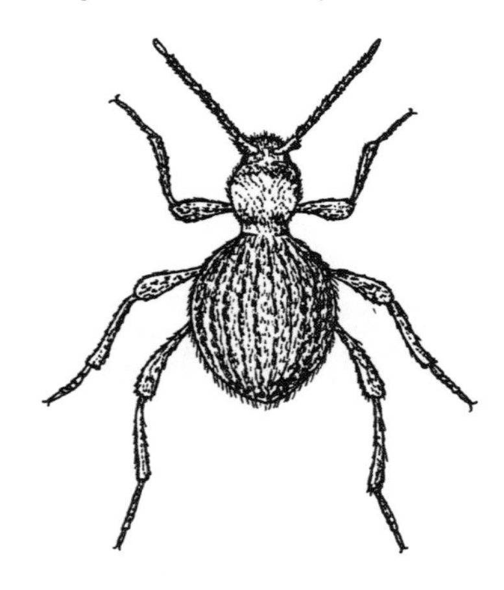

(b) Elytra and prothorax uniform gold-yellow hairs, prothorax without depression, 3.0–4.5 mm (Figure A1.31)
Golden spider beetle, *Niptus hololeucus*

26 (a) 1.5–1.8 mm — Mycaetidae 27
(b) 2–10 mm — 28

27 **Mycaetidae**
This family of small beetles is dependent upon decaying vegetable matter. *Mycetaea hirta* occurs in dung and vegetable refuse but may occur in buildings, particularly in wine cellars. It is 1.5–1.8 mm long,

body convex and hairy, antennae less than half body length and progressively broadening to form club (Figure A1.32). This family also includes the black and red *Endomychus coccineus.*

Figure A1.32 *Mycetaea hirta*

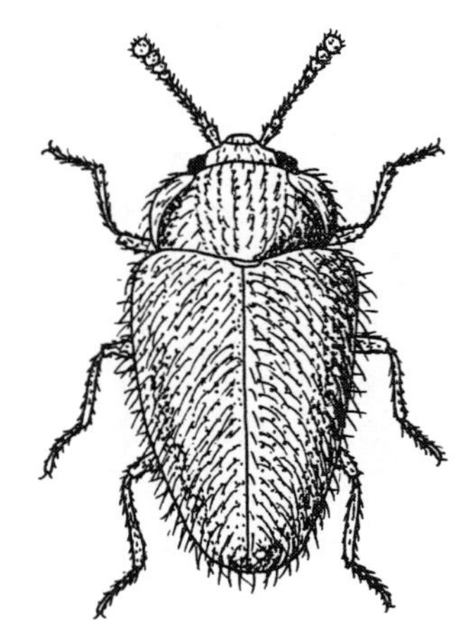

28 (a) Tarsi apparently three segmented (fourth is concealed)
Coccinellidae 29

(b) Tarsi four segments but front in male only three segments
Mycetophagidae 30

(c) Tarsi five segments, antenna short with large club
Dermestidae 31

29 **Coccinellidae**

These are the ladybirds which feed on aphids and other plant pests. The bright red species are well known and are usually found in buildings in hibernating masses.

30 **Mycetophagidae**

Species of this family are superficially similar to the Cryptophagus genus (34) but the latter has a distinct 'tooth' in the centre of each side of the prothorax. The Mycetophagidae and the Cryptophagidae are both fungus feeders and thus found in similar situations. The two most common Mycetophagids in buildings can be readily identified as follows:

30.1 (a) Antenna club four segments, prothorax with distinct pit on each side adjacent to elytra, elytra striate with light spots on each side at base and in each side towards apex, 3.5–4.0 mm (Figure A1.33)
Mycetophagus quadriguttatus

Figure A1.33 *Mycetophagus quadriguttatus*

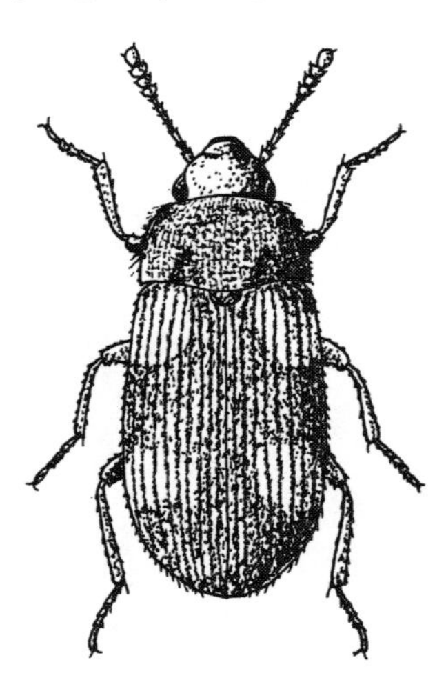

Figure A1.34 *Typhaea stercorea*

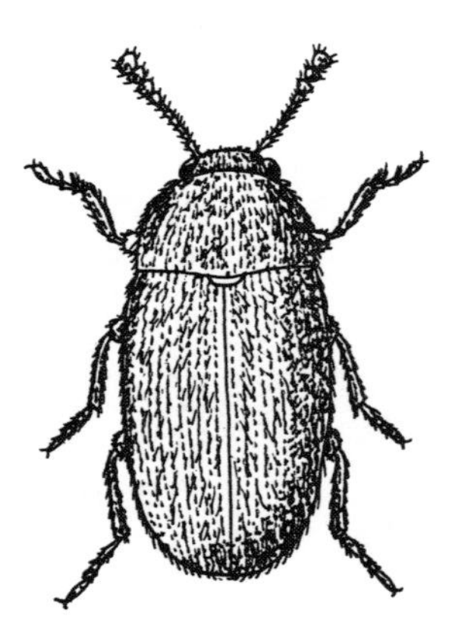

(b) Antenna club three segments, prothorax without pits, elytra hairy and non-striate but hairs in distinct longitudinal rows, 2.5–3.0 mm (Figure A1.34) *Typhaea stercorea (fumata)*

31 **Dermestidae**

The Dermestids feed on dried animal matter such as skins, furs, wool products, dried fish etc., and some species also attack dried vegetable products. The most important species can be identified using the following key:

31.1 (a) 5.5–10 mm 31.2
(b) 1.5–5.5 mm 31.6
31.2 (a) Black perhaps with brown hairs but never with white 31.3
(b) Edges of prothorax white 31.4
31.3 (a) Base half of each elytron with brown patch with three black spots, 7–9 mm Larder beetle, *Dermestes lardarius*
(b) Elytra colour uniform, 7–9 mm *Dermestes ater (cadaverinus)*
31.4 (a) Elytra formed at rear mid-dorsal line into sharp point, 5.5–10 mm *Dermestes maculatua (vulpinus)*
This species is known as the Hide or Leather beetle. It is sometimes found in hen-house litter and the larvae can bore into structural wood in order to pupate.
(b) Elytra without point 31.5

Figure A1.35 Underside of abdomen of (a) *Dermestes frischia* **and (b)** *D. carnivorus*

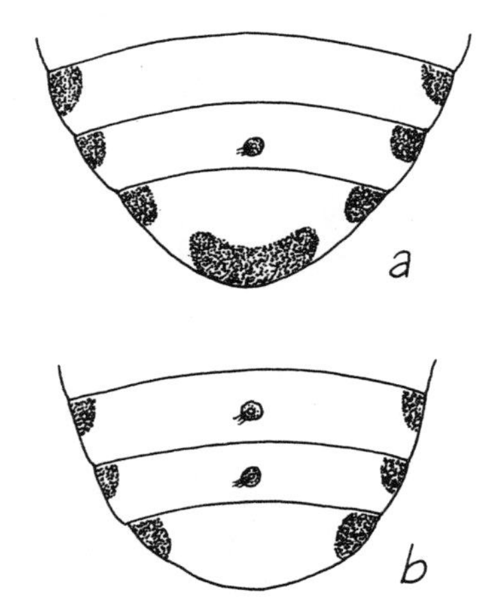

31.5 (a) Last segment of abdomen on ventral (lower) surface with dark patch each side and third patch in centre on rear edge (Figure A1.35), 6–9 mm *Dermestes frischia*
(b) Last segment of abdomen on ventral surface with dark patch each side but without central patch (Figure A1.35), 6.5-8.5 mm *Dermestes carnivorus*
31.6 (a) Body almost round, distinct whitish patches on body, legs short with grooves for housing on underside of body, 2–4 mm *Anthrenus* spp.
One of the most common beetles of this genus is *Anthrenus verbasci* (Figure A1.36), known as the Varied Carpet beetle.
(b) Body oblong 31.7

Figure A1.36 Varied Carpet beetle, *Anthrenus verbasci*

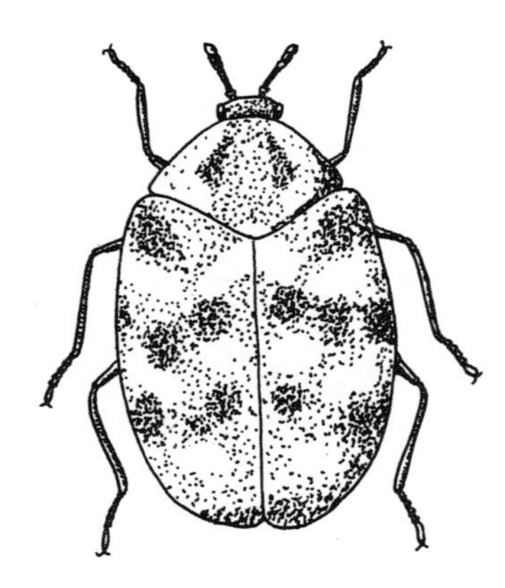

31.7 (a) Antenna club three segment, first segment of tarsi half length of second 31.8

(b) Antenna club four segment, first segment of tarsi much longer than second 31.9

Figure A1.37 *Attagenus pellio*

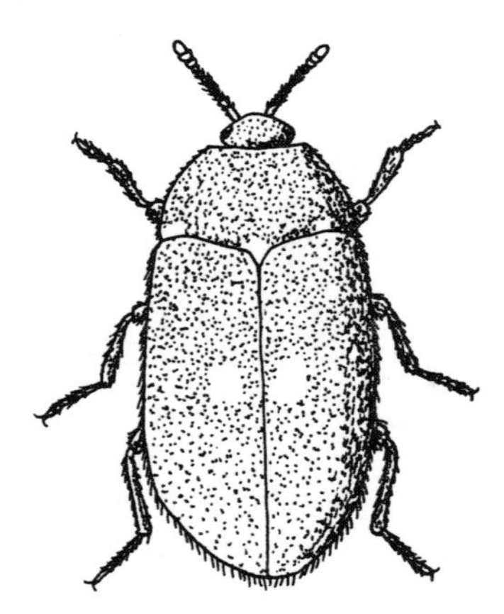

31.8 (a) Elytra black with white patch near centre of each elytron, 4.5–5.0 mm (Figure A1.37) Fur beetle, *Attagenus pellio*

(b) Elytra entirely black, 3.3—5.0 mm Black Carpet beetle, *Attagenus piceus (megatoma)*

31.9 (a) 1.5–3.0 mm, elytra uniform coloration Khapra beetle, *Trogoderma granarius (khapra)*

31.9 (b) 2.5–5.0 mm, elytra black mottled with brown, the brown areas with white hairs — Larger Cabinet beetle, *Trogoderma versicolor*

32 (a) Distinctly flat form — Silvanidae 33
(b) Distinctly cylindrical — 36
(c) Moderately elongate and hairy — Cryptophagidae 34
(d) Not distinctly flat, cylindrical or elongate, antennae without club but perhaps combed or serrated — Anobiidae 35

33 **Silvanidae**

A family of small narrow and distinctly flattened beetles, with eleven-segment antennae with distinct clubs. Tarsi are all five segment. Most species are predacious. Three species may be found in buildings in north-west Europe and can be identified using the following key:

33.1 (a) Prothorax with single tooth forming front angles on each side, 2–3 mm — *Ahasverus (Cathartus) advena*
This species is known as the Foreign Grain beetle and is found on refuse, feeding on moulds.
(b) Prothorax with six large teeth on either side and three distinct longitudinal ridges — 33.2

33.2 (a) Distinct constriction of head adjacent to prothorax, 2.5–3.5 mm — Saw-toothed Grain beetle, *Oryzaephilus (Silvanus) surinamensis*
(b) No distinct constriction of head adjacent to prothorax, 2.5–3.5 mm — Merchant Grain beetle, *Orzaephilus (Silvanus) mercator*

34 **Cryptophagidae**

These very small insects are generally found in association with fungi and decaying organic matter. They are pale brown and hairy. They are sometimes known, together with the Lathridiidae (44), as Plaster beetles as they are frequently observed feeding on mould on damp plaster.

34.1 (a) Prothorax with distinct tooth in centre of each side (Figure A1.38), 1.5–3.5 mm — *Cryptophagus dentatus*
(b) Prothorax with distinct tooth at front angle of each side (Figure A1.38), 3 mm — *Cryptophagus acutangulus*
(c) Prothorax without tooth (Figure A1.38) — *Atomeria linearis*
(d) Prothorax with nine or ten small teeth on each side (Figure A1.38), 1.7– 2.1 mm — *Henoticus californicus*

35 **Anobiidae**

The Anobiid beetles are principally wood-borers of considerable importance.

35.1 (a) Elytra distinctly striate — 35.2
(b) Elytra not striate — 35.3

Figure A1.38 Prothorax details of (a) *Cryptophagus dentatus,* **(b)** ***Cryptophagus acutangulus,*** **(c)** *Atomeria linearis* **and (d)** *Henoticus californicus*

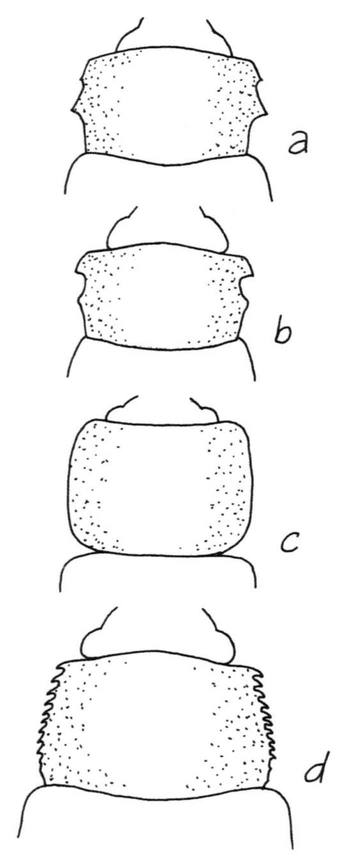

35.2 (a) Prothorax with distinct hump when viewed from side (Figure A1.39), 2.5–5.0 mm (Figure A1.40) *Anobium punctatum (domesticum, striatum)*

This species is known as the Common Furniture beetle and damages sapwood of dry hardwood and softwood. The related *Anobium pertinax* is found in buildings in north Europe but only in association with decayed wood.

(b) Prothorax not humped when viewed from side (Figure A1.39), 2–3 mm *Stegobium (Sitodrepa) paniceum*

This species is known as the Drug Store beetle or Biscuit beetle and is often found in biscuits, particularly dog food, but it can also infest flour.

Figure A1.39 Side views of (a) *Anobium punctatum,* **(b)** *Xestobium rufovillosum,* **(c)** *Stegobium paniceum* **and (d)** *Lasioderma sernicornia*

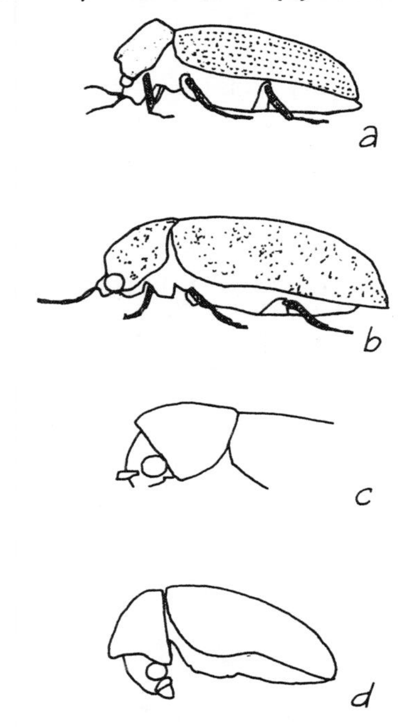

Figure A1.40 Common Furniture beetle, *Anobium punctatum*

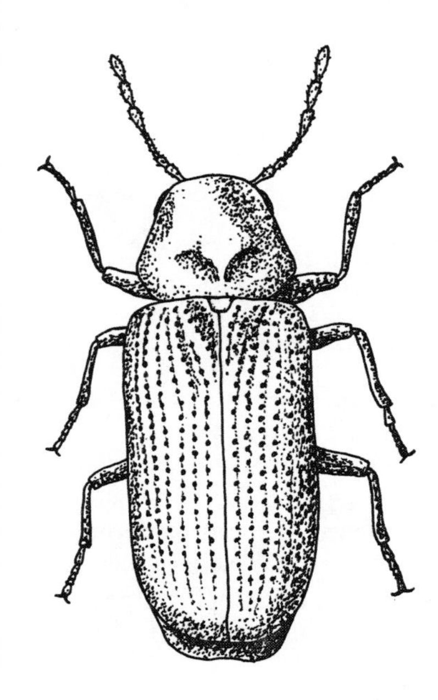

35.3 (a) Antennae comb- or saw-like — 35.4
(b) Antennae simple — 35.5

35.4 (a) 2.0–2.5 mm (Figure A1.39), antennae serrate or saw-like — *Lasioderma serricornia*
This species is known as the Cigarette beetle and is usually found in tobacco.

(b) 3.0–6.0 mm, antennae comb-like in male and serrate or saw-like in female (Figure A1.41) — *Ptilinus pectinicornis*
This species causes damage similar to that of the Common Furniture beetle in beech, maple, sycamore and elm.

Figure A1.41 *Ptilinus pectinicornis*

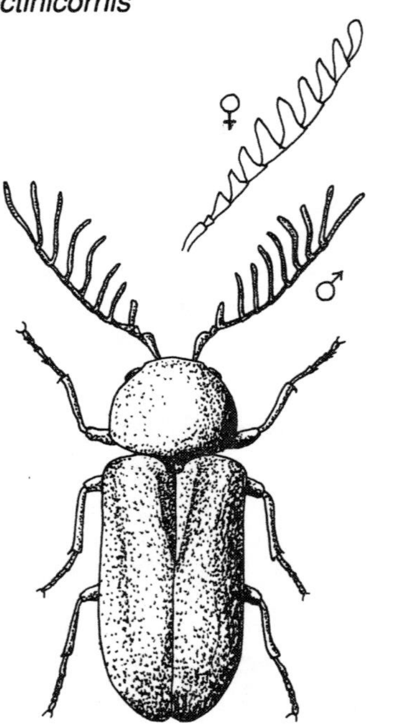

35.5 (a) 3.0–6.0 mm, large eyes can be seen from above projecting on either side of prothorax (Figure A1.42) — *Ernobius mollis*
This species is a borer of freshly felled softwood, attacking the bark and adjacent sapwood.

(b) 6.0–9.0 mm, head and eyes completely concealed from above by prothorax which is very wide at its base; irregular mottled appearance through patchy hair (Figure A1.43) — *Xestobium rufovillosum (tesselatum)*

Figure A1.42 *Ernobius mollis*

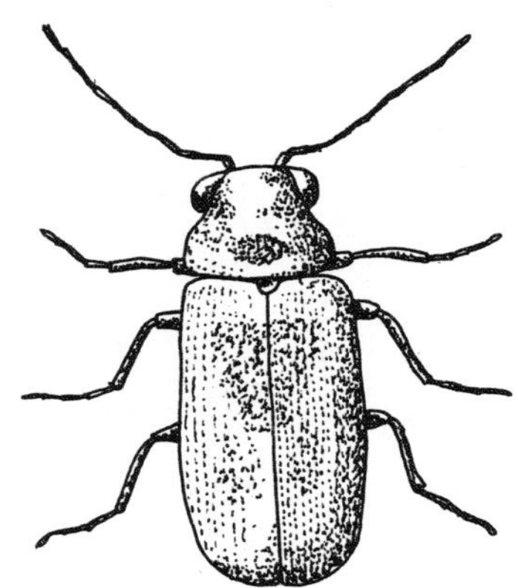

Figure A1.43 Death Watch beetle, *Xestobium rufovillosum*

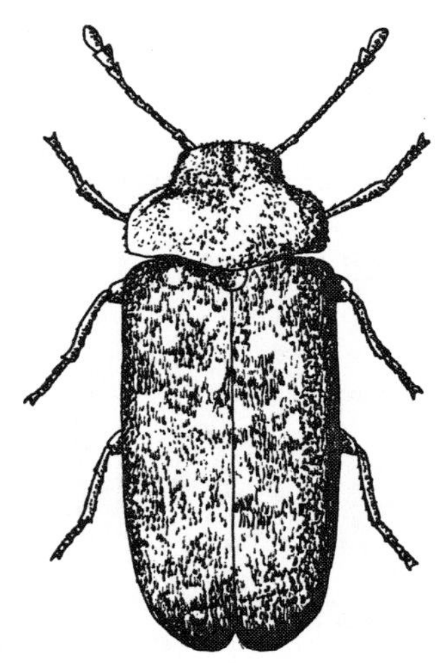

This species is known as the Death Watch beetle and attacks hardwoods such as oak affected by limited fungal decay.

36 (a) Head projecting in front of prothorax, elongate beetle, 3–5 mm, antenna with 2 two-segment club (Figure A1.44) Lyctidae 37
(b) Prothorax, hooded and very rough and concealing head from above, short beetle, antenna with three-segment club (Figure 5 A1.46 and 47) Bostrychidae 38
(c) Head projecting in front of prothorax, elongate beetle, antenna with large one-segment club (Figure A1.48) Platypodidae 39
(d) Prothorax hooded and usually concealing head from above, short beetle, antenna with one-segment club (Figure A1.49) Scolytidae 40

37 Lyctidae

Figure A1.44 *Lyctus brunneus,* **a Lyctid**

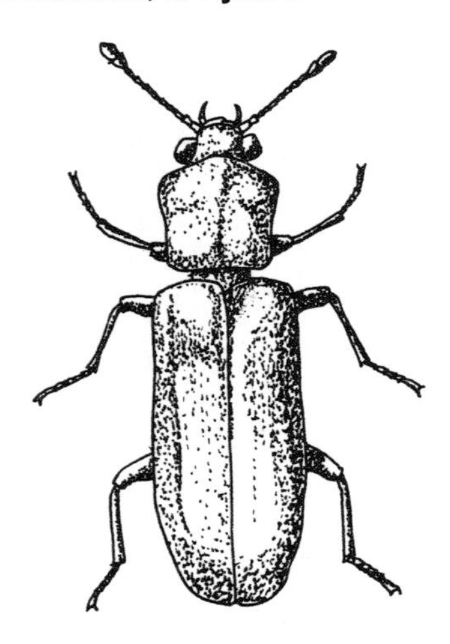

Figure A1.45 Prothorax details of (a) *Lyctus brunneus,* **(b)** *L. linearis,* **(c)** *L. parallelopipedus* **and (d)** *L. planicollis*

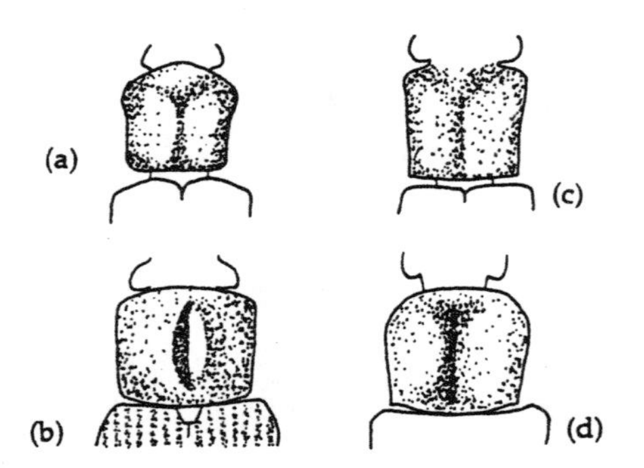

These are wood-borers known, with the Bostrychidae, as Powder Post beetles. All species are 3.0–5.0 mm long and the most common species is *Lyctus brunneus* (Figure A1.44). Identification of species can be mostly simply achieved by examining the shape of the prothorax (Figure A1.45); *L. brunneus* has a wider front end and concave sides, *L. parallelopipedus* is rather similar but with straight sides, *L. planicollis* has rounded front corners, and *L. linearis* has a distinct identation in the centre as well as distinct rows of hair on the elytra.

Figure A1.46 Lesser Grain borer, *Rhizopertha dominica*

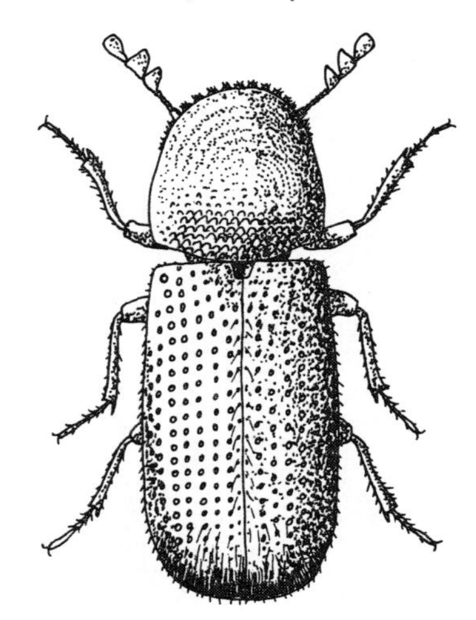

Figure A1.47 *Apate capucina,* **a Bostrychid**

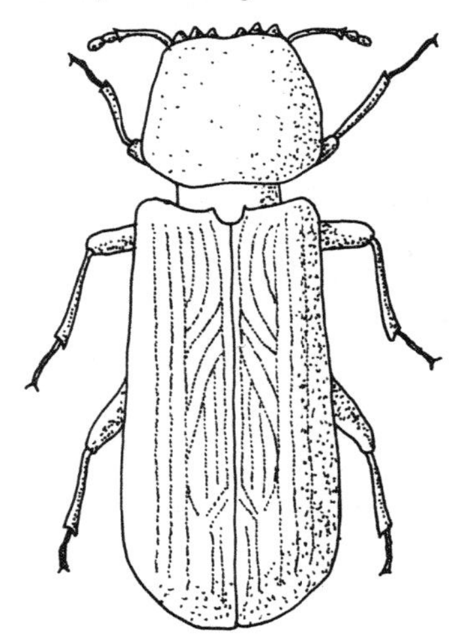

38 **Bostrychidae**

A large and important family of starch feeding beetles but rare in buildings in north-west Europe. Some attack grain, such as the Lesser Grain borer, *Rhizopertha dominica,* 2.5–3.0 mm (Figure A1.46). Others are Powder Post wood-boring beetles, some tropical species being quite large, but species likely to be found in north-west Europe are very few and small, 3–6 mm, such as *Apate capucina* (Figure A1.47), which has distinctive red or red-brown elytra. Larger species may be found in imported wood.

Figure A1.48 *Platypus cylindrus*, **a Platypodid**

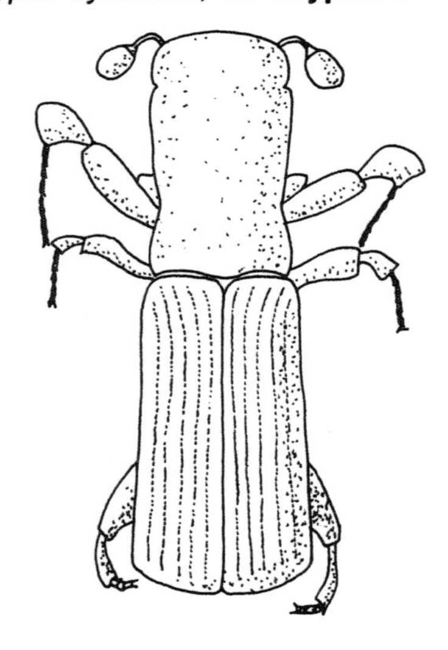

39 **Platypodidae**

This family is known, with the Scolytidae (40), as the Ambrosia beetles, or Pinhole borers. They are 3–6 mm long and their form is very characteristic (Figure A1.48). They bore in the bark of freshly felled green wood, causing characteristic galleries and sometimes penetrating more deeply into the sapwood. They are essentially forest pests but may be found in buildings, emerging from fire logs, for example.

Figure A1.49 *Xyleborus fraxesini*, **a Scolytid**

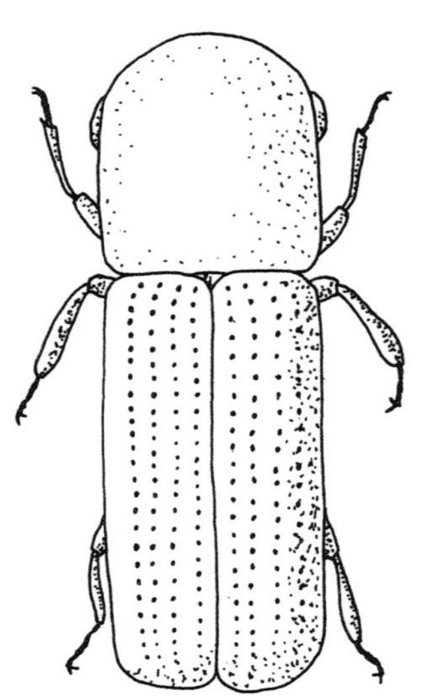

40 **Scolytidae**

These are similar in habits to the Platypodidae (39). They are also 3–6 mm long but their form is distinctly different (Figure A1.49) from that of the Platypodidae.

41 (a) Very flat, antenna long and greater than half body length but without club — Cucujidae 42
(b) Very flat, antenna with three-segment club — Ostomidae 43
(c) Not distinctly flat, antenna with three-segment club — Lathridiidae 44

42 **Cucujidae**
Species of this family are very flat to enable them to explore beneath loose bark. They are sometimes found in old borer galleries in decayed wood. They are small. Only *Laemophloecus* species, 1.3–5.0 mm long, are likely to be found in buildings. They can be readily identified by the longitudinal ridge close to each side of the prothorax.

Figure A1.50 Cadelle, ***Tenebroides mauritanicus***

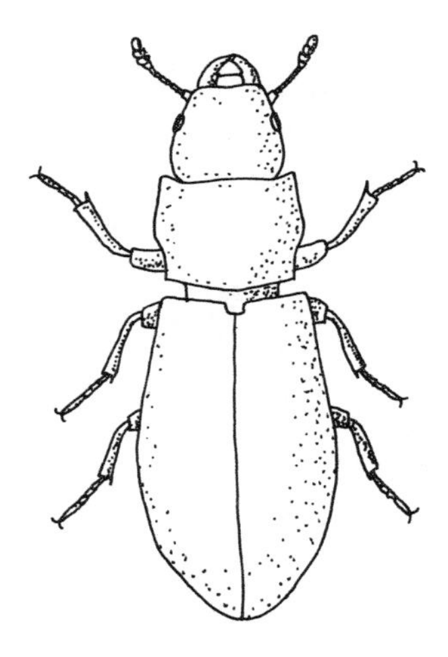

43 **Ostomidae**
Species of this family are distinctly flattened. Only two species are likely to be encountered in buildings. The Cadelle, *Tenebroides mauritanicus*, 5–11 mm (Figure A1.50), has smooth elytra and the prothorax distinctly separated from the elytra. The Siamese Grain beetle,

Figure A1.51 Siamese Grain beetle, *Loptocateres pusillus*

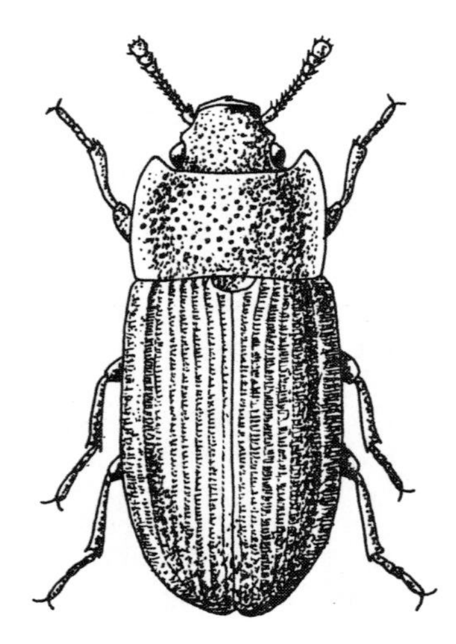

Loptocateres pusillus, 2.7–3.2 mm (Figure A1.51), has distinct ridges on the elytra and the prothorax is in close contact with the elytra.

44 **Lathridiidae**

These are very small beetles which feed on fungi. They are found in buildings browsing on mould on food, plaster or wallpaper. They are sometimes known, together with the Cryptophagidae (34), as Plaster beetles. They are pale brown to black with eleven-segment antennae. The tarsi have three-segments. The most common species can be identified using the following key:

Figure A1.52 *Corticaria pubescens*

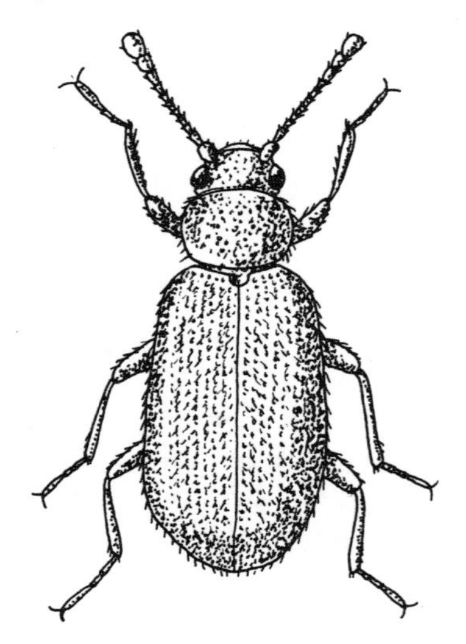

44.1 (a) Elytra not ridged and lines of punctures shallow and largely concealed by hair, 2.3–3.0 mm (Figure A1.52) *Corticaria pubescens*
(b) Elytra punctures deep with no apparent hair 44.2
44.2 (a) Prothorax with ridge on either side of centre, two rows of punctures between elytra ridges 44.3
(b) Prothorax without ridges, elytra without distinct ridges 44.5

Figure A1.53 *Lathridius bergrothi*

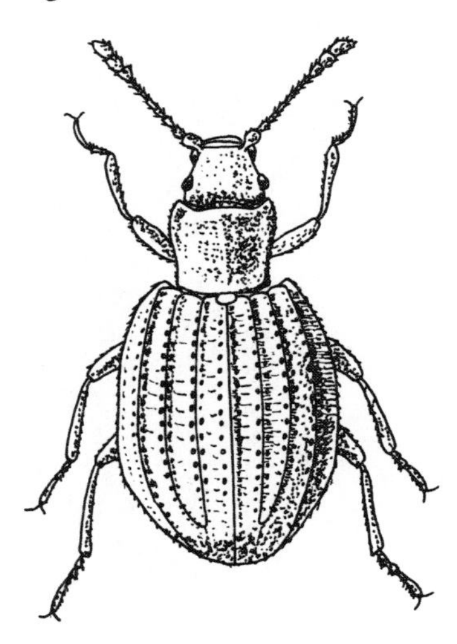

44.3 (a) Tarsi with first segment distinctly shorter than second, 1.8–2.2 mm (Figure A1.53) *Lathridius bergrothi*
(b) Tarsi first and second segments equal in length 44.4
44.4 (a) Antenna with three-segment club, 1.5–2.1 mm *Coninomous (Aridius) nodifer*
(b) Antenna with two-segment club, 1.2–1.7 mm *Coninomous constrictus*

Figure A1.54 *Enicmus minutus*

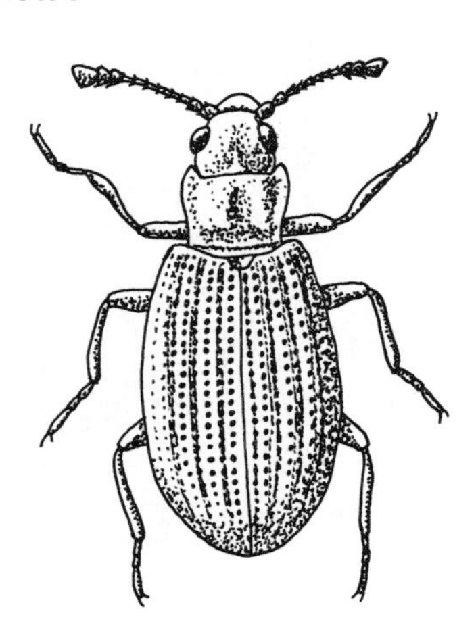

44.5 (a) Eyes larger than distance between eye and base of antenna, prothorax distinctly narrower than base of elytra, 1.2–2.4 mm (Figure A1.54) *Enicmus minutus* (*Lathridius pseudminutus*)

(b) Eye much smaller than distance between eye and base of antenna, prothorax front as wide as base of elytra, antenna with two-segment club, 1.2–1.6 mm *Cartodere filiformis*

Figure A1.55 *Cartodere filiformis*

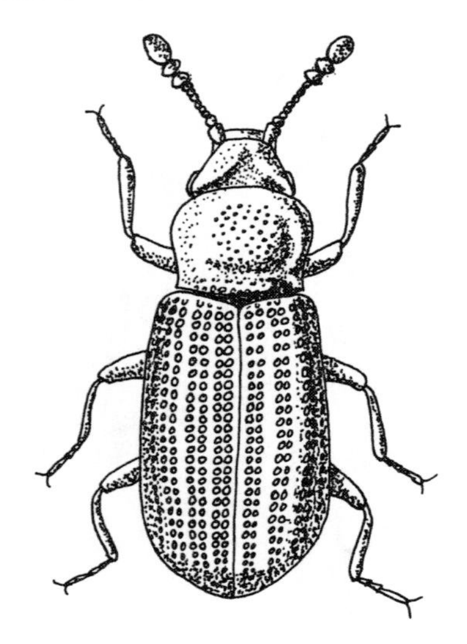

Order Hymenoptera (ants, bees and wasps)

This is a very large order, many of its members having highly developed social habits. Most Hymenoptera have two pairs of membranous wings but some, such as the ants and parasitic wasps, have no wings. Common ants, bees and wasps are well known and this section therefore describes only the wood-boring Hymenoptera, those that are predators on other wood-borers and those that are sometimes confused with wood-borers.

Figure A1.56 Worker of Carpenter ant, *Camponotus herculeanus.* **Reproductive castes are winged**

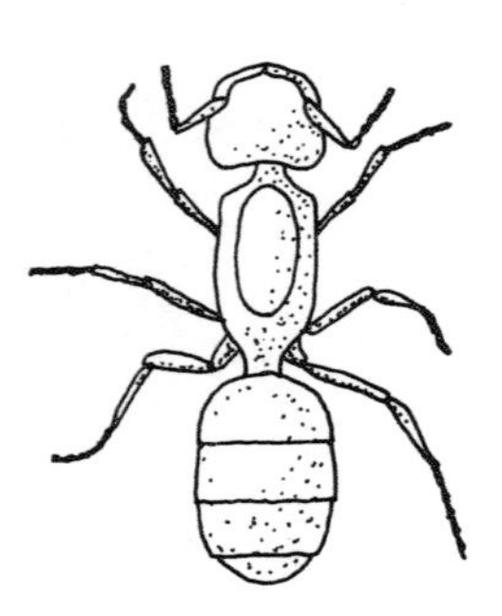

The Carpenter ants, *Componotus herculeanus* (Figure A1.56) and *C. ligniperda*, have been causing increasing damage in buildings in Scandinavia in recent years. In nature these insects tunnel into old trees affected by interior decay in order to establish nests but buildings in forest areas are being attacked to an increasing extent. Carpenter ant damage is occasionally seen in wood imported into the British Isles from Scandinavia or North America. Most species attack only decayed wood but there have been recent reports that some can attack dry and even preservative-treated wood.

Figure A1.57 (a) Worker and (b) queen of Pharaoh's ant, ***Monomorium pharaonis***

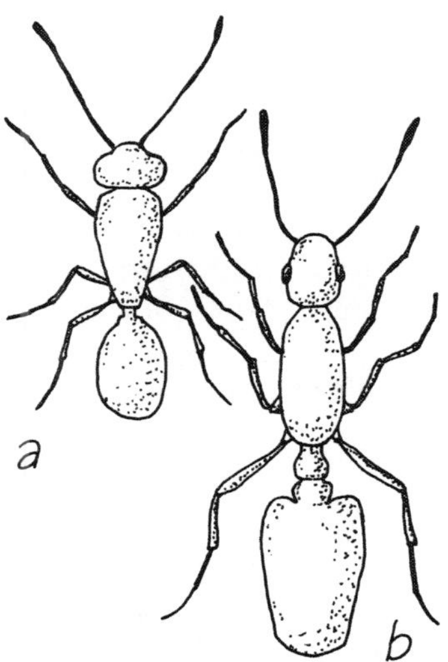

Pharaoh's ant, *Monomorium pharaonis* (Figure A1.57), is a tropical species which has become established in England in some permanently heated buildings such as hospitals and bakeries, but it is also now found occasionally in domestic buildings. It is a small reddish-yellow ant, the workers and males being about 3 mm long but the queens somewhat larger. Nests are usually rather inaccessible in heating pipe ducts and other permanently warm cavities, infestation spreading through groups of workers carrying broods to new sites to establish new colonies. These multiple inaccessible colonies make eradication rather difficult once an infestation has become established. The Argentine ant, *Iridomyrmex humilis*, is similar in size but slightly darker and more rare in Great Britain. A slightly larger black ant, *Paratrechnia longicornis*, sometimes known as the Cheltenham ant, is also rare. These three species are unusual and confined to high-temperature conditions, unlike the well-known garden ants which are easily recognised by their dark blackish-brown colour and their size; they are at least twice the size of Pharaoh's ant.

The family Xylocopidae are the Carpenter-bees. These are the largest known bees, black with dark and often irridescent wings, the fine hairs over the body frequently being yellow, white or brown. They generally resemble the bumble-bees, *Bombus* species, but they are more flattened and less hairy. Some species have been reported in France boring deeply in beams, rafters and other structural timbers but there have been no reports of similar damage by Carpenter-bees in buildings in the British Isles.

Species of bees, and occasionally wasps, are sometimes found boring in soft mortar in buildings. They are not usually of economic importance as their presence invariably indicates defective mortar requiring repointing.

The family Siricidae are the Wood-wasps, which bore through bark with their ovipositors, laying eggs which hatch to release larvae which bore into the wood. Adult Wood-wasps occasionally emerge from timber in buildings following infestation of logs in the forest. The Giant wood-wasp, *Urocerus (Sirex) gigas*, may be up to 50 mm long, striped yellow and black, and is often confused with hornets.

The family Cephidae are the saw-flies, slender-bodied flying insects which frequently cause damage to standing crops. However, one species, *Ametastegia glabrata*, is occasionally found boring in wood.

Figure A1.58 *Eubadizon pallipes*, **a predator on Lyctid Powder Post beetles**

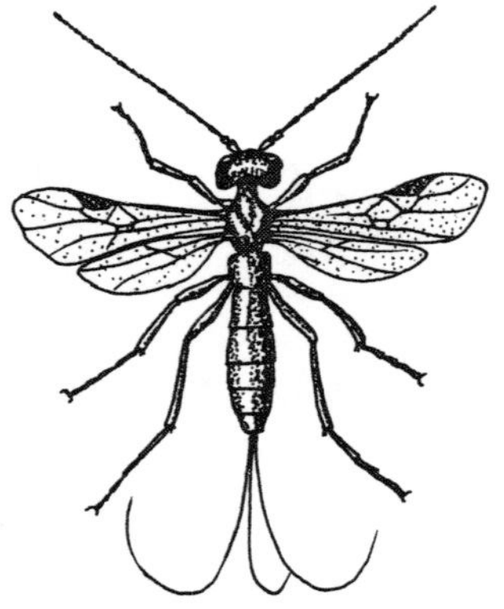

There are a number of Hymenoptera which are predators on wood-borers, so their presence generally indicates a widespread concealed wood-borer infestation which may, of course, be controlled by the predators! Lyctid Powder Post beetle (order Coleoptera, 37) infestations attract minute, ant-like, wingless *Sclerodermus domesticus* and *S. macrogaster*, and the minute winged flies *Eubadizon pallipes* (Figure A1.58). *Calosoter vernalis*, a similar winged fly but which lacks the abdomen or tail appendages, is a predator on the fairly rare Anobiid wood-boring beetle *Ptilinus pectinicornis* (order Coleoptera, 35).

Figure A1.59 *Theocolax formiciformis*, **a predator on the Common Furniture beetle**

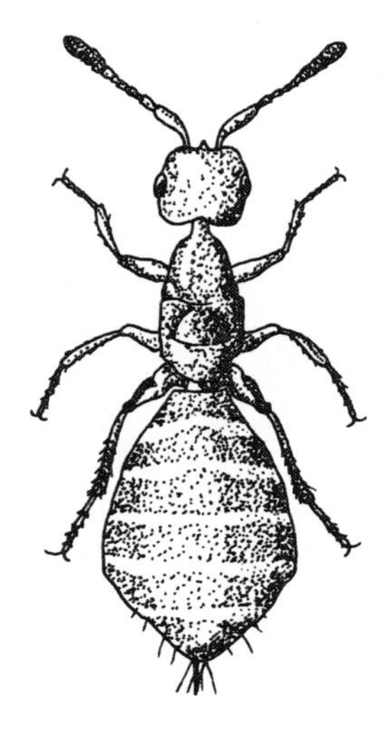

Other Hymenoptera are associated with the Common Furniture beetle *Anobium punctatum* (order Coleoptera 35), for example the small, wingless, ant-like *Theocolax formiciformis* (Figure A1.59) and *Spathius exarata*. The latter, which is sometimes seen walking rapidly over window panes, also attacks larvae of other insects, including those of the clothes-moth (order Lepidoptera).

Order Diptera (flies)

The Diptera are the true flies and can be readily identified as they have only a single pair of wings. Many of the flies are well known and easily identified but it is not usually appreciated that this order also includes the crane-flies or daddy-long-legs, mosquitoes, gnats and midges.

Several small species of *Drosophilia* may be encountered in buildings, the fruit-fly, *D. melanogaster*, on fruit which is beginning to deteriorate and ferment, and the vinegar-fly, *D. funebris*, on vinegar or decaying detritus, often becoming troublesome in warm conditions as they breed very rapidly. The window-fly, *Scenopinus fenestralis*, is a small fly 3–6 mm long with a narrow oblong body; its larvae are sometimes found in grain or other foodstuffs as predators on the larvae and pupae of beetles and moths. The cluster-fly, *Pollenia rudis*, is larger, up to 10 mm long and typically fly-shaped, with a dark grey thorax covered with golden hairs. The larvae are parasites on a species of earthworm but enormous numbers of adults swarm together for hibernation in dark nooks and crevices, such as corners of attics and window sash boxes, sometimes causing considerable annoyance. The green cluster-fly, *Dasyphora cyanella*, has similar swarming habits but has the appearance of a fairly small greenbottle and its larvae develop in decaying matter such as dung. The yellow swarming-fly, *Thaumatomyia notata*, also has similar habits but it is a smaller yellow fly with black stripes, longitudinal on the thorax but transverse on the abdomen.

Figure A1.60 The dog flea, ***Ctenocephalus canis***

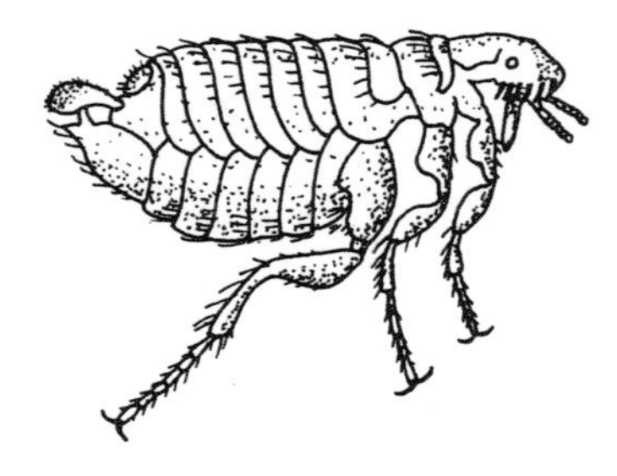

Order Siphonaptera (fleas)

At one time the human flea, *Pulex irritans*, was a serious problem in buildings but improved personal hygiene and the use of powerful modern insecticides have largely eliminated it. Animal fleas are, however, still a problem. Whilst care can ensure that dogs and cats are kept free from their specific fleas, *Ctenocephalus canis* (Figure A1.60) and *C. felis*, they may temporarily acquire rodent fleas through hunting or general exploring, and these will temporarily transfer to humans, as will the fleas found on domestic fowl, causing considerable irritation to sensitive persons.

Class Arachnida (spiders, scorpions and mites)

This description and identification key to pests in buildings has been confined to insects, but mention should also be made of the Arachnids. Spiders are, of course, regularly encountered in buildings and are easily recognised by their eight legs, compared with six for insects. The mites are minute pests which can also occur widely in buildings and which are again recognisable by their eight legs. The red spider mites are well known on fruit trees and are occasionally found in buildings. However, there are several other mites which often occur but which are not so easily recognised. They are all very small, usually about 0.5 mm long, the flour, sugar and house mites occurring on damp furnishings and in larders. In addition there are a number of parasitic mites which can be very irritating. Poultry, rat and harvest mites can be picked up by humans and domestic animals, resulting in severe itching, but the itch mite, recognised by its very short stumpy legs, is a true parasite on man, burrowing into the skin and causing the itching disease known as scabies.

Class Crustacea (woodlice, gribbles)

The Crustaceans include the lobsters, crabs and shrimps, and also all the well-known wood-lice which sometimes cause concern when they are found in buildings. Wood-lice do not, in fact, cause damage to wood and their name arises only through their association with damp decayed wood; they occur in buildings, often with silver-fish and firebrats, where there are suitable damp conditions. The Crustaceans also include the gribbles, which are marine borers that cause damage in marine piles and boats.

Phylum Mollusca (snails, ship-worms)

The only Molluscs that are found in buildings are snails, but this phylum is included here to draw attention to the ship-worms that cause damage in marine piles and boats; some related marine species are also found boring in rocks.

Identification key to wood-borer damage

Borer damage is usually caused by larvae tunnelling invisibly within the wood and the infestation only becomes apparent when the larvae pupate and then emerge from the wood as adult insects, leaving flight holes in the wood surface. All borer damage of economic significance in buildings in the British Isles is caused by beetles (order Coleoptera) but wood damaged by other insects, and even marine borers, is sometimes found in buildings, so descriptions of some of these other forms of damage are also given. In order to use this key fully it is necessary to have a small transparent scale to measure flight hole diameters and an x10 or x15 hand lens, preferably illuminated, to examine bore dust. Larvae are described briefly as they are often found in seriously damaged wood and can be useful for confirming an identification. However, it must be appreciated that, if recent flight holes are present, adult insects will have emerged and they may be found in dust deposits or cobwebs, providing the most positive method for identifying damage. Care must be taken to ensure that old galleries, exposed by sawing wood before use in the building, are recognised; flight holes normally pass straight through the surface whereas sawing usually exposes galleries parallel or diagonal to the surface. The only certain method for confirming current activity is the discovery of a live larva within the damaged wood but clean flight holes with fresh bore dust discharges are usually evidence of continuing activity, although it must be appreciated that disturbance during an inspection, such as lifting floorboards, may cause discharge from old holes.

Throughout this key the terms 'softwood' and 'hardwood' are used in the botanical sense, that is, softwoods are coniferous woods such as pine, fir and spruce, whilst hardwoods are deciduous woods such as oak, elm, beech and mahogany.

1 (a) Oval flight holes — Cerambycidae 2
(b) Round flight holes, more than 5 mm — Siricidae 3
(c) Round flight holes, less than 4 mm — 4
(d) Rare borers — 9

2 **Cerambycidae (Longhorn beetles)**
(a) *Damage in softwoods*
House Longhorn beetle, *Hylotrupes bajulus*, damage occurs in sapwood of dry softwood (Plate 2.5). The oval flight holes are 5–9 mm across and galleries are packed with bore dust containing cylindrical pellets. There is often extensive damage throughout the sapwood, leaving only a thin surface veneer distorted where galleries run beneath. Larvae are straight bodied with a slight taper, up to 30 mm long and distinctly segmented (Figure A1.61). House Longhorn beetle is also known as the Camberly beetle and is generally confined to south-central and south-east England, where it is of considerable economic significance. Remedial treatment is essential.

Figure A1.61 Typical wood-borer larvae: (a) Death Watch beetle, *Xestobium rufovillosum*, **a typical curved larva; (b) House Longhorn beetle,** *Hylotrupes bajules*; **(c) Wharf borer,** *Nacerdes melanura*

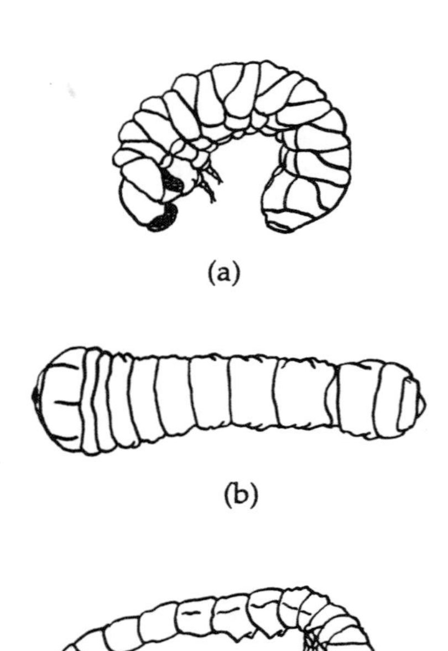

Some damage found in building timbers may have been caused in the forest before conversion, particularly in wood that has been slightly decayed. Some forest species may survive in wood after conversion. *Ergates spiculatum*, a North American species, is sometimes found in Douglas fir up to 30 years after installation in buildings. The Two-toothed Longhorn beetle, *Ambeodontus tristus*, has been reported in England in pine imported from New Zealand but it is not known whether this beetle has become established. It produces an oval flight hole about 5 mm across.

(b) *Damage in hardwoods*
Oak Longhorn beetle, *Phymatodes testaceus*, damage is sometimes found in sapwood of oak. Oval flight holes may be found but galleries are principally under the bark and extend into the sapwood, where pupation occurs. Egg-laying occurs through the bark after wood has been felled in the forest or during seasoning; however, adults may emerge for a few years following installation in buildings. Remedial treatment is not necessary. Some damage found in building timbers may have been caused in the forest before conversion, particularly in wood that has been slightly decayed; oak may be attacked in this way by *Rhagium mordax* and *Leiopus nebulosus*. Some forest species may survive in wood after conversion. *Eburia quadrigeminata*, a North American species, is sometimes found in American oak, usually furniture, up to 25 years after conversion. Damage in beech, similar to that due to the Oak Longhorn beetle, has been observed in England, sometimes caused by an Indian Longhorn beetle *Trinophylum cribratum*, but it is not known whether this insect is widely established.

(c) *Damage in wickerwork*
The two smallest British Longhorn beetles, *Gracilia minuta* and *Leptidea brevipennis*, the Osier Longhorn beetles, are sometimes found causing oval flight holes 2 mm across in wickerwork. Round Common Furniture beetle flight holes may also be present.

3 **Siricidae (wood-wasps)**
Wood-wasps such as *Sirex noctilis* and the Giant wood-wasp, *Urocerus (Sirex) gigas*, attack sickly or felled softwoods in the forest but may emerge after installation in buildings. Flight holes are large and round and galleries are tightly packed with coarse shredded bore dust. Remedial treatment is not necessary.

4 (a) Damage in tropical or temperate hardwoods. Galleries stained brown or black with stain spreading along grain on either side of

hole; branched galleries beneath bark, if present (Plate A1.1). Scolytidae or Platypodidae 5

(b) Damage in tropical or temperate hardwoods. Galleries packed with soft, fine bore dust Bostrychidae or Lyctidae 6

(c) Damage in damp or decayed structural timbers. Flight holes round with ragged edges; galleries forming laminar pattern with fine bore dust Curculionidae 7

(d) Bore dust containing distinct elongate oval, rod, or bun-shaped pellets Anobiidae 8

Plate A1.1 Typical branched galleries under bark produced by Ambrosia beetles *(Cementone-Beaver Ltd)*

5 **Scolytidae or Platypodidae (Ambrosia beetles)**

The Ambrosia beetles bore into the bark of sickly or felled trees in the forest, forming characteristic branched galleries under the bark in which they lay eggs. The galleries are inoculated with the Ambrosia fungus on which the hatching larvae feed. Larval galleries often penetrate deep into sapwood and sometimes heartwood in light-coloured tropical hardwoods, and when the wood is cut a characteristic stain is seen spreading along the grain on either side of each hole. Small holes are described as pinholes and larger holes as shotholes. Ambrosia beetles are a forest problem; remedial treatment is not necessary.

6 Bostrychidae or Lyctidae (Powder Post beetles)

The adult Bostrychidae, sometimes known as Auger beetles, tunnel in the bark of sickly trees or freshly felled logs in the forest, producing dust-free galleries in which they lay their eggs. Hatching larvae penetrate into the sapwood, forming galleries packed with soft, fine bore dust. Larvae can survive after wood has dried as they are starch feeders (cf. Scolytidae and Platypodidae, 5). Larvae are curved, with three pairs of four-jointed legs.

The adult Lyctidae, the most common Powder Post beetles, lay eggs in the large vessels exposed on the end-grain of ring-porous tropical and temperate hardwoods, such as African 'mahoganies', oak, ash, walnut, elm and hickory. Hatching larvae are at first straight and work along the vessel but then become curved and form extensive galleries within the sapwood and these may eventually merge causing complete collapse. Galleries are packed with soft, fine bore dust. Larvae are curved, with three pairs of minute three-jointed legs (Figure A1.61) The flight hole is round, 0.8-1.5 mm in diameter. Powder Post beetle infestation can occur only in fresh hardwood. If flight holes are observed in, for example, furniture or panelling, immediate injection treatment may eradicate the infestation and avoid the collapse of the sapwood that is otherwise inevitable. In older wood remedial treatment is unnecessary, but in valuable items resin reinforcement treatment may be considered.

Continuing activity in sapwood of beech, sycamore, maple and elm may be due to the Anobiid *Ptilinus pectinicornis*, which creates damage and bore dust similar to that produced by the Powder Post beetle, but the Common Furniture beetle, *Anobium punctatum*, is also often present, (see 8 (c)).

7 Curculiondae (Wood weevils)

The Wood weevils attack damp and decayed wood, but visible fungal decay is not always present. The adults with their distinct protruding snouts (Figure A1.11) can be found on the wood surface or boring in galleries at any time of the year. The larvae, which are curved but legless, cause most boring damage. Galleries contain fine bore dust consisting entirely of small elongate pellets and tend to form a distinct laminar pattern, following spring wood only. They are usually in sapwood but extend into heartwood if distinct fungal decay has occurred. Flight holes have ragged edges and may appear round or oval, 2 mm across. *Euophryum confine* will attack relatively dry wood which may have little evidence of decay, whilst *Pentarthrum huttoni* will attack damp wood and *Caulotrupis aeneopiceus* is confined to wet and decayed wood, such as plates and doorposts in cellars.

8 **Anobiidae (Furniture beetles)**

(a) Bore dust containing elongate ovoid or pointed rod-shaped pellets Anobiinae 8 (c)

(b) Bore dust containing distinct bun-shaped pellets Ernobiinae 8 (d)

(c) *Anobiinae*

The Common Furniture beetle, *Anobium punctatum*, will attack sapwood in hardwoods and softwoods, as well as heartwood in light-coloured hardwoods; attack may spread into heartwood if activity is encouraged by the presence of decay. Galleries always remain separated by a very thin membrane of wood (cf. Powder Post damage, 6) and are loosely packed with gritty bore dust containing granular debris and characteristic elongate ovoid or pointed rod-shaped pellets. Flight holes are round, 1.5 mm in diameter. Larvae are curved, with three pairs of five-jointed legs (Figure A1.61). Damage extends relatively slowly but remedial treatment is essential.

Anobium pertinax is generally similar but attacks only more severely decayed wood in cellars or subject to roof or plumbing leaks; it is common in Scandinavia but relatively rare in the British Isles. Remedial treatment for decay is more essential than eradication of the borer.

(d) *Ernobiinae*

The Bark borer, *Ernobius mollis*, attacks only softwoods with bark adhering. Galleries are concentrated between the bark and sapwood, usually causing the bark to become detached, but extend into the bark and for about 1 cm into the sapwood; bore dust is a brown and white mixture from the bark and wood, granular but with characteristic bun-shaped pellets. Flight holes are round and 3 mm in diameter. Remedial treatment is unnecessary as infestation is confined to waney edges with bark adhering and dies out naturally only about two years after wood is felled. It is important that damage by this beetle should not be confused with Common Furniture beetle damage.

The Death Watch beetle, *Xestobium rufovillosum*, attacks only hardwoods that are affected by fungal decay. Visible decay may not be present but the wood usually has at least a characteristic brown colour. Attacks are most common in oak, probably because it is often used structurally in old buildings, but chestnut, elm, walnut, alder and beech can also be attacked, and infestation can sometimes spread from hardwoods into adjacent softwood timbers affected by fungal decay. Attack is usually concentrated in sapwood but may extend into heartwood if decayed. Only relatively strong wood is attacked and where decay is too severe other insects may occur, such as *Helops coeruleus*, a Tenebrionid sometimes found in south and east England, which has a long straight larva with two spines on the last segment (Figure A1.61). Death Watch beetle is confined to England, Wales and part of southern Ireland. Its flight holes are round, 2.5– 4 mm in diameter and the galleries contain fairly coarse bore dust, usually brown, with characteristic bun-shaped pellets.

9 Rare wood-borer damage

Several examples of comparatively rare wood-borer damage have been given where appropriate in the key, such as damage caused by Forest Longhorn beetles, wood-wasps, the Anobiid beetles *Anobium pertinax* and *Ptilinus pectinicornis*, and the Tenebrionid beetle *Helops coeruleus*, which is sometimes found in decayed hardwood in association with the Death Watch beetle. Whilst the key covers the borer damage normally observed in buildings it will be appreciated that wood damaged elsewhere is likely to be moved into buildings where identification of the damage can present considerable difficulty.

Fire logs are often obtained from dead trees or allowed to remain in the open, where they may be attacked by a variety of wood-borers, particularly the Ambrosia beetles (5), which produce extensive branched galleries under the bark. Wood-borers in logs do not generally represent a risk to structural or furnishing woodwork; fungal decay introduced on logs can be a far more serious problem. Near coasts old ships' timbers are sometimes used in buildings, perhaps showing old marine borer damage. Gribble damage consists of galleries which riddle the wood to a depth of about 1 or 2 cm, the tunnels being of uniform diameter throughout, usually about 2.5 mm, in contrast with damage caused by insect larvae, which is characterised by galleries of progressively increasing diameter as the larvae grow. Ship-worm damage consists of galleries, perhaps quite large, usually along the grain and with a characteristic calcareous lining. Wood in marine situations but above water level is sometimes attacked by an Oedermerid beetle, the Wharf borer, *Nacerdes melanura*, particularly in concealed zones such as where a beam rests on top of a pile. Galleries can be relatively large and the damage is usually laminar in form. Such damage, which can also occur in boat houses and even wet cellars, usually becomes apparent when the distinctive adult beetles emerge in large numbers but the elongate larva is also very distinctive (Figure A1.61)

There are two forms of borer attack that have not been observed in buildings in the British Isles but which would have considerable economic significance should they become established. Termite damage can be readily identified by the 'white ants' that will be seen if the damaged wood or associated nests and walkways are disturbed.

Carpenter ants occur in North America and Scandinavia, where they usually cause damage to posts in ground contact, but there have been reports that some species can also attack dry wood. A few examples of damage, typically wide galleries or cavities spreading from an end-grain surface, have been seen in imported wood and there is thus a danger that these borers may become established in the British Isles.

There are many other insects that can bore into wood, such as carpenter-bees, various moths and Dermestid beetles, but they generally form only pupal chambers of limited extent and the damage they cause is therefore of little structural significance.

Appendix 2 *Identification of wood rots and other micro-organisms infecting building materials*

Most householders are familiar with mould, the most common fungal infection found within buildings, and the moss and lichen on external surfaces. Similarly most remedial treatment surveyors are familiar with common wood rots such as Dry rot and Cellar rot. The problems arise when infections or damage are discovered which cannot be readily identified or explained.

The following key will enable micro-organisms found infecting building materials to be readily identified. In the case of important infections such as the wood-destroying fungi a reasonably comprehensive account is given. An infection can generally be visually identified on site using this key without the need for magnification or laboratory examination, but only major groups can be identified rather than individual species in some cases.

This key covers micro-organisms likely to be found infecting building materials, that is fungi, algae and lichens; mosses and higher plants are also mentioned when appropriate. Bacteria are ignored as they are invisible and their identification would serve no useful purpose in connection with normal remedial treatment. The significance of bacteria in relation to stone deterioration is considered in Chapter 4.

1 (a) Organic substrates (wood, textiles, paper, also painted surfaces and organic deposits on inorganic surfaces) 2

(b) Inorganic substrates (masonry, brickwork, concrete, rendering, tiles, paving) 6

2 (a) Superficial growth causing patchy surface discoloration usually green, grey or black; always associated with high humidity; can occur on painted and inorganic surfaces if organic contamination is present (some paints support growth); serious problem in breweries, bakeries, kitchens, larders, bathrooms, laundries and other situations where high humidity affects an organic substrate or contaminated inorganic substrate. Typical examples are growth on bread and other foodstuffs and superficial growth on wood cladding.

Mould fungi

(b) Staining of wood, often throughout sapwood, usually blue or black; occurs on freshly felled wood with high moisture content but damage persists after wood dries; usually no significance when found in wood in buildings Sapstain (Bluestain)

(c) Staining of wood under varnish (or invisibly under paint), usually blue or black, often accompanied by small black surface growths through the coating; sometimes accompanied by surface mould; similar fungi to those described in 2(b) but occurring under different conditions through moisture accumulations under coating systems Sapstain ('stain in service')

(d) Deterioration of wood, or fungal growth showing distinct spread over wood (emanating from a distinct point or area) rather than the more uniform and structurally indistinct growth associated with mould 2(a); white fluffy growth or strands or sheet, sometimes darker strands, sometimes spreading over adjacent masonry or brickwork 3

3 (a) Sapwood and heartwood softened to a consistent depth when wet, and showing small cracks along and across the grain when dry. Only associated with waterlogged wood and therefore present in boats or groynes rather than buildings; however, quays, jetties, mills and boat houses may have affected wood components, and affected wood may be reused in buildings (if wood is painted, as in a boat hull, only areas from which the paint has been lost are usually affected). If deterioration is deep the wood will break with a typical 'brash' or clean cross-grain fracture; no visible fungal infection Soft rot

(b) Wood distinctly brown with cracks along and across the grain, cracks along the grain sometimes more prominent; fungal growth often visible Brown rots 4

(c) Wood approximately normal colour but soft and stringy; fungal growth often visible White rots 5

4 Brown rots

(a) Wood brown with cracks along and across the grain tending to form large cubes, and forming soft powder when crushed between fingers; fluffy masses of white hyphae like cotton wool present with active growth, compacting to form surface film of greyish mycelium when older and developing distinct strands or rhizomorphs up to 6 mm in diameter which are brittle when dry. Mycelium may become yellowish with lilac tinges when exposed to light and often

Plate A2.1 Typical Dry rot, *Serpula (merulius) lacrimans*

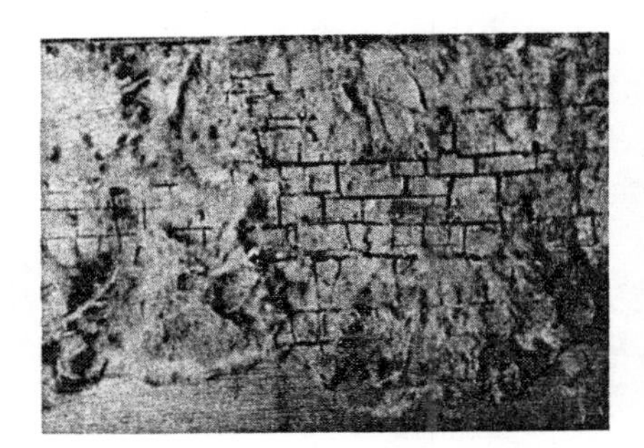

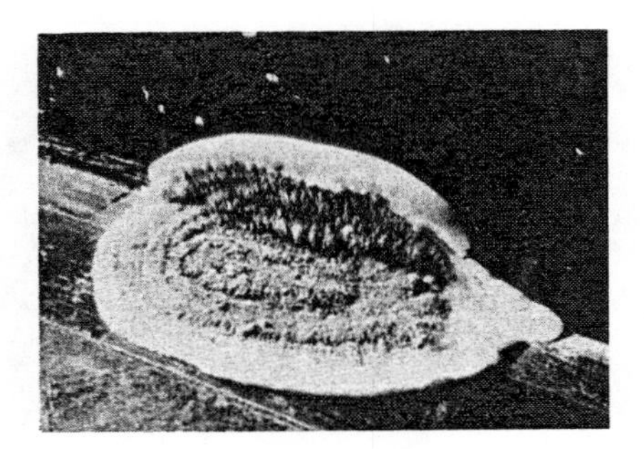

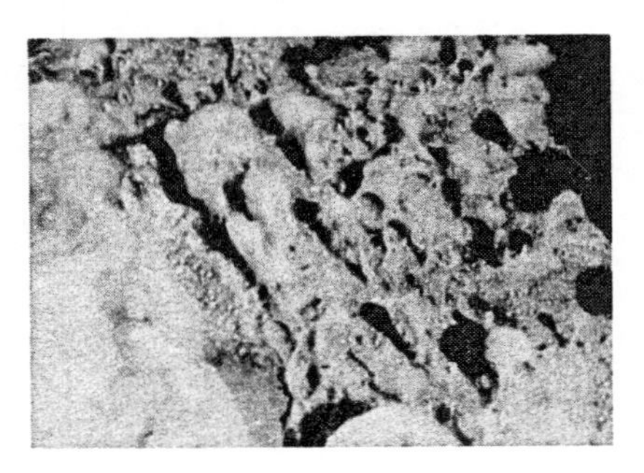

greenish through development of surface mould; rhizomorphs commonly spread across and through masonry or brickwork; fruiting bodies or sporophores may develop, perhaps emerging through otherwise unaffected plaster surfaces from concealed timbers beneath, sporophores being flat plates or brackets from a few centimetres to a metre across, grey at first with a white margin, later the corrugated hymenium or spore-bearing surface releasing rust-red spores covering surroundings with red dust; associated with poorly ventilated or concealed timbers.

Dry rot, *Serpula (Merulius) lacrimans*

(b) Wood dark brown with cracks along and across the grain, the longitudinal cracks more dominant. Some white hyphae may be present with active growth but no distinct mycelium formation, rhizomorphs fine and becoming brown or black with only limited spread over masonry and brickwork, sporophore rare and on wood itself (cf. 4(a)), consisting of thin skin with irregular lumps, yellowing and then darkening to brown as spores released; associated with wood in contact with continuous dampness, such as ground floor plates or skirtings in contact with damp walls (a Wet rot)

Cellar rot, *Coniophora puteana (cerebella)*

Plate A2.2 Typical Cellar rot, *Coniophora puteana (cerebella)*

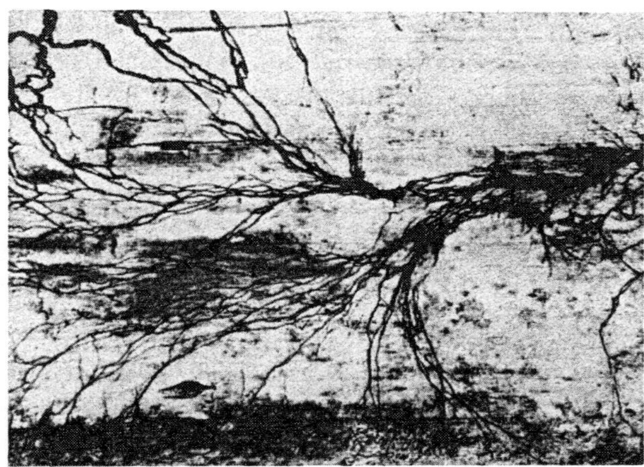

(c) Wood brown with cracks along and across the grain but smaller cubes than 4(a) and wood gritty when crushed; fluffy white hyphae present with active growth, no distinct mycelium formation but rhizomorphs white, often flattened, with adjacent margin on wood surface and rarely spreading far over masonry or brickwork, sporophore rare and on wood itself, consisting of white irregular plate up to 12 mm thick with distinct pores; requires high moisture content but will tolerate occasional drying and therefore found in roofs and greenhouses in conjunction with roof leaks (a Wet rot)
White Pore fungus, *Antrodia sinuosa (Poria vaporaria)*

(d) As 4(c) but fungal growth confined to cracks between cubing; common in greenhouses *Antrodia (Poria) xantha*

(e) Wood bright yellow in initial stages but darkening to deep reddish-brown with shallow cracking along and across grain; hyphae yellow fine distinctly branching strands, fibrous with limited mycelium which may develop lilac coloration, no distinct rhizomorphs, no spread onto masonry or brickwork; sporophore attached at one distinct point but without stalk, tending to curl around edges and becoming irregular in shape and fleshy, branching gills on upper surface radiating from point of attachment, dingy yellow but darkening as spores develop (a Wet rot)
Tapinella (Paxillus) panuoides

(f) Small pockets of brown rot with cracking along and across grain; pockets may be filled with fluffy white hyphae; in early stages only a brown stain may be present but rot can be confirmed by brashness detected by lifting fibres with the point of a knife. Infection occurs in logs or sickly trees in the forest but can extend as a normal Brown rot under suitable conditions. Pipe or Pocket rot; Dote

5 **White rots**

(a) Wood normal colour or bleached in dark species but much softer than sound wood with longitudinal fibrous texture, yellow or brown mycelium sometimes present, sporophore thick fawn-coloured plate or bracket darkening as spores develop Stringy Oak rot, *Donkioporia expansa* (*Phellinus megaloporus* or *cryptarum*)

(b) Wood white flecked initially but eventually bleached, attack often confined to sapwood in hardwoods, sporophore rare but thin bracket up to 75 mm across, grey and brown on top with concentric hairy zones and cream pore surface underneath from which spores are released; often found in decayed hardwood joinery, sometimes in softwood joinery *Trametes (Polystictus) versicolor*

6 (a) Superficial growth causing patchy surface discoloration, usually green, grey or black, usually in building interiors; see 2(a)
Mould fungi

(b) Uniform coloration on surfaces, usually green, usually exterior surfaces but sometimes in light interiors, always associated with moisture, disappearing on drying and reappearing on wetting; no visible structure; occur on any surface Algae

(c) Clumps of green growth, superficially grass-like, associated with deposits of organic matter (humus) Mosses

(d) Growth on masonry, brickwork, concrete, tiles, wood etc., with distinct structure; encrustations in close contact with surface varying from minute to large, or distinctly branched and attached

Figure A2.1 A crustose lichen, *Caloplaca aurantia*

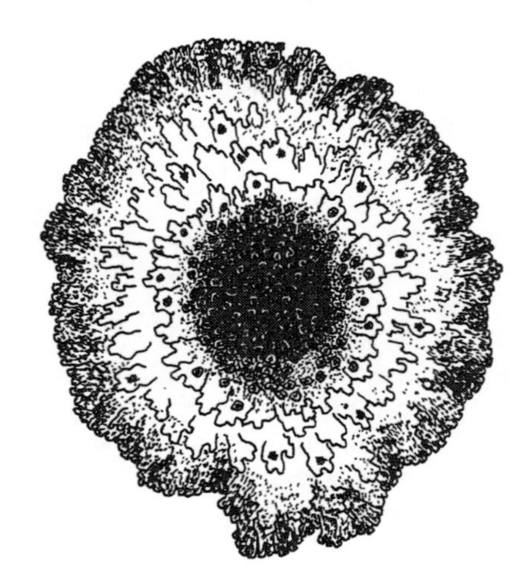

Figure A2.2 A fruticose lichen, *Ramalina siliquosa*

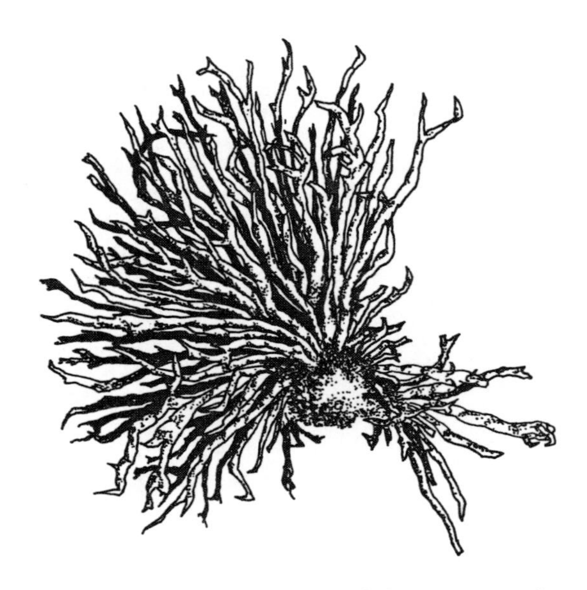

firmly to surface at base, or with leaves or scales attached loosely to surface at base — Lichens 7

(e) Uniform coating, usually dark green, brown or black, usually glossy surface, occurring on interior and exterior surfaces in suitable damp conditions — Slime fungi

Figure A2.3 A foliose lichen, *Parmelia perlara*

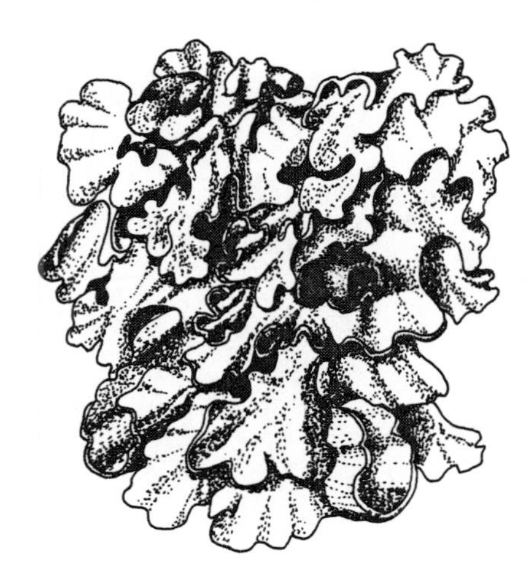

(f) Grasses, ferns, liverworts and many higher plants up to small trees can grow on masonry if deposits of organic matter can first accumulate in which seeds can germinate.

7 **Lichens**

(a) Thallus (i.e. growth) forming a crust firmly attached to the substrate and removed only with difficulty; varying colours and sizes from minute, perhaps within pores, to a diameter of 300 mm or more. Black spots or localised black areas on light-coloured stones are often ascocarps or spores from minute crustose lichens within pores (Figures A2.1)

(b) Thallus distinctly branched, usually erect and bushy, attached at base (Figure A2.2) Fruticose lichens

(c) Thallus leaf-like, sometimes gelatinous (Figure A2.3) Foliose lichens

Identification of individual lichens is usually unnecessary in relation to remedial treatment. However, the occurrence of a particular species can give an interesting guide to the nature of the substrate, climatic conditions and degree of atmospheric pollution. *The Observer's Book of Lichens* by K. L. Alvin (Frederick Warne) is a useful starting point for identification and gives a list of books for further reading.

Appendix 3 *Diagnosis and measurement of dampness in buildings*

The amount of water in air is known as the moisture content or humidity, typically 5 to 10 grams of water per kilogram of air (g/kg). Air has a limited capacity to carry water and when this capacity is reached the air is said to be saturated.

The humidity of air at a particular temperature relative to its saturated humidity at the same temperature is known as relative humidity, normally expressed as a percentage, which gives a direct indication of the dryness or wetness of the air. If the temperature increases, the water-carrying capacity of the air also increases so that air which is warmed will not change in humidity but the relative humidity will progressively fall and the air will appear to be drier; this is the principle of drying by heating. However, if the temperature is reduced the moisture-carrying capacity is also reduced so that the relative humidity increases progressively and the air appears to become damper.

Eventually the relative humidity reaches 100% and the air is saturated.

This saturation occurs at the dew point, a temperature that depends only on the humidity of the air, and any further reduction in temperature results in condensation as either formation of mist within the air or dew on cold surfaces.

The properties of air in relation to moisture content are illustrated in the psychometric diagram A3.1.

Air with a moisture content of 7.5 g/kg will have a relative humidity of about 70% at 15°C (point A) but if it is heated to 25°C the relative humidity is reduced to about 37.5% (point B), indicating that it 'feels' drier although the moisture content is actually unchanged. Conversely, if the same sample of air is cooled to about 10°C, the relative humidity will increase to about 100% or saturation (point C); any further cooling will result in condensation, and the temperature at which the air reaches saturation is therefore known as the dew point. An increase in moisture content will also increase the dew point, so that if the moisture of the same air sample is increased to 13 g/kg, the dew point is increased to about 18°C (point D).

Vapour pressure is directly proportional to moisture content and does not vary with temperature. The saturated moisture content of air depends on the temperature. These two factors are very important in relation to ambient moisture contents in buildings. The low temperature of the exterior air means that it is often saturated but, as the temperature is low, saturation involves a relatively low moisture content and the water vapour pressure is also low. In an unoccupied building the water vapour pressure will ensure that the moisture content of the air is the same both internally and externally but, whilst the air may be saturated with a relative humidity of 100% at the

Figure A3.1 Properties of air in relation to moisture content and temperature. This diagram enables the dew point to be determined by measuring the temperature and relative humidity. The average relative humidity is indicated best by the moisture content of wood in contact with the air, and this moisture content can be used with an estimate of average air temperature to give a more accurate assessment of dew point and condensation risk

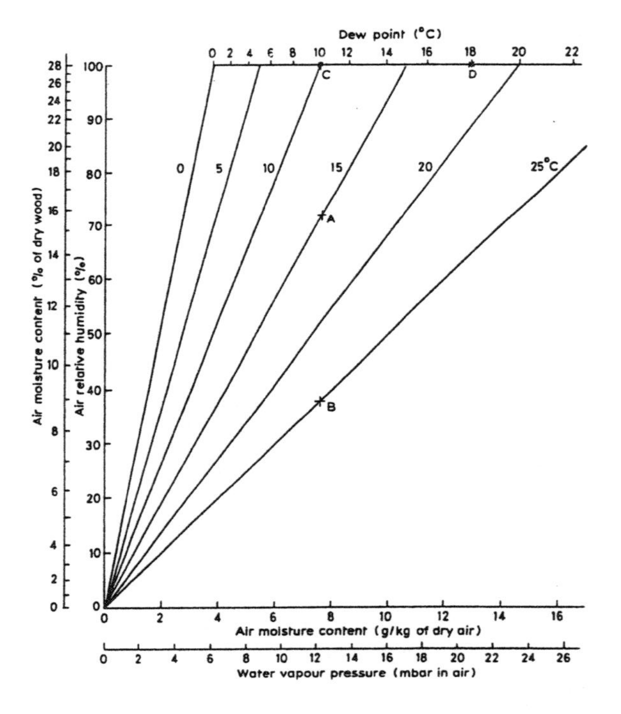

low external temperature, the higher internal temperature will result in a much lower internal relative humidity. For example, if the external air is saturated at 10°C (point C). it will have a relative humidity of only 70% in cooler parts of the buildings at 15°C (point A) but an even lower relative humidity of only about 37.5% in the hottest parts of the building at 25°C (point B). In all cases the moisture content of the air is the same at about 7.5 g/kg but, if this moisture content is increased by breathing, cooking or washing, the extra moisture will increase the vapour pressure of the air and there will therefore be an outward movement of this moisture towards the exterior, the natural process which ensures that the interior of a warm building is also dry.

Measurements

Air moisture content, vapour pressure and dew point cannot be measured conveniently on site but they can be easily determined from temperature and relative humidity measurements using Figure A3.1. For example, if the air in a building has a relative humidity of about 70% at a temperature of 15°C (point A), it can be seen that it has a moisture content of about 7.5 g/kg and a dew point of about 10°C. Dew point is usually the most important factor, at least in winter conditions, and this information indicates that, if the air diffuses to a cooler part of the accommodation with a temperature below 10°C, condensation will form as mist; this situation often occurs when warm humid air diffuses from a kitchen, bathroom or utility room into an adjacent cooler area such as a corridor or larder, warm air from such sources often being virtually saturated so that only slight cooling is necessary to cause condensation. If warm humid air is in contact with a cold surface with a temperature below the dew point, condensation will develop as dew, but this effect is not confined to impervious surfaces; humid air diffusing into an external wall or roof will cause condensation within the structural material as it approaches the cool exterior and reaches the dew point, an effect known as interstitial condensation.

Various instruments can be used to determine the temperature and relative humidity of air with varying degrees of accuracy. Hygrometers directly measure relative humidity, usually using the change in length of a piece of hair or strip of paper, but generally the best technique is to use the wet and dry temperature method. This involves using a thermometer or other temperature measuring instrument in the normal way to determine the dry air temperature and then fitting a porous material wetted with distilled or deionised water to the thermometer bulb or sensing element to determine the wet temperature. Water evaporation from the wet material causes cooling so that the 'wet' temperature is lower than the 'dry' temperature except when the air is already saturated with a relative humidity of 100%. The dry and wet temperatures can therefore be used in conjunction with Figure A3.2 to determine the relative humidity. The main advantage of this method is that the same temperature measuring system is used to determine both temperatures so that any errors in the measurement virtually cancel out in normal relative humidity measurements; in conventional wet-and-dry bulb thermometer instruments the two thermometers can be compared for a dry air measurement and a temperature correction applied if one is reading higher or lower than the other. The method can be very accurate if there is a reasonable flow of air around the sensing element.

Temperature and relative humidity measurements are only an indication of the condition of air at the time of measurement and they must be carefully interpreted. If central heating is controlled by a clock which switches it off at night and also perhaps when occupants are at work during weekdays, or if other heating is used intermittently, the temperature fluctuations will affect relative humidity to a greater extent than any fluctuations in the moisture content in the air. The temperature and relative humidity measurements must therefore be used, in conjunction with Figure A3.1, to determine the

Figure A3.2 Determination of relative humidity of air using wet and dry temperatures

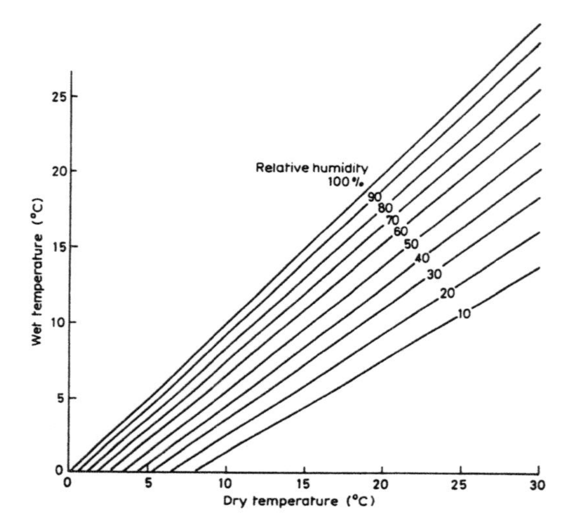

moisture content of the air, and thus the dew point and the relative humidites that will occur at various other room temperatures.

If a room is well ventilated at the time of the measurements, the results may be very misleading as they will not indicate normal ambient conditions. Average relative humidity can be determined much more accurately and more easily by using an electrical probe moisture meter, such as one of the Protimeter instruments, to determine the average moisture content of the softwood joinery components in the accommodation area. These measurements must preferably be made on wood that is painted so that the moisture content does not change rapidly with fluctuations in atmospheric relative humidity. The most suitable sample point is usually the top surface of the architrave across the head of a doorway in an internal partition wall, well away from possible sources of penetrating, rising or condensing moisture; the top surface is used so that the probe holes do not show! The wood moisture content determined in this way is then used, in conjunction with the ambient room temperature and Figure A3.1, to determine the average air moisture content and thus the dew point and the relative humidity of the air at various temperatures. In a living room it can be assumed that the temperature during occupation usually averages about 20°C but there may be long unheated periods which reduce the average temperature. Generally an average of about 15°C is appropriate for living rooms in winter and 17°C in summer, with 12.5°C in winter and 15°C in summer for bedrooms, but it may be possible to estimate average temperatures more accurately by observing the apparent conditions in the building. Thus if the wood moisture content is about 15%, this indicates an ambient relative humidity of about 70% and, if the measurement is made in the living room in winter with an average temperature of about 15°C (point A), the ambient air moisture content is evidently about 7.5 g/kg so that relative humidities at various temperatures can be determined, as well as the dew point which is about 10°C for this particular example. Dew or surface condensation will develop on internal surfaces if their temperatures are below the dew point for the accommodation air; the surface temperatures will depend on the interior and exterior air temperatures in relation to the thermal properties of the roof, wall, window or door structure involved; Figure 5.2 enables *U* and *R* values to be determined from air and surface temperature measurements, but calculated dew points can be used in conjunction with this diagram to determine the external temperatures below which condensation is likely to develop on internal surfaces.

Most porous structural materials are hygroscopic to a greater or lesser extent and will absorb water from the surrounding air. If these materials are in equilibrium with the air and there is no other source of moisture, their moisture content will depend on the atmospheric relative humidity. Wood is perhaps the best example; its moisture content in equilibrium with saturated air with a relative humidity of 100% is known as the fibre saturation point and is usually 25–30%. The amount of water absorbed will depend on the amount of wood present so that the moisture content must be expressed as percentage of the dry weight of wood, a system that is normally used to express moisture content for all materials.

However, some instruments such as the Speedy carbide moisture detectors give moisture content as a percentage of the wet weight, always a lower figure. Other instruments express moisture content as a percentage of volume but these are not generally used in dampness inspections. Moisture contents arising through equilibrium with surrounding air are generally low and insignificant as they are not sufficient to cause darkening on structural surfaces, or even sufficient to allow salts to migrate in diffusing moisture, but wood and wood-based products such as fibreboard and wallpaper can all achieve equilibrium moisture contents when exposed to high relative humidities which are sufficient for biological growth and spoilage by mould, with a danger of more severe Dry rot in some cases.

Treatment cannot dry structural materials below their equilibrium moisture contents and, if they are excessive, they can be corrected only by reducing ambient relative humidity, usually by improving ventilation. However, it is necessary to check that the moisture content is solely due to natural hygroscopicity and atmospheric relative humidity and not to some other cause. There are two other principal explanations for a high moisture content: the presence of hygroscopic salts which increase the equilibrium moisture content, and water derived from other sources such as penetrating and rising dampness.

Dampness diagnosis

It must be emphasised that dampness is generally insignificant if it is not visually apparent but, on the other hand, recent decorations may conceal dampness which can be detected only by means of suitable instruments. If dampness is detected, either visually or by means of instruments, it must first be assumed that it is caused by rain penetration or water absorbed from the soil, the diagnosis depending on the distribution:

1 Driving rain penetration: uniform dampness on solid external walls, perhaps less severe at upper levels protected by eaves, no dampness on internal partition walls.
2 Roof leaks at eaves, gutter defects and damaged downpipes: patches at tops of walls or on ceilings, vertical areas of dampness down solid external walls.
3 Condensation under flat roofs: decay in roof boarding and joists if untreated, uniform damp staining on ceilings.
4 Accumulating rain penetration at base of solid external walls: dampness largely confined to base of external walls and most severe on most exposed elevations, no dampness in internal partition walls.
5 Rising dampness in walls: dampness largely confined to bases of both external and partition walls (may be confused or combined with 4; check whether damp-proof course is absent or has been bridged by pointing, rendering, plaster, soil levels or floor levels).
6 Rising dampness in inner skins of cavity walls: dampness confined to external walls (check whether incorrect damp-proof course heights or slovens accumulated at base of cavity).
7 Mortar slovens on cavity wall ties: damp patches on internal surfaces of external walls, particularly towards base of wall.
8 Flue defects: patches on surfaces concealing flues such as chimney breasts, dampness following line of flue with intense patches opposite bends or shelves in flue which catch rain penetration through pot or allow condensation to accumulate (hygroscopic salts may be present).
9 Hygroscopic salts incorporated during construction: uniform dampness on all walls, perhaps through the use of marine sand or bricks containing sulphates.

Brickwork with stretcher bond normally indicates cavity construction, provided the wall is more than about 275 mm (11 in) thick, but other bonds are also used in cavity wall construction for aesthetic reasons. If there is any doubt about the form of construction it can usually be resolved by measurement; solid walls have thicknesses in units of 112 mm ($4^1/_2$ in) plus perhaps 12 mm ($^1/_2$ in) for a plaster or render coat, whereas cavity walls will generally have a similar modular thickness plus 50 mm (2 in) for the cavity. Only modern masonry walls incorporate cavities. Rubble walls can present particular problems as driving rain penetration through a face can percolate through the wall, accumulating at distant points to cause local damp patches.

Whilst hygroscopic salts deposited by rising dampness may have only limited significance, salts from other sources can be particularly troublesome. Hygroscopic salt deposits have been identified in the past by chemical analysis but it is far better to assess the actual hygroscopicity of the sample and compare it with the moisture content. A slow-speed cordless electric drill can be used with a masonry bit of about 9 mm ($^3/_8$ in) diameter to obtain samples. Generally material from about the first 10 mm ($^1/_2$ in) should be retained and tested separately from deeper material. A suitable watch glass is weighed (*Wo*); the sample is added and the watch glass is weighed again (*Ww*). The watch glass with the sample evenly distributed is then placed overnight in a container at a relative humidity of 75% (most easily achieved using a saturated solution of common salt (sodium chloride) in the bottom of the container) and the watch glass is weighed again (*Wh*). Finally the watch glass is placed in an oven at a temperature above 100°C for more than one hour, giving a final weight (*Wd*) The calculation proceeds as follows:

$$Ww - Wo = \text{wet weight of sample}$$
$$Wd - Wo = \text{dry weight of sample}$$
$$Wh - Wo = \text{hygroscopic weight of sample}$$

$$\text{Thus: } 100\ \frac{Ww - Wd}{Ww - Wo}\% = \text{moisture content of sample}$$

$$100\ \frac{Wh - Wd}{Wh - Wo}\% = \text{hygroscopic moisture content of sample}$$

The oven drying method for determining moisture contents is not essential and a carbide moisture detector can be used as an alternative. A rather larger sample may be necessary which must then be divided into two parts; the moisture content of the first part is determined on site whilst the moisture content of the second part is determined after the sample has been exposed to the relative humidity of 75%, the latter sample determining the hygroscopic moisture content. If the normal site moisture content is much greater than the hygroscopic moisture content, the hygroscopic salt alone cannot be responsible for the dampness. However, if the moisture content is approximately equal to the hygroscopic moisture content it would seem that there is no other source of moisture.

It is recommended by the Building Research Establishment that moisture

Plate A3.1 Dampness at the base of a wall. (a) Normal rising dampness. (b) Isolated dampness: the dry area at the bottom of the wall results from patching with waterproof cement rendering

(a)

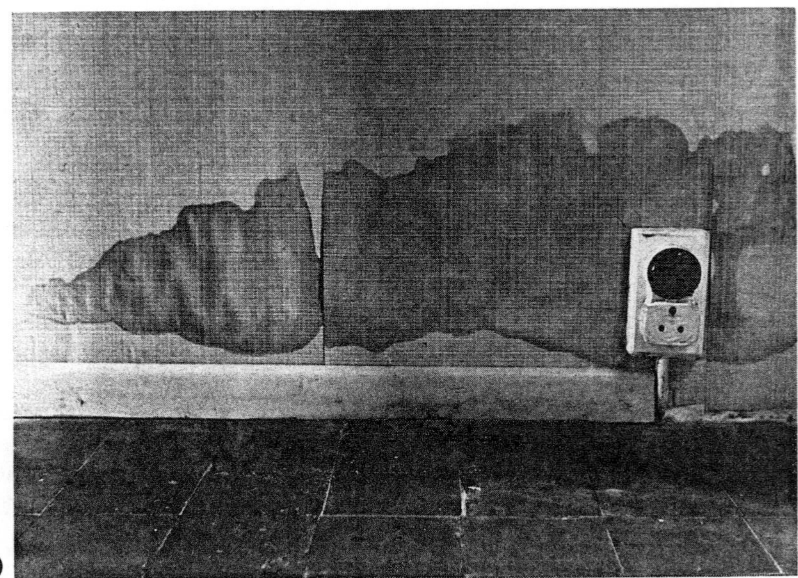

(b)

content and hygroscopic moisture contents should be determined in this way for the diagnosis of rising dampness, perhaps sampling every two or three courses from the base of a wall to the limit of the rising dampness. It is suggested that, if the moisture content exceeds the hygroscopic moisture content at any point, rising dampness is clearly present. Unfortunately this is not necessarily true for several reasons; this conclusion fails to take account of the fact that penetrating rainfall may be accumulating at the base of an external wall whether or not a damp-proof course is present, and salts are only rarely detected in significant quantities within walls as they are only usually present in rising dampness at low concentrations and are only significant where they become concentrated at the evaporation surfaces, precisely the sample zone that is rejected according to the recommended method.

It has already been explained that hygroscopic salts have little significance when they are found in conjunction with rising dampness, provided the surface plaster in which they have accumulated can be removed, but the above method has great value in monitoring hygroscopic salt deposits arising from other sources. For example, dampness following the line of a flue may be due to rain penetration into the pot or stack, condensation, or hygroscopic salt deposits, and this sampling method can be used to show that only accumulated salts are responsible for persisting dampness when a flue has been sealed. The method can be similarly used to confirm the presence of hygroscopic salts introduced in building materials such as marine sand, bricks and clinker blocks.

In this connection it is perhaps worth mentioning that the origin of hygroscopic salt is not always apparent. The reaction of flue gases and atmospheric pollution with mortar and other building materials has already been mentioned in Chapter 3 as an indirect route by which hygroscopic salts can be introduced but there are several more obscure routes. For example, many water-repellent, consolidant and biocidal treatments contain active components solubilised as sodium, potassium or sometimes magnesium salts. In the presence of gypsum plasters these may all result in the formation of limited quantities of hygroscopic sulphates. Sea salt is itself hygroscopic to a certain extent, probably through the magnesium chloride content rather than the dominant sodium chloride, but sodium chloride solution in the presence of gypsum plaster can result in the formation of both calcium chloride and sodium sulphate. Whilst the determination of hygroscopic moisture content may confirm the significance of salt deposits, chemical analysis will still be necessary to confirm their source.

Dryness or dampness

It will be observed that no comment has been made on the actual moisture content of material suffering from rainwater penetration or rising dampness. Dampness is significant only if it is apparent to the occupants of a building and requires remedial treatment only if this is the case, so that the detection and measurement of dampness in building materials is necessary only if, for example, recent decorations have concealed a defect or it is necessary to check whether remedial treatment has been effective. Thus the actual moisture content of building materials is of little significance and it is more important to consider whether they appear to be damp or dry. It will be recalled that the actual humidity of air is similarly unimportant, and the dryness or dampness of the air as expressed by relative humidity is far more significant. Unfortunately a similar relative scale is rather difficult to establish for solid materials whose moisture contents can depend on so many different factors. With the single exception of wood and wood-based products, solid materials have insignificant moisture contents in normal dry buildings. However, if a measurable moisture content is found, say 5%, the significance of this figure is difficult to decide. If the total porosity of the material is 5% it is clearly saturated, but if the total porosity is 40% it is relatively dry. Unfortunately it is necessary to distinguish between total porosity and absorption, the latter indicating the amount of water that can be absorbed under normal circumstances. Absorption over total porosity gives saturation coefficient, as explained in Chapter 4. For these various reasons it is best to avoid reference to particular moisture contents and to concentrate instead on relative dampness, that is the use of simple moisture measurements to show the highest and lowest moisture contents across a wall to indicate the probable source of the dampness.

Moisture content of wood

Whilst 'dampness' is more important than moisture content for brick, plaster and mortar, moisture content is very important for wood which is liable to decay at moisture contents in excess of about 23%. The moisture content of wood is consistent and measurable, depending directly on the 'moisture tension' in the surrounding air and contacting materials, so that measurements on wood skirtings and architraves will give a realistic assessment of the 'dampness' of contacting plaster and masonry. Measurements on wood in the same room or roof space but well away from a possible source of moisture will indicate the average moisture content due to atmospheric relative hmidity; 100% relative humidity is equivalent to about 28% moisture content in wood. If the wood moisture content exceeds 12–20% it must be suspected that atmospheric relative humidity is excessive through lack of ventilation or perhaps through a water leak or penetration, and if the wood moisture content is even higher where it is in contact with brickwork or masonry it is affected by a source of water such as penetrating rainfall or rising dampness.

Measuring moisture content

Various methods have been used for measuring the moisture content of materials. The most accurate is to take a sample, weigh it, dry it, weigh it again and calculate the dampness from the weight loss in comparison with the dry weight. Whilst this method is appropriate for checking the moisture content of wood in connection with kilning, as explained in *Wood in Construction* by the same author, it is far too cumbersome for use in connection with building inspections and can be considered realistic only where a special investigation is being made into possible hygroscopicity, as previously described.

The most convenient portable instruments for determining moisture content in buildings rely on the fact that electrical conductivity increases with moisture content. Generally these instruments are calibrated to indicate moisture content as a percentage of dry weight but different scales are required for each different material and if the material varies in any way the scales become inaccurate. At first, direct current instruments were used, but it was found that salts in brickwork, masonry and plaster tended to cause polarisation, resulting in a progressive fall in the apparent moisture content. Whilst this polarisation was distinctive and thus assisted in the identification of salt deposits, it was considered to be a disadvantage and most instruments now operate using high-frequency alternating current, although the presence of salts still interferes by exaggerating the apparent moisture content.

Most conductivity instruments rely on needle probes but only limited penetration can be obtained in many harder building materials and readings tend to indicate surface rather than deeper conditions; the surface may be particularly dry due to good ventilation or may be particularly wet due to condensation. Hammer probes can be used with some instruments such as the Protimeter Timbermaster to obtain deeper readings in wood. In some instruments the needle probes are replaced by flat plates which have the particular advantage that no damage is caused to decorations. These instruments operate at high frequencies, in theory using capacitive coupling through paint and other surface decorations to detect the conductivity of the moisture beneath, but these systems are particularly susceptible to surface effects and are normally measuring the change in dielectric constant due to the presence of moisture quite apart from the conductance, an important point as dielectric constant varies with moisture content per unit volume rather than with moisture content per unit weight. For these various reasons the plate-type electric moisture meters tend to give rather inconsistent results and normal probe-type conductivity meters are generally preferred. Whilst most electrical instruments for measuring moisture show the results as a meter deflection or numerical display, several other methods have been developed such as the flash rate of a neon lamp and the tone of a sound signal.

If a weighed sample of material is mixed with calcium carbide, a reaction with water occurs which liberates acetylene gas and, if the reaction occurs in a 'bomb' or sealed pressure vessel, the pressure gives a measure of the amount of water present in the sample. This is the principle of all carbide moisture detectors. Samples are normally obtained by drilling, usually with a tungsten-tipped masonry bit with a diameter of about 9 mm ($^3/_8$ in). A slow-speed drill must be used to avoid the generation of excessive heat which might reduce the moisture content of the sample, a cordless drill being most convenient. Obviously blunt drills must never be used as they may result in over-heating, and the first 10 mm ($^1/_2$ in) of each drilling should be discarded if the purpose of the investigation is to determine the true moisture content of the wall at depth, free from abnormal surface drying, condensation or hygroscopic salt deposits. Carbide instruments give very accurate measurements of moisture content but that does not mean that they are more reliable than electrical moisture meters; in all cases the actual moisture content may be irrelevant. One warning is necessary; some instruments such as the Speedy give moisture content as a percentage of wet weight, thus giving a rather low result in comparison with the more normal method of percentage of dry weight. However, the readings can be easily corrected. If the moisture content as a percentage of wet weight is x, then the moisture content as a percentage of dry weight is $x/100 - x$). Thus, if the moisture content is 20% of wet weight, this is equivalent to 25% of dry weight.

Monitoring damp-proofing treatments

Carbide moisture detectors are often used to monitor the efficacy of rising damp treatment. Moisture contents are recorded for samples taken during the inspection, and further measurements after damp-proof course installation confirm whether the structure is drying satisfactorily. Samples must be obtained by drilling for this purpose and some damage will therefore be caused. Probe-type electrical moisture meters normally measure surface effects, and the surfaces are often contaminated and are generally removed and replaced as part of the rising damp treatment. However, stainless steel or brass electrodes can be permanently installed in a wall at the commencement of treatment so that the situation can be monitored by use of a suitable resistance meter. Whilst such systems are generally entirely satisfactory, difficulties are sometimes encountered with obtaining a tight fit within holes drilled in the wall and it is sometimes found to be more satisfactory to drill a larger hole in which the electrodes can be bedded in a standard plaster, thus enabling standard moisture contents to be determined. Whilst this is certainly the most satisfactory method for research purposes it is not essential for monitoring the performance of commercial treatments where it is only really necessary to show that the moisture content decreases steadily until it becomes insignificant. One method of monitoring is simply to provide permanent holes with a diameter of perhaps 6 mm ($^1/_4$ in) and a depth of about 75 mm (3 in). These holes can be fitted with plugs which can be opened at a later date, providing access to the sample points at the end of the drillings within the wall. Contact is made with these sample points by using probes fitted with cotton wool or gauze moistened with a conductive gel. This method enables a consistent contact to be established with the brickwork or masonry at the end of the holes, the moisture content being determined by the electrical conductivity between the holes in the normal way.

Plastic patches

It has already been explained that surface effects, particularly ventilation drying, condensation and hygroscopic salt deposits, interfere with electrical moisture meter readings unless the surface is first removed so that the readings can be taken well below these surface effects. An alternative is to fix a piece of impermeable plastic to the wall so that the surface comes into equilibrium with the moisture content deeper within the wall, enabling a probe-type moisture meter to be used through the plastic to measure the apparent moisture content at depth. Whilst this method removes the uncertainties due to ventilation drying and condensation, surface salt deposit problems still remain and the method is rather expensive to use because of the time involved in reaching equilibrium after the plastic patches are fixed. It has also been suggested that moisture 'tension' or the wetness of a wall can be determined most reliably by measuring relative humidity within a drilling, but if there is any free water present the relative humidity will be 100% and no measurement will be possible of the degree of moisture introduced by rain penetration or rising dampness.

Methods selection

Whilst moisture measurement techniques are not entirely satisfactory, they are certainly adequate if the most appropriate methods are selected and properly used. Sound and electro-magnetic radiations have been proposed as methods for measuring moisture content but, as with electrical conductivity, the variation in the substrate material is often greater than the variation due to the moisture content changes. In summary, the most important requirements are to be able to detect the wettest and driest areas to locate the source of dampness. If surface effects are suspected then the surface must be removed to enable determinations to be made at a greater depth, and here the carbide meter may be the most convenient instrument at present available, pending the development of a probe which will take electrical conductivity measurements at a reasonable depth down a relatively small-diameter hole. If condensation is suspected it is necessary to measure surface temperatures and air humidities. If hygroscopic salts are suspected it is necessary to check the hygroscopicity of the material in some way, either by direct hygroscopicity determination or by chemical analysis. If a surveyor carrying out inspections is capable of making all these measurements and interpreting them, reliable diagnosis is possible!

Appendix 4 *Drying buildings*

Large quantities of water may be introduced into buildings during construction, through roof and plumbing leaks, through flooding or through fire-fighting. Drying can present problems, either through the large quantities of water involved or because the water is relatively inaccessible.

Evaporation

Drying involves the evaporation of water from the structural materials and this can be divided into stages. The superficial surface moisture can be lost comparatively rapidly but the deeper moisture is more difficult to evaporate.

Water in wood

In wood the removal of free moisture from the pores is followed below the fibre saturation point of about 25–30% by the removal of combined water from the cell walls. The loss of this combined water will lead to shrinkage but the decay danger remains above 20%. Consideration should always be given to preservation treatment if wood has moisture contents in excess of 18%, but *in situ* preservation treatment of wood normally involves the application of organic solvents which will only penetrate reliably at moisture contents below about 22%. Clearly drying may be necessary to reduce the moisture content to below 22% to permit preservation treatment or to reduce the moisture content to ambient levels of about 12–15%, or 8–12% in centrally heated buildings.

Energy equivalent

In theory heat is required at a rate of 0.694 kW h per litre (3.15 kW h per gallon) to evaporate water. If heat is not supplied evaporation causes a decrease in the temperature, the reason why a damp building feels cold. If the structure is heated to encourage evaporation and drying, the consumption will exceed this theoretical value as the structure itself will be heated and some losses will occur. The rate of evaporation depends on the temperature which in turn depends on the rate of supply of heat. The vapour pressure of water or its tendency to evaporate increases with temperature, and heating of a structure is essential if evaporation is to be encouraged; if heating is not used the evaporation will be very slow. A further problem is the drying of the deepest parts of the structural members; the rate slows as the evaporating surface recedes into the material so that the evaporating moisture must diffuse to the surface, and it is thus essential to expose all structural members as far as possible to encourage drying.

Vapour removal

The next process is to remove the evaporating water. As water is released into the atmosphere the humidity increases and this inhibits further evaporation. However, the water-carrying capacity of the atmosphere increases as the temperature increases (see Appendix 3) so that again it is important to maintain an elevated temperature to encouraged evaporation. The humidity of the air compared with its water-carrying capacity is its relative humidity so that at 100% relative humidity the air is completely saturated with water. If the humidity of air remains constant but the temperature falls the water-carrying capacity decreases and the relative humidity rises until it reaches 100% and condensation then occurs. For these reasons the temperature must be as high as possible to ensure that the relative humidity is as low as possible, this latter requirement achieving the actual removal of water from the building.

Free ventilation

The most efficient method of drying a building is to permit free circulation of external air. External air always has a low humidity. In warm weather this results in a low relative humidity and, if this air is circulated within the building, it will be able to accept a considerable volume of moisture, the energy for evaporation being introduced in the large volumes of air. In cool weather the relative humidity of the external air may be very high but it will be warmed when it comes into contact with the structural materials so that its relative humidity will be reduced and drying can still occur. If the external air temperature remains below freezing for a significant period the relative humidity will fall dramatically and the ventilating air can then achieve quite rapid drying. Whilst freely ventilated parts of a building such as unlined roof areas are generally much drier than areas with restricted ventilation, it is important to appreciate that this drying has occurred progressively over a period of years and may not occur so rapidly when drying is urgently required. The important limitation is the energy required for evaporation from the structural material; whilst this can be provided by warming incoming air by passing it through propane blower heaters, for example, it is usually far more efficient to arrange for night-time heating to be alternated with day-time ventilation.

Dehumidifiers

Dehumidifiers are often suggested as a means for reducing the relative humidity of the atmosphere and thus for drying structural materials. Hygroscopic absorbants have been used on a small scale but mechanical dehumidifiers are essential if structural drying is to be attempted in this way. Mechanical dehumidifiers involve a circulator fan which passes the air over a low-temperature condenser. As the temperature is reduced the relative humidity rises and the moisture in the air condenses as dew which is then collected. Dehumidifiers must operate in an unventilated space as they otherwise dehumidify the ventilation air rather than drying the structural materials. Air leaving mechanical dehumidifiers has been cooled but, whilst this cooling is compensated by the heat produced by the compressor which cools the condenser, the entire system progressively cools owing to the evaporation of moisture from the structure, and eventually evaporation ceases through this cooling. This can be corrected by heating the air leaving the dehumidifier, but the energy required for direct evaporation of water from structural materials is enormous, as previously indicated, and dehumidifiers operating in an unventilated space are therefore extremely inefficient and very expensive to operate.

Ventilation represents the most efficient method for drying structural materials. Mechanical dehumidifiers should be used only in spaces that cannot be readily ventilated, such as for drying an isolated room within an occupied building, and only where a limited amount of water is involved, perhaps through replastering or laying a floor screed. In all cases the material to be dried must be freely exposed. This normally means that floorboards must be lifted, all sound and thermal insulation materials removed and all parts of the structure made accessible to the air flow. Problems may be encountered in some structures where, for example, plaster on laths and battens is fixed to walls, perhaps containing timber framing, so that substantial parts of the property are largely inaccessible. It is generally essential for any concealed wood to be exposed for drying and, if necessary, preservation treatment, or at least for its moisture content to be reliably checked to ensure that decay is unlikely. This may mean the entire stripping of plaster and other dry-lining materials in properties affected by fire-fighting water or severe plumbing leaks, but it is sometimes possible to use unconventional methods to treat wood when the concealing plaster-work is considered to be of special value; examples are given in Chapter 2.

Finally, it should be appreciated that the rate of drying is usually restricted through cooling induced by the water evaporation itself, and it is thus essential to attempt to introduce heat energy to encourage evaporation. Heaters should always be installed when dehumidifiers are in use and, whilst incoming air can be heated when drying by ventilation, it is usually more efficient to heat the unventilated structure overnight and ventilate freely only during the day when the relative humidity of incoming air will tend to be lower. Additional heating may not be necessary during the summer months but drying problems are usually encountered only during the winter.

Appendix 5 *Assessing the exposure of buildings*

The insulation of a roof, wall, window or door will depend to a certain extent on the ventilation to which it is subject at its external surface. For example, with a wall of normal high emissivity the surface resistance to heat loss (units $m^2°C/W$) is reduced from 0.08 for sheltered conditions to 0.03 for severe conditions, and it is thus apparent that the insulation required for a particular building will depend on its exposure. Whilst the Building Regulations for England and Wales (and Scotland) have defined maximum thermal transmittance or *U* values for many years, they fail to emphasise the need to take account of the degree of exposure. Exposure significantly affects the thermal properties of a building and must be considered when designing to meet certain performance requirements.

Degrees of exposure

For most normal purposes, such as assessment of external surface resistances in thermal insulation calculations or assessment of driving rain indexes for roof tile lap calculations, only three degrees of exposure need to be defined:

1 Sheltered exposure: buildings in city centres up to third floor.
2 Normal (standard) exposure: buildings in city centres from fourth to eighth floors and most suburban or country premises.
3 Severe exposure: buildings in city centres above ninth floor, buildings in suburban or country districts above fifth floor, buildings within 5 miles (8 kilometres) of coasts or exposed on hill sites.

Very severe exposed conditions may occur in some cases, such as for tall buildings on mountainous sites close to coasts, and these may require special consideration.

Rainfall is not itself a special problem in buildings, even if it is particularly heavy at times or the total rainfall is high during the year, but the degree of exposure to driving rain is of considerable importance in relation to rain penetration through windows, doors and walls. In central southern England around London it usually rains for about 5% of the time but on the western coasts of England and Wales this figure rises to 7% or 8% and, in hilly districts, to over 10%. Figures are even higher for the west coast of Ireland, and in north-west Scotland it normally rains for more than 15% of the time.

Annual rainfall figures for these various zones generally show that annual precipitation is even heavier than indicated by these rainfall periods. However, for most purposes it is sufficient to divide the British Isles into three rainfall zones:

1 Light rainfall: England east of a line drawn through the Pennines to Poole on the south coast.
2 Average rainfall: all areas not classified as light or heavy.
3 Heavy rainfall: all western coastal areas of England, Wales, Scotland and Ireland; all mountain areas.

An index of exposure to driving rain can be derived from these categories of exposure and rainfall by adding the classification numbers and then referring to the following:

Sheltered	2
Moderate	3, 4
Severe	5, 6

Clearly in this calculation it is necessary to take account of very severe exposure situations which might be classified as 4 rather than 3 and which can, in effect, increase these indices by an extra point. An alternative index of exposure to driving rain has been prepared by the Building Research Establishment by multiplying annual rainfall by average wind speed. The results are quoted in BRE Digest 127 (1971) but have now been revised as shown in Figure A5.1 which is from the BRE Report *Driving Rain Index* (1976) by R. E. Lacy.

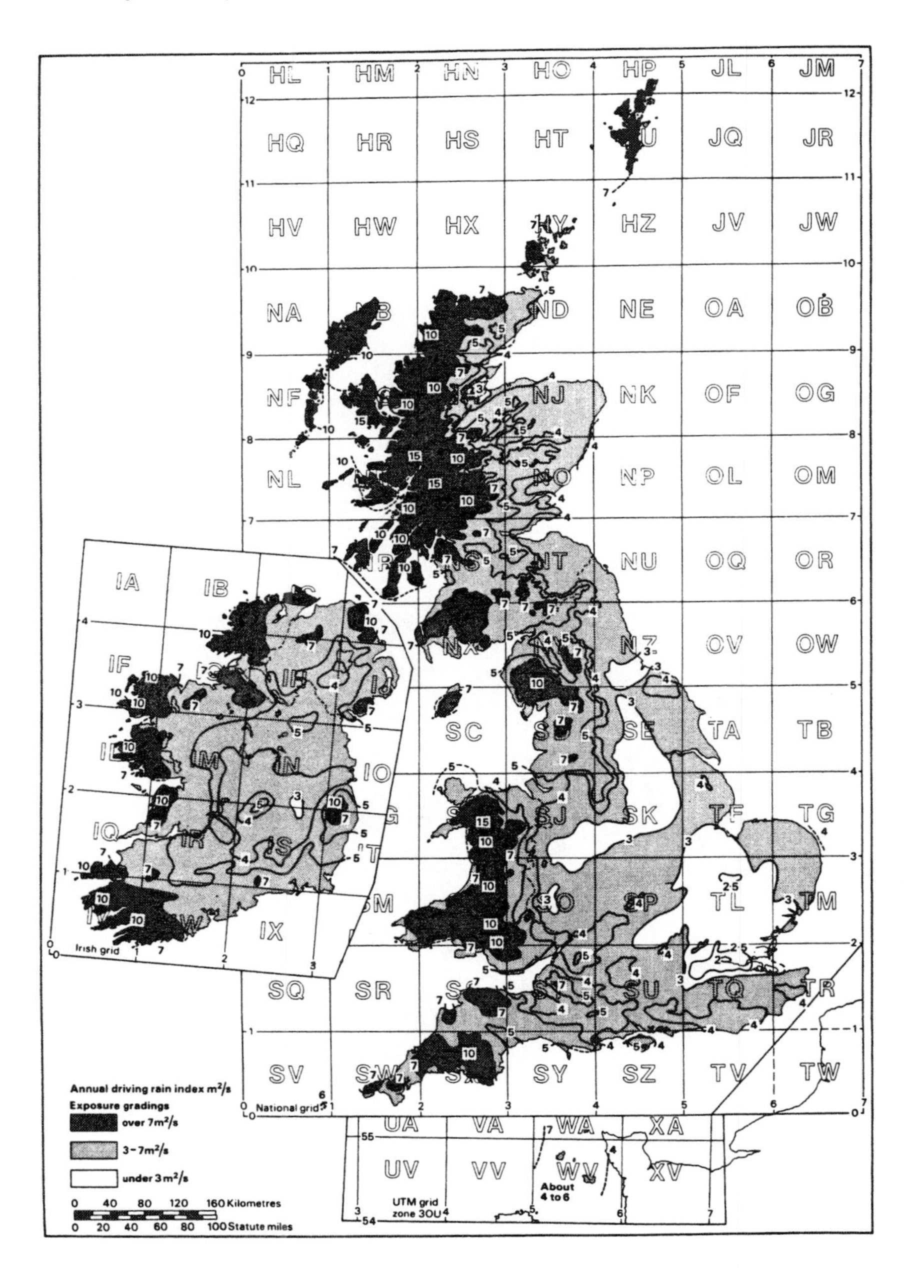

Figure A5.1 Map of annual mean driving rain index at a standard height of 10m above open level country (*from BRE Report* Driving Rain Index *by R. E. Lacy, reproduced by permission of the Controller of Her Majesty's Stationery Office.*)

Appendix 6 *Thermal properties of buildings*

The function of a building is to isolate its occupants from the surrounding environment. Good thermal insulation is essential if the internal conditions are to remain constant despite fluctuating external temperatures. Thermal insulation affects only the rate at which internal temperature changes following an increase or decrease externally, so that a well-insulated building is largely unaffected by daily fluctuations but cannot resist seasonal fluctuations. If heating is used in cold climates, or cooling in hot climates, good insulation will reduce the energy consumption necessary to maintain a particular temperature, or in unusually cold weather it will be possible to achieve a higher temperature if the available heat energy is limited.

Thermal transmittance

The rate of heat loss from a building element is usually specified in terms of the thermal transmittance or *U* value, defined as the rate of heat transfer (watts) through unit area (one square metre) of the building element for unit temperature difference (one kelvin or one degree Celsius or centigrade), so that the units for *U* value are W m^2 K or W m^2°C; kelvins and degrees centigrade are identical, but the kelvin notation, in which the freezing point for water is 273 K and the boiling point is 373 K, is used in the current Building Regulations, presumably to confuse architects, engineers and contractors who have only recently absorbed the transition from degrees Fahrenheit to degrees centigrade! For convenience the Celsius or centigrade notation will be used throughout the text as it is more widely understood.

Energy loss

The advantage of defining the thermal properties of a building element in terms of thermal transmittance or *U* value is that the energy loss through the element can be calculated by multiplying the *U* value by the area and the temperature gradient. For example, if an exposed wall has a *U* value of 0.6 and an area of 25 m^2, the total energy loss through the wall will be 15 W for a difference in temperature of 1°C, so that if the average temperature difference over the year between the interior accommodation and the exterior air is 10°C, the rate of heat loss is 150 W, representing a total energy loss over the year of 1314 kW. It is obviously in the national interest to limit energy consumption for heating buildings, and the current Building Regulations therefore limit *U* values; the requirements for dwellings are summarised in Table A6.1. As the purpose of these requirements is energy conservation it is essential in an element of mixed construction such as a wall with windows to achieve the overall requirement, even if individual parts cannot comply for some reason. If the proportion of windows is excessive, the *U* value of the windows and/or the walls must be reduced in compensation, as explained in the notes to Table A6.1. For example, a house with 100 m^2 of wall meeting the critical requirements will have 12 m^2 of windows with a *U* value of 5.7 W/m^2C and a heat loss of 68.4 W/°C, and 88 m^2 of wall with a *U* value of 0.6 W/m^2°C an a heat loss of 52.8 W/°C, giving a total heat loss of 121.2 W/°C which must not be exceeded. If the window area is increased to 20 m^2 with single-glazed windows with a *U* value of 5.7 W/m^2°C and a heat loss of 114.0 W/°C, the wall *U* value must be reduced to below 0.09 W/m^2°C to ensure that total heat loss still does not exceed 121.2 W/°C. In fact, a *U* value of 0.09 W/m^2°C cannot be achieved with conventional construction so that single-glazed windows cannot be used if the window area is 20% of the wall area. The solution to this dilemma is to use double glazing. A 20 mm air gap and a well-insulated wood, uPVC or thermal break aluminium frame will give a *U* value of 2.5 W/m^2°C and a heat loss for 20 m^2 of 50 W/°C, the wall requirement then being a heat loss not exceeding 71.2 W/°C which represents for 80 m^2 a *U* value not exceeding 0.89 which is easily achievable.

Table A6.1 Maximum thermal transmittance on *U* values for elements of dwellings.*

Building element	*Maximum U value†* *(W/m^2°C)*
Exposed walls and floors	0.6
Roof	0.35
Windows and roof-lights‡	5.7

*These requirements are based on the Building Regulations 1985 for England and Wales.
† Kelvins (K) are used in the Building Regulations but they are identical to degrees Celsius or centigrade (°C).
‡ The area of the windows and roof-lights shall not in effect exceed 12% of the external walls; if their area is greater, the maximum *U* values of one or more of the elements must be reduced in compensation.

Although *U* values are the most convenient way in which to express heat loss from a building, they cannot be directly determined from the properties of the various components or layers of a structural element. The converse of thermal transmittance or *U* value is thermal resistance or *R* value, these two values being reciprocals, and it is the *R* value that can be calculated more easily as the thermal resistance of each succeeding layer can be added to give the total *R* value of the element. The simplest situation is represented by a single structural layer, such as single glazing, but in addition to the resistance of the material itself there are also external and internal surface resistances which represent the resistance to transfer of heat from air to solid material.

Surface resistance

Typical surface resistances are shown in Table A6.2 from which it can be seen that surface resistance depends upon the surface emissivity, or the efficiency of the surface in transferring heat, high emissivity representing normal surfaces and low emissivity representing reflective surfaces such as polished aluminium window frames and aluminium foil backing on plasterboard. Surface resistance also depends upon the direction of heat flow, that is horizontal through walls and windows, upwards through ceilings and

Table A6.2 Surface resistances*

Internal surface resistance R_{Si}

Building element	*Surface emissivity*†	*Heat flow*	*Surface resistance (m^2 °C/W)*
Walls	High	Horizontal	0.123
	Low	Horizontal	0.304
Roofs, ceilings, floors	High	Upward	0.106
	Low	Upward	0.218
	High	Downward	0.150
	Low	Downward	0.562

External surface resistance R_{So}

Building element	*Surface emissivity*†	*Surface resistance*‡ *(m^2 °C/W)*		
		Sheltered	*Normal*	*Exposed*
Walls, windows, doors	High	0.08	0.055	0.03
	Low	0.11	0.067	0.03
Roofs, floors over ventilated areas	High	0.07	0.045	0.02
	Low	0.09	0.053	0.02

Values from Building Research Digest 108 (*IHVE Guide*, Book A).

† Emissivity is high for all normal building materials including glass, but unpainted metal and other reflective surfaces have low emissivity.

‡ External surface resistance depends upon exposure (see Appendix 5):

1 Sheltered: up to third floor in city centres
2 Normal: most sites, fourth to eighth floors in city centres
3 Exposed: coastal or hill sites, ninth floor and above in city centres, fifth floor and above elsewhere.

roofs (except summer conditions or warm climates when heat flow is downwards), and downwards through ground floors. The resistance of external surfaces also depends upon the degree of exposure, normal exposure being adopted except in special cases. As far as the example of single glazing is concerned, the internal surface resistance involves high emissivity and horizontal heat flow, and thus a surface resistance of 0.123 $m^2°C/W$. The external surface resistance for high emissivity and normal exposure is 0.055 $m^2°C/W$.

Figure A6.1 Variation of thermal conductivity *k* with density.

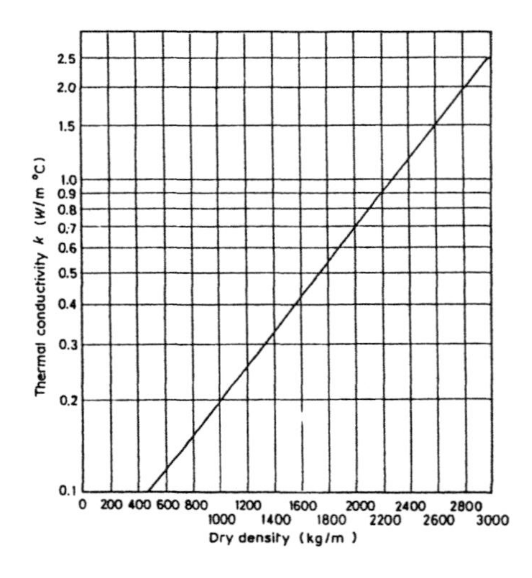

The graph shows the thermal conductivity for typical dry brickwork, masonry, concrete and plaster, but thermal conductivity increases with moisture content and a correction must be applied by multiplying by the appropriate factor, as follows:

Exposure conditions and materials	*Moisture content* (% by volume)	*Correction factor*
Protected brickwork, granite, sandstone	1	1.3
Protected concrete, limestone, plaster	3	1.6
Exposed brickwork, concrete, masonry	5	1.75
Severe exposure to driving rain, porous masonry	10	2.1

The thermal resistance of a layer of material depends upon thickness divided by thermal conductivity *k*; Table A6.3 lists typical *k* values for building materials. Continuing the example of single glazing, the glass is perhaps 3 mm thick (0.003 m) and, as the *k* value from Table A6.3 is 1.05, the resistance of the glass itself is about 0.003. For brickwork, concrete, masonry, mortar, plaster etc. the thermal conductivity depends upon bulk density and moisture content, as shown in Figure A6.1.

Table A6.3. Typical thermal conductivity *k* values for building materials. For brickwork, concrete, masonry, mortar, plaster etc. the thermal conductivity depends upon bulk density and moisture content, as shown in Figure A6.1.

	Material	*Moisture content (% dry weight)*	*Bulk density (kg/m³)*	*Thermal conductivity k(W/m °C)*
Asbestos	Cement sheet	5	1,600	0.4
	Insulation board	5	750	0.12
Asphalt	Roofing		1,920	0.58
Brick	(see Figure A6.1)			
	Common, dry	0	1,760	0.81
	Conditioned at 17.8°C and 65% RH	6	1,870	1.21
	Wet	16	2,034	1.67
Building board	Asbestos insulator	2	720–900	0.11–0.21
	Fibreboard		280–420	0.05-0.08
	Hardboard, medium		560	0.08
	Plasterboard, gypsum		1,120	0.16
	Wood/chip board		350–1,360	0.07–0.21
	Wood/wool slab	5	400–800	0.08–0.13
Carpeting	Wilton type			0.058
	Wool felt underlay		160	0.045
	Cellular rubber underlay		270	0.065
			400	0.10
Concrete	(see Figure A6.1)			
	Gravel 1:2:4		2,240–2,480	1.4
	No fines, gravel 1:10		1,840	0.94
	Clinker aggregate	4	1,680	0.40
	Expanded clay aggregate	5	800–1,280	0.29-0.48
	Pumice aggregate	4.6	770	0.19
	Vermiculite aggregate		400–880	0.11-0.26
	Cellular		320–1,600	0.08-0.65
Cork	Granulated, raw	7	115	0.046
	Slab, raw	7	160	0.050
	Slab, baked	3–5	130	0.040
Felt	Under-carpet felt		120	0.045
	Asbestos felt		144	0.078
	Roofing felt		960–1,120	0.19–0.20
Glass	Sheet, window		2,500	1.05
	Wool, lightweight mat		25	0.04
Metals	Aluminium alloy, typical		2,800	160
	Brass		8,400	130
	Copper 99.9%		8.900	200
	Iron, cast		7,000	40
	Lead		11,340	35
	Steel, mild		7,850	47
	Steel, high alloy		8,000	15
	Zinc 99.99%		7,130	113

	Material	Moisture content (% dry weight)	Bulk density (kg/m³)	Thermal conductivity k(W/m °C)
Plaster	(see Figure A6.1)			
	Gypsum		1,120–1,280	0.38–0.46
	Perlite aggregate		400–610	0.079-0.19
	Vermiculite aggregate		480–960	0.14–0.30
	Sand cement		1,570	0.53
Plastics, cellular	Polystyrene, expanded board		15	0.037
	Polyurethane foam		30	0.026
	Polyvinyl chloride, rigid foam		25–80	0.035–0.041
	Urea formaldehyde foam		8–30	0.032–0.038
Plastics, solid sheet	Acrylic resin		1,440	0.20
	Nylon		1,100	0.30
	Polycarbonate		1,150	0.23
	Polyethylene, low density		920	0.35
	Polyethylene, high density		960	050
	Polypropylene		915	0.24
	Polystyrene		10.50	0.17
	PTFE		2,200	0.24
	PVC rigid		1,350	0.16
Roofing felt			960–1,120	0.19–0.20
Stone	(See Figure A6.1)			
	Granite		2,650	2.9
	Limestone		2,180	1.5
	Marble		2,700	2.5
	Sandstone		2,000	1.3
	Slate		2,700	1.9
Tiles	Burnt clay		1,900	0.85
	Concrete		2,100	1.10
	Cork		530	0.085
	PVC asbestos		2,000	0.85
	Rubber		1,600–1,800	0.30–0.50
Vermiculite	Loose granules		100	0.065
	Plastering		480–960	0.144–0.303
Wood	Across grain:			
	Beech	15	700	0.165
	Deal	12	610	0.125
	Mahogany	10	700	0.155
	Oak	14	770	0.160
	Pitch pine	15	660	0.140
	Spruce	12	420	0.105
	Teak	10	700	0.170
	Along grain:			
	Deal	12	610	0.215
	Oak	14	770	0.290

Single glazing thermal calculation

The calculations for single glazing can be conveniently based on the formula $R = R_{So} + R_G + R_{Si}$

Where R_G is the glass resistance. For example:

External surface resistance R_{So} (high, normal)	0.055
Glass resistance R_G. (3 mm, $k = 1.05$) 0.003/1.05 =	0.003
Internal surface resistance R_{Si} (high)	0.123
Total thermal resistance R	0.181
Thermal transmittance $U = {}^1/_R$	5.53

It is apparent from this calculation that the surface resistances are far more significant than the insulation properties of the glass, indicating that it is the envelope around the accommodation that provides the most important isolation from the surroundings. The reciprocal of the R value is the U value which is 5.53 in this case for single glazing alone. In fact, windows are multi-component elements consisting of frames as well as glazing.

Complete windows

The U value for the frame, which usually occupies 20 to 30% of the window area, depends upon its construction; the U value for a normal uninsulated aluminium or painted steel frame is about 5.5, the same as for the glass. Examples of U values for complete frames are given in Table A6.4. An interesting point is the relatively high U values for proprietary metal frames with factory-sealed double glazing with a relatively small air gap, but U values are greatly reduced, indicating improved thermal insulation, with wider air gaps and more efficient wood, uPVC or thermal break aluminium frames. The importance of a sufficient air gap is illustrated in Figure A6.2.

Figure A6.2 Thermal transmittance or *U* values for double glazing

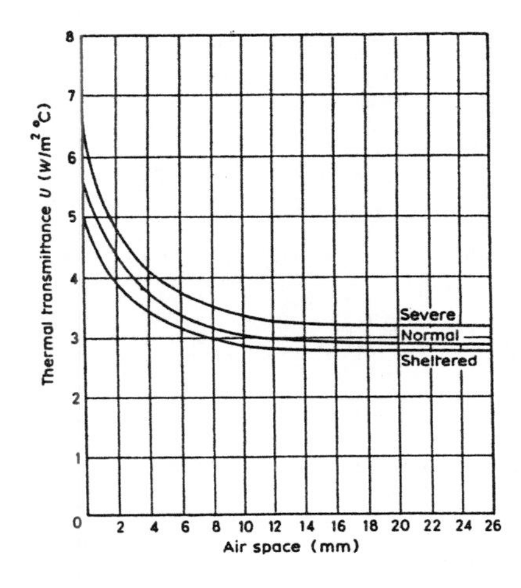

Table A6.4 Thermal transmittance or *U* values for windows

Construction	*Frame proportion (%)*	*U Value (W/m² °C)* Sheltered	Normal	Exposed
Metal, single glazed	20	5.0	5.6	6.7
Wood or UPVC, single glazed	30	3.8	4.3	4.9
Metal, double glazed (6 mm)	20	3.6	3.8	4.4
Metal thermal break, double glazed (20 mm)	20	3.0	3.2	3.5
Wood or UPVC, double glazed (20 mm)	30	2.3	2.5	2.7

Cavity resistance

In more complex forms of construction exactly the same principles are involved in which the total thermal resistance or R value is determined by adding the resistances for the surfaces and each successive layer of the structural element. In some cases cavities occur, their thermal resistances depending upon the emissivities of the surfaces on either side of the cavity, as well as the ventilation, as summarised in Table A6.5.

Cavity wall, thermal calculation

The R value for a composite building element can be calculated using the general formula

$$R = R_{So} + R_{Si} + R_C + R_1 + R_2 + \ldots$$

where R_C is the cavity resistance and R_1, R_2 etc. are the resistances of the structural elements. For example, the R value and its reciprocal the U value for a typical external cavity wall can be calculated as follows:

External surface resistance R_{So} (wall, high, normal)	0.055
Internal surface resistance R_{Si} (wall, high)	0.123
Outer brickwork R_1 (0.1 m, 1700 kg/m³, 5% MC) 0.1/(0.47x1.75)	0.122
Cavity resistance R_C (20 mm +, high, horizontal)	0.180
Inner blockwork R_2 (0.1 m, 600 kg/m³, 3% MC) 0.1/(0.13x1.6)	0.481
Plaster R_3 (0.01 m, 600 kg/m³, 3% MC) 0.01/(0.13x1.6)	0.048
Total thermal resistance R	1.009
Thermal transmittance $U = 1/R$	0.991

It can be seen from these calculations that the lightweight inner blockwork, with a density of 600 kg/m³, compared with about 1700 kg/m³ for normal brickwork or dense concrete blocks, provides the main contribution to thermal resistance and is thus most significant in minimising the thermal transmittance or *U* value. The use of a lightweight block inner skin and lightweight plaster was sufficient for many years to keep the *U* value below 1.0 and thus meet the requirements of the Building Regulations, but the 1985 Regulations required the *U* value to be less than 0.6. This can be achieved only by the use of cavity fill insulation, built in as sheets or installed by injection in standing walls.

Table A6.5 Cavity resistances*

Building element	*Surface emissivity*†	*Heat flow*	*Thermal resistance*‡ *(m^2 °C/W)*
Unventilated air spaces			
5 mm air space	High		0.11
	Low		0.18
20 mm minimum air space	High	Horizontal/upwards	0.18
		Downwards	0.21
	Low	Horizontal/upwards	0.35
		Downwards	1.06
Ventilated air spaces, minimum 20 mm			
Air space between asbestos-cement or dark-painted metal cladding with unsealed joints over high-emissivity surface such as brickwork			0.16
As above but over low-emissivity surface such as foil-backed plasterboard			0.30
Loft space between flat ceiling and unsealed asbestos-cement or dark metal cladding pitched roof			0.14
As above but aluminium cladding instead of black metal or with low emissivity upper surface on ceiling such as foil-backed plasterboard			0.25
Loft space between flat ceiling and unsealed tiled pitched roof			0.11
Air space between tiles and sarking felt on pitched roof			0.12
Air space behind tiles on tile-hung wall			0.12

* Values from Building Research Digest 108 (*IHVE Guide*, Book A).
† Emissivity is high for all normal building materials, but unpainted metal and other reflective surfaces have low emissivity.
‡ The thermal resistance values include the surface resistances on either side of the cavity.

Cavity fill

Cavity fill insulation such as granulated expanded polystyrene and pelletised glass fibre will increase the cavity resistance R_C from 0.180 to about 1.30, increasing total resistance *R* to 2.129, giving a thermal transmittance or *U* value of 0.47, well within the current limit of 0.6. Mineral wool is less effective but polyurethane foam is more effective. When the insulation is incorporated in new buildings as thick sheets or bats, the sheets are not usually as thick as the cavity and are secured to the inner leaf, leaving a narrow cavity between the insulation and the outer leaf, slightly reducing the thermal insulation but keeping it within the necessary limits, and also largely avoiding the moisture penetration problems that can arise when the cavity is fully filled with insulation.

Tile-clad timber-frame wall, thermal calculation

Thermal resistance R and transmittance U values can be calculated for any other structural elements in the same way. For example, the following calculation is for a timber-frame wall, finished internally with plasterboard on a polythene vapour barrier over glass-fibre quilt insulation, with a cavity, then sheathing plywood covered externally with breather paper, and finally counter-battens, battens and vertical tiling. In thermal calculations the polythene vapour barrier and the breather paper are ignored as their effect is insignificant, and the calculation is made through the panels which comprise most of the wall area as the wood frames or studs have good insulating properties, comparable with the cavity and glass-fibre quilt. The vertical tiling cavity resistance is taken from Table A6.5.

External surface resistance R_{So} (wall, high, normal)	0.055
Internal surface resistance R_{Si} (wall, high)	0.123
Vertical tiles R_1 (0.015 m, $k = 0.85$) $0.015/0.85 =$	0.018
Tile cavity R_{C1}	0.120
Plywood R_2 (0.01 m, $k = 0.125$) $0.012/0.125 =$	0.096
Frame cavity R_{C2}	0.180
Glass-fibre insulation quilt R_3 (0.050 m, $k = 0.033$) $0.050/0.033 =$	1.515
Plasterboard R_4 (0.010 m, $k = 0.16$) $0.010/0.16 =$	0.063
Total thermal resistance R	2.170
Thermal transmittance $U = 1/R$	0.461

Temperature gradient

The exceptionally high thermal resistance of the glass-fibre insulation quilt is very apparent from these calculations, but is particularly striking if the thermal resistances of the various layers of the wall are plotted graphically, as in Figure A6.3. The temperature gradient is in proportion to the thermal resistance so that the temperature loss across the building element can be plotted on this graph as a straight line, thus indicating the temperatures at the interfaces between various layers of the structural elements. If these temperatures are then transferred to the interfaces on a diagram of a section through the wall and the points joined up to show the temperatures through the wall, the steep slopes indicate the best insulation which is, in this case, across the glass-fibre insulation quilt.

Figure A6.3 (a) Section through brickwork-clad timber frame wall showing temperature gradients through each component. (b) Temperature plotted against thermal resistance to determine temperatures at component interfaces

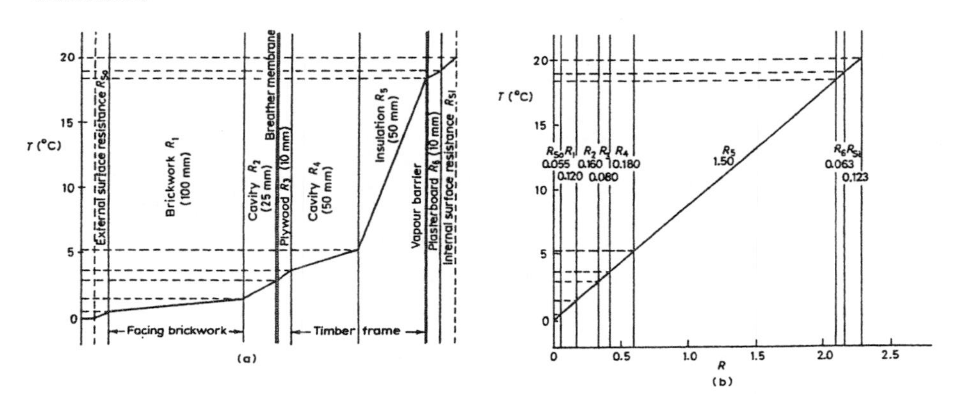

Interstitial condensation

Diagrams of temperature drop across a building element are thus useful as an indication of the thermal significance of the various components, but they are also useful in assessing the dangers of interstitial condensation, as discussed in greater detail in Chapter 3. In this particular example, illustrated in Figure A6.3, in which the internal temperature is 20°C and the external temperature is 0°C, it is clear that if the internal air has a humidity equivalent to the dew point of 11°C, interstitial condensation will occur where the structure is at lower temperatures, that is from within the insulation to the exterior of the element. Precautions are therefore necessary to prevent this condensation from accumulating and causing fungal decay in the sheathing plywood and stud frames.

Vapour barrier

The usual precautions are to provide an impermeable vapour barrier on the warm side of the main insulation to prevent diffusion of warm humid air from the accommodation into the building element, coupled with a permeable breather paper on the outside of the sheathing plywood which will allow any condensation to disperse to the exterior. If the internal vapour barrier is omitted there is still a danger of condensation damage as the rate of condensation formation will be too great in comparison with the rate at which it can be dispersed by diffusion through the sheathing plywood and breather paper, and in such circumstances the danger can only be avoided by introducing some limited ventilation into the cavity between the glass-fibre insulation quilt and the sheathing plywood. The importance of such studies on building elements can be appreciated from the fact that the vapour barriers are often damaged during construction, particularly by electricians. It is therefore safer, if possible, to avoid dependence upon a vapour barrier and to rely instead upon ventilation on the cold side of the insulation.

Pitched roof

This is precisely the situation that exists in a conventional roof structure where the roof space is ventilated and condensation problems are never encountered. However, in order to meet modern requirements, sarking felt must be introduced beneath the slating or tiling battens and the thermal insulation properties must be improved. The insulation of the roof space must be maintained in order to avoid interstitial condensation dangers, usually by providing ventilators in eaves soffits, but the function of this ventilation in terms of interstitial condensation control will only be properly maintained if the main insulation is provided on the ceiling so that the roof space is cold and freely ventilated.

A normal pitched tile roof with sarking felt, and where the ceiling beneath is formed with foil-backed plasterboard, has a typical R value of about 0.66, or a U value of about 1.5. The addition of 50 mm of glass-fibre quilt between the ceiling joists will increase the R value to about 2.0, reducing the U value to about 0.5, which was sufficient for many years to meet the requirements in the Building Regulations. However, the 1985 Regulations reduced the U value limit to 0.35, equivalent to an R value of about 2.86. Assuming that glass-fibre quilt has a k value of about 0.033, these requirements can be met by using glass-fibre quilt more than about 72 mm thick; in fact, it is normal now to use a 100 mm quilt. The critical thicknesses of other insulations, compared with 72 mm of glass-fibre quilt, are about 82 mm for granulated expanded polystyrene, 88 mm for pelletised glass-fibre, 99 mm for mineral wool, and 142 mm for Vermiculite; obviously vermiculite and mineral wool are not really suitable to meet modern requirements in view of the excessive depths of materials that are required.

Flat roof

The same principles apply to flat roofs, whether they are constructed from timber, steel or concrete. Their properties are calculated in precisely the same way as for the earlier cavity wall and timber-frame wall examples, although with suitable corrections to the surface resistances, and diagrams similar to those in Figure A6.3 can be prepared to show the temperature gradients and the condensation dangers associated with dew point positions under appropriate conditions. However, the importance of designing to avoid interstitial condensation is not widely understood. In timber roofs it is possible to construct a normal 'cold' roof, that is a roof in which the main insulation is on the ceiling and there is a cold ventilated space above the insulation. The alternative is a 'warm' roof in which the main insulation is on the roof deck, immediately beneath the external roof covering. In a warm roof interstitial condensation within the insulation is inevitable, unless warm humid air frm the accommodation is prevented from diffusing into the insulation by provision of a reliable vapour barrier on the warm side of the insulation. It is virtually impossible to provide a sufficiently reliable vapour barrier at ceiling level as there is always a danger that air can diffuse round the edges of the barrier and there is also a danger that electricians will damage the barrier when installing light fittings. A far more reliable arrangement is to provide the vapour barrier on the roof deck, but it would be apparent from plotting temperature curves as in Figure A6.3 that the ceiling, warm cavity and deck itself make a significant contribution to the insulation, and there is a danger that interstitial condensation will occur in the deck if insufficient insulation is provided on top of the vapour barrier. In 'warm' roof construction of this type problems generally arise because the insulation beneath the roof covering is designed to conform with Building Regulations thermal requirements and the risk of interstitial condensation has not been assessed.

Floors

Calculations on the thermal properties of floors can be as complex as for roofs and it is best to use information for typical structures, as in Table A6.6, adjusted to incorporate differences in the structure. The technique is always to work in thermal resistance *R* values, so that when a transmittance or *U* value is known, the reciprocal must be calculated in order to provide the *R* value before commencing the calculation. The change in *R* value due to a difference in material or thickness is then simply incorporated to calculate a revised *R* value of the element, the reciprocal being the *U* value. For example, one important feature of a floor is the covering; carpet in particular is an excellent insulation. Carpet has a *k* value of about 0.05 so that a thickness of 10 mm will have a thermal resistance *R* of about 0.02. A 3 m wide suspended floor above the ground will have a *U* value of about 1.05 according to Table A6.6, that is an *R* value of about 0.95. If carpet is laid on top this *R* value increases to about 1.15, reducing the *U* value to about 0.87. Carpet and other floor coverings also have other thermal advantages, particularly reducing draughts by sealing gaps between boards and under skirtings, a point discussed more fully in Chapter 5.

Carpet is not the only furnishing of importance in relation to the thermal properties of a building. For example, a normal wood-framed single-glazed window will have a *U* value of about 4.3, or an *R* value of 0.23. Closed curtains will introduce a cavity with an additional *R* value of about 0.12, increasing the total *R* value to 0.35 and thus reducing the U value to about 2.86, similar to the *U* value for an efficient double-glazed window. In the past thick solid walls, heavy drapes and sawdust or shavings between joists were often used to improve thermal insulation, not necessarily achieving efficiencies comparable with modern dwellings but certainly similar to those in modern commercial buildings!

Table A6.6 Thermal transmittance or *U* values for typical floors

Solid floors in contact with earth

Width (*m*)	*U value* (W/m^2 °C)
3	1.47
7.5	0.76
15	0.45
30	0.26
60	0.15

This table assumes that floors are square, with four exposed edges. For oblong floors the *U* values can be estimated from the average edge length, but for widths of less than 10 m the narrow width becomes of increasing importance and should be considered alone at widths of 3 m or less. If one or more sides are insulated or within the building the *U* value is reduced by about 22% per insulated side. An adjustment must be made to appropriate sides before averaging for oblong floors, but the reduction is less for widths of less than 10 m and becomes only 14% per side at widths of 3 m.

Suspended wood floors above ground

Width (*m*)	*U value* (W/m^2 °C)
3	1.05
7.5	0.68
15	0.45
30	0.28
60	0.16

This table assumes that floors are square and constructed with tongued and grooved boarding or similar precautions to prevent draughts from the sub-floor ventilation. For oblong floors the *U* values can be estimated from the average edge length, but for widths of less than 15 m the narrow width becomes of increasing importance and should be considered alone at widths of 7.5 m or less. Carpet, parquet and cork tiles can reduce the *U* value slightly for all suspended floors, but this reduction becomes significant at narrow widths, the reduction being 2% at 15 m, 4% at 7.5 m and 6% at 3 m.

Intermediate floors

Construction	*U value* (W/m^2 °C) *Heat flow downwards*	*Heat flow upwards*
Wood		
20 mm wood floor on 100 mm x 50 mm joists, 10 mm plasterboard ceiling	1.5	1.7
Allowing for 10% bridging by joists	1.4	1.6
Concrete		
150 mm concrete with 50 mm screed	2.2	2.7
With 20 mm wood flooring	1.7	2.0

Figure A6.4 Thermal resistance R for typical window-frame components

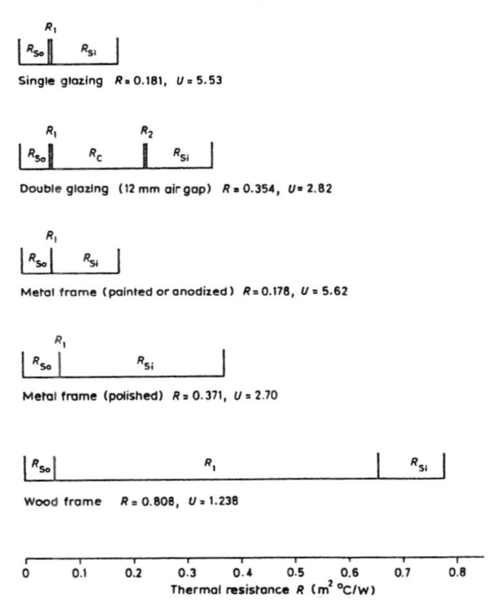

Site measurements

Thermal resistance R and transmittance U values can be calculated as explained previously, but calculations are difficult for complex structures and it is always more reliable to determine U values experimentally in the laboratory under standard conditions. Values can also be determined with reasonable accuracy on site, provided that there is a reasonable temperature difference across the building element that is being tested, and this technique is particularly useful when it is necessary to check the U value of a structure as part of an investigation. Heat flow through an element is proportional to the temperature difference and inversely proportional to the resistance, the thermal equivalent of the well-known Ohm's law for electric current. When an electric current passes through resistances in series, the potential between any two points is proportional to the resistance between them, or in thermal terms the temperature is proportional to the resistance. This is, of course, the origin of the graph in Figure A6.3 in which temperature is plotted against resistance, enabling the temperature to be determined at any interface in a building element. Conversely the same types of graph or calculation can be used in conjunction with temperature measurements at interfaces to calculate resistance values, including the total resistance or R value and thus the U value of the complete element.

For example, Figure A6.4 shows temperature and thermal resistance plots for simple wood and aluminium frames. T_i is the temperature of the interior air whilst t it is the temperature of the inner surface of the element. Similarly T_o is the temperature of the outside air and t_o is the temperature of the outside surface of the element. It is apparent from Figure A6.4 that the temperature difference $T_i - T_o$ is proportional to the total thermal resistance R of the element, that is the reciprocal of the U value. However, it is also apparent that the temperature differences $T_i - t_i$ and $t_o - T_o$ are proportional to the internal and external resistances respectively, or

$$(T_i - T_o)/R = (T_i - t_i)/R_{Si} = (t_o - T_o)/R_{So}$$

Therefore

$$R = R_{Si}\,(T_i - T_o)/(T_i - T_i)$$
$$= R_{So}\,(T_i - T_o)/(t_o - T_o)$$

or

$$U = (T_i - t_i)/(R_{Si}\,(T_i - T_o)$$
$$= (t_o - T_o)/(R_{So}\,(T_i - T_o)$$

These formulae enable the R or U values to be calculated from simple temperature measurements related to the internal or external surface resistance. The surface resistances quoted in Table A6.2 are the standard values used for design calculations. In particular, the external resistances are average values for sheltered, normal or exposed conditions which will only apply at the time of test if the wind strength happens to be the average value for such conditions. It is therefore best to avoid the use of external surface resistance and to use only the following formulae based on internal surface resistance, assessed in accordance with Table A6.7:

$$R = R_{Si}\,(T_i - T_o)/(T_i - t_i)$$

$$U = (T_i - t_i)/(R_{Si}\,(T_i - T_o)$$

These formulae involve external air temperature measurements T_o but it is assumed that the external surface temperature of the element t_o depends upon the external air temperature, which is not always the case as the external surface may be heated by the sun or cooled by evaporation of rainwater. It is therefore better to measure the actual external surface temperature t_o, but this then involves a slightly more complex calculation for R, with U determined most simply as the reciprocal of R:

$$R = R_{So} + R_{Si}\,(T_i - t_o)/(T_i - t_i)$$

$$U = 1/R$$

The external surface resistance value R_{So} in this formula is the appropriate design value from Table A6.2. This formula can be used for checking the R and U values of any building element and is particularly useful for checking thermal insulation values in individual components, such as double glazing and frame components in windows, and local features in structural elements which may form thermal bridges, such as concrete lintels over openings, and posts and beams in major buildings. In order to minimise errors, the same temperature device should be used for both surface and air temperatures. For surface measurements the probe should have a large flat collecting surface, a particularly important point when making measurements on good insulators as there is otherwise a danger that the measured temperature will be that of the probe rather than the surface! In order to minimise this effect the probe disc must be moved gently over the surface until a constant temperature reading is obtained, although care is necessary with some probes to avoid temperature increases through holding the probe.

Table A6.7 Internal surface resistances R_{Si} for use in R and U value measurements

Building element	*Surface emissivity*	*Surface resistance (m^2°C/W)*
Walls windows (horizontal flow):		
bare masonry, dark paint	High	0.12
white paint		0.15
anodised aluminium		0.20
polished aluminium	Low	0.30
Roofs, ceilings, floors (upward flow):		
normal surfaces	High	0.11
reflective surfaces*	Low	0.22
Roofs, ceilings, floors (downward flow):		
normal surfaces	High	0.15
reflective surfaces*	Low	0.56

* Reflective surfaces include aluminium paint on roof surfaces and foil backing on plasterboard, even though the foil is concealed by the plaster ceiling.

Appendix 7 *Glossary of useful terms*

Abutment
The junction between a roof and a wall which rises above it; the walls which carry the thrust of an arch.

Aggregate
Sand, gravel, crushed stone etc. used in concrete and plaster mixes.

Air brick
A perforated block incorporated in a wall to provide ventilation to a floor space, small room or cavity.

Air shaft/Light well
An open well or area within a building to provide some light or ventilation to rooms opening onto it.

Aisle
Part of a church parallel to the nave and divided from it by an arcade; a passage between rows of pews.

Alcove
A recess in a room.

Almery/Aumbry
A cupboard within the thickness of a wall. In some churches there is a very tall one near the west end for housing the processional cross.

Altar
A table or raised structure for offering sacrifices or dedicated to certain religious ceremonies in churches.

Ambulatory
A covered promenade, arcade or cloister; an aisle round an apse.

Anchor
A metal tie built into a wall to secure an overhanging cornice etc.

Angel beam
A hammer beam with an angel carved on its end.

Angle block/Glue block
A small piece of wood usually of triangular section, glued into an angle to stiffen it, as under the step of a stair.

Angle post/Corner post
The corner post of a timber-framed building.

Angle tie/Dragon tie
A horizontal timber which acts as a tie to the wall plates across the corner of a building and also carries the inner end of the dragon beam.

Apron
A panel on the internal face of a wall under the window board.

Apron flashing
The flashing along the face of the lowest part of a chimney stack at its junction with the roof.

Apron lining/Breast plate (Scots)
A board used to finish the edge of a floor round the stair well.

Apse
A semicircular or polygonal wing with an arched or domed roof as in some churches.

Arcade
A line of arches supported on columns.

Arch
A curved structure bearing the weight of a wall over an opening and transferring it to the abutments.

Arch brace
The arch-shaped brace which supports the hammer beam in a hammer-beam roof.

Architrave
A strip or moulding used to cover the joint between a frame and a wall, as round a door or window frame; the lowest of the three sections of an entablature in classical architecture.

Architrave block/Plinth block
The block at the foot of an architrave.

Area
A space separating the external walls of a basement from the surrounding higher ground.

Arris
The sharp angle along the edge of a piece of wood or other material.

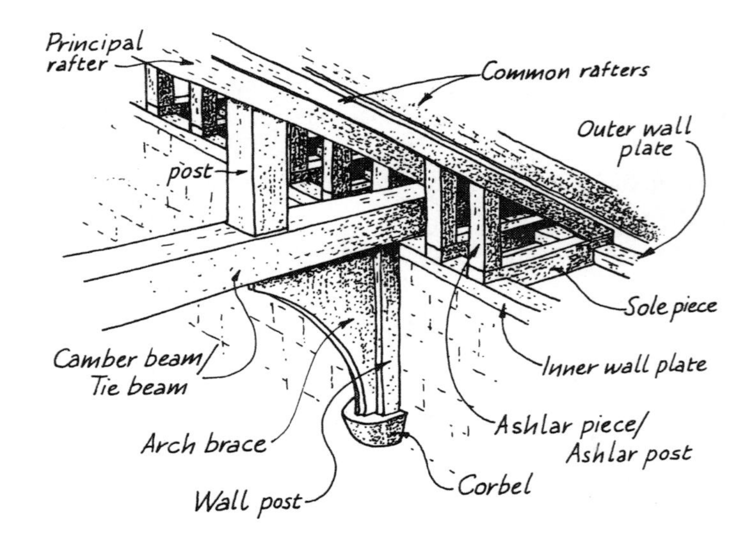

Ashlar
Square hewn stones and masonry built from them.

Ashlar areas
The spaces between the rafters and the wall tops.

Ashlar piece/Ashlar post
The post set between the sole piece and the common rafter or between the tie beam and principal rafter in the ashlar area of a roof.

Astragal
Scots for a **Glazing bar**.

Attic
The uppermost storey in a building formed immediately or partially under the roof and usually with sloping ceilings.

Aumbry
See **Almery**.

Back-fill
Earth, rubble etc. used for filling excavations, usually after foundations have been formed in them.

Back gutter
The lining between the back of a chimney and the roof slope, usually made of lead.

Backing coat/Render coat
The first layer applied to a wall or ceiling in the process of plastering.

Back lining
A thin piece of wood closing the jamb of a cased frame and giving protection to the balancing weights of a sash window.

Balcony
An external platform with access from a window and with a balustrade around it.

Baluster/Banister (Scots)
An upright or shaped pillar supporting the horizontal rail of a balustrade.

Balustrade
A row of balusters supporting a rail or coping forming the surround to a balcony or a parapet and providing a guard to it; the infilling between the handrail and outer string or floor of a staircase.

Baptistery
Part of a church used for baptism.

Barge board
A board fixed to the projecting end of a roof over a gable, usually in pairs, one to each slope.

Barrel roof/Wag(g)on roof
A ceiling of semicircular section often under a medieval trussed-rafter church roof.

Barrel vault
A vault of semicircular section carried on parallel walls throughout its length.

Bartisan
A small turret projecting from the angle of a wall and sometimes from the top of a tower.

Base
The foot of a column under the shaft; the lowest visible course of a masonry wall.

Basement
A storey partly or wholly below ground level, usually providing living accommodation; in larger buildings it may serve as the pedestal to the main part of the building, *q.v.* **Cellar**.

Base plate
Scots for a **Skirting board**.

Batten
A piece of softwood usually only a few inches in breadth.

Batter
Usually applied to the slope of a wall built intentionally out of vertical.

Battlement
A parapet wall with rectangular notches or embrasures spaced regularly along it, originally for defensive purposes.

Bay
The space between uniform divisions of a building, as between the trusses of a roof or the columns of an arcade.

Bay window
A window formed in a projection forward of the main face of the building and which rises from the ground or from a basement, *q.v.* **Oriel window**.

Bead
A half or three-quarter round moulding used to mask a joint.

Beam filling/Wind filling
Brick and mortar filling between the joists and rafters in a roof near the eaves, or between floor joists at their bearing ends.

Bearer
A horizontal timber or joist which supports other timbers.

Bed mould
A moulding under a window board, shelf or cornice.

Belfry
The part of a church tower in which the bells are hung.

Bellcast eaves
Scots for eaves with sprockets or cocking pieces.

Bell cote/Bell gable/Bell turret
The arch or turret rising from the apex of the gable of small churches and usually housing a single bell.

Belvedere
A raised turret or room from which the surrounding scenery is viewed.

Bevel
The angled surface remaining when the square arris is cut off the edge of a piece of wood or stone, etc. to an angle *other than* 45°.

Binder
A wooden or steel beam spanning between opposite walls and supporting the joists.

Birdsmouth
A V cut as in the end of a rafter to house it over the edge of a wall plate.

Blocking course
One or more courses of brickwork or masonry over a cornice to stabilise the cornice and form a low parapet.

Bonding timbers
Long timbers incorporated in brick walls at regular intervals to strengthen the brickwork.

Bonnet
A roof over a bay window.

Bonnet tile
A round-topped tile for covering the angle of a hip, each tile bedded in mortar over the other.

Boot
A projection from a concrete lintel or floor slab to carry face brickwork.

Borrowed light
A window in an internal partition.

Boss
A projecting carving in wood or stone usually at the intersection of ribs or groynes in a ceiling.

Bottle nosing
Scots for a half-round nosing.

Bottom rail
The lowest horizontal member of the frame of a door, sash or casement.

Bow window
Similar to a bay window but curved in plan.

Box gutter
A rectangular sectioned trough usually lined with lead or asphalt to form a gutter in valleys or behind parapets.

Boxing
The hollow box structure at the side of some windows into which internal wooden shutters fold and are housed.

Braces/Struts
Components in a truss, usually set at an angle into the foot of a king or queen post and supporting the principal rafters above.

Bracket
A projecting support. In brickwork or masonry it would be called a **Corbel**.

Breast/Chimney breast
A section of wall projecting into a room and containing a fireplace and flue(s): Scots for the riser of a step.

Breastplate
Scots for an **Apron lining.**

Bressummer/Breastsummer
A wood or steel or flitched beam which carries a wall over a large opening, as over a shopfront or a bay window.

Bridle
Scots for a **Trimmer joist**.

Broach
A church spire which rises from a tower without a parapet.

Buttress
A brick or masonry pier, usually battered, built against a wall to give support to that wall.

Camber beam
A timber tie beam with a slightly arched shape.

Came
An H-sectioned strip of lead or copper in which the separate pieces of glass are held in lattice or leaded-light windows.

Canopy
A roof-like projection over a door or window; a hood.

Cantilever
A term applied to a beam or structure which projects without struts, brackets or other external support.

Capillary groove
A groove formed between two surfaces in close contact to prevent the movement of water between them by capillary action.

Capital
The head or crowning feature of a column.

Carcase
The load-bearing part of a structure.

Carriage piece
A timber fixed longitudinally to the underside of a stair to give support to the structure.

Cased frame/Case (Scots)
The built-up hollow-sectioned fixed outer frame of a sash window.

Casement
The hinged opening part of a casement window.

Casement door/French door
A hinged door or pair of doors without a central mullion and usually fully glazed.

Castellated
Having battlements.

Catslide roof/Hall-span roof/Lean-to roof
A roof with a single slope, *q.v.* **Penthouse roof**.

Caulking
The packing in a spigot and socket joint in a pipe, to make it watertight.

Cavity wall/Hollow wall
A wall built in two leaves with a continuous space between them and across which they are jointly stabilised with metal wall ties.

Ceiling joist
A joist which carries a ceiling but which does not support a floor above it.

Cellar
Part of a building wholly or partially below ground level and used for storage, not living accommodation, *q.v.* **Basement**.

Cement fillet
A cement strip sometimes used to weather the joint between the tiles or slates of a roof and an abutting wall or chimney stack.

Chair rail/Dado rail
A wooden moulding fixed horizontally to a wall at a height that will prevent damage to the plaster from chair backs.

Chamfer
The shape of the edge of a piece of wood or other material when the square arris is cut off to an angle of 45°. If the angle is other than 45°, it is known as a **Bevel,** *q.v.*

Chancel
The part of a church at the east end reserved for the clergy and choir and usually separated from the nave by a ***Chancel rail.***

Chancel arch
The arch at the west end of the chancel, separating it from the nave in a church.

Chapter house
A building or chamber used for the meetings of the canons and clergy of a cathedral or collegiate church.

Charnel house
A building in or near a church for the keeping of bones from graves which have been disturbed.

Chase
A groove cut or formed in a wall or floor to receive pipes, cables, etc.

Check
Scots for **Rabbet or Rebate.**

Cheek
The side of a dormer window.

Chimney breast
See **Breast**.

Church
The principal parts of a church are shown in the diagram.

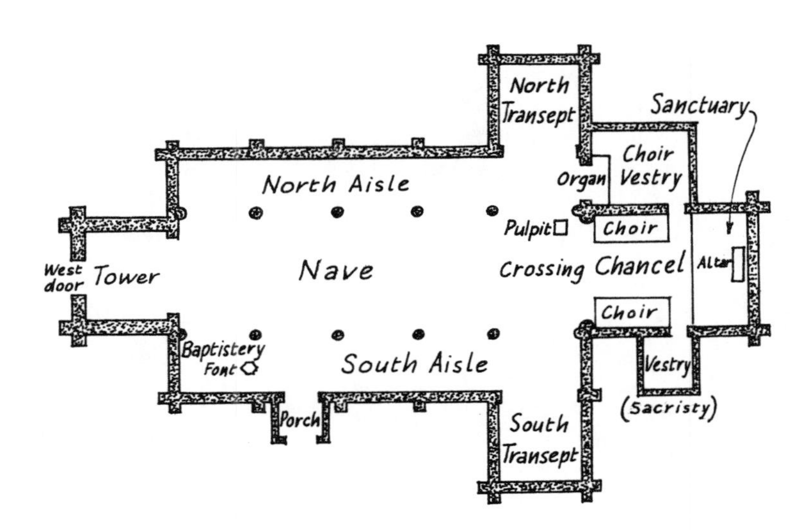

Chimney stack
That part of the masonry or brickwork containing several vertical flues which stands above the level of a roof.

Choir
The part of a church where the choristers are gathered, usually part of the chancel.

Chute
A slope for conveying things to a lower level as with a **Coal chute** to a coal cellar.

Cill/Sill
A slab of stone or wood at the base of a window or door opening, giving protection to the wall beneath.

Cistern
A rectangular open-topped vessel for the storage of water.

Cladding
The sheathing of a wall or roof. It is not load-bearing.

Clapboard/Weatherboarding
Feather-edged boarding fixed horizontally to walls with the lower part of each board covering the upper part of the board below.

Classical architecture
Some of the principal features are shown in the diagram.

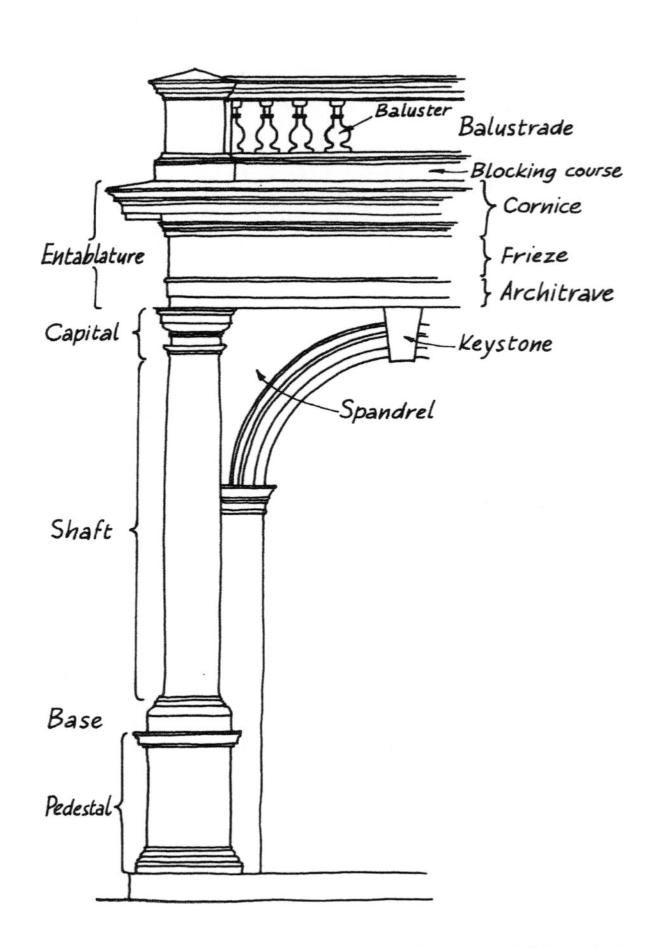

Cleaning eye/Rodding eye
A hole in a pipe with a removable cover giving access for cleaning or clearing with rods.

Cleat
A small piece of wood used to reinforce another.

Cleft timber
Timber which is reduced to size by splitting longitudinally along the grain, e.g. cleft battens for tiling and cleft chestnut fencing.

Clerestorey/Clearstorey
The upper part of a wall containing windows which are above the level of an adjoining roof as in the nave of a church where the windows are above the roof of the side aisle.

Cloister
A covered walk often round a quadrangle, with windows or an open colonnade on the inner side.

Close-boarding
Boarded covering of a roof or wall where the boards touch each other on their long edges.

Close-couple
The term applied to a simple roof of common rafters, connected at wall plate level with a tie-beam.

Clunch
Chalk used for building in medieval times.

Cob
Compacted mud used for walling; sometimes clay, marl or chalk mixed with gravel and straw to bind it.

Cocking piece/Sprocket piece
A short tapered length of wood fixed to the lower end of a common rafter to lift the angle of the roof at the eaves. In Scotland this makes **Bellcast eaves.**

Coin/Quoin
The outer corner of a wall; sometimes the large stone set in such a corner.

Collar beam/Collar tie
A horizontal tie beam connecting the common rafters of a roof, usually about mid-way up their length.

Colonnade
A series of columns.

Column
An upright shaft supporting the entablature of an arch or carrying an axial load in compression.

Common rafter
One of a series of sloping timbers spanning between the ridge and wall plate of a roof and carrying the tile battens and roof fabric.

Compass rafter
A rafter which is curved in elevation, so giving a curved roof or **Compass roof**.

Console
An ornamental bracket of double curvature used to support a canopy, balcony, etc.

Coping
The capping, usually of stone or concrete slabs, which gives weather protection to the top of a wall.

Corbie step gable (Scots)
A stepped stone coping to an upstanding gable wall.

Corbel
A projection from a wall, often in brick, iron, wood or stone, which provides support for a beam or roof truss. Sometimes decorated.

Corner post
See **Angle post**.

Cornice
A moulding, usually horizontal and projecting from the top of the outside face of a wall or at the top of an internal wall against a ceiling.

Counter-battens
Battens nailed parallel to the rafters over a boarded or felted roof with the tile battens nailed over and across them to allow free drainage of penetrating water.

Counter-floor
The lower of two sets of floorboards, often laid at right angles or diagonally to the upper layer.

Course
A line of bricks, stone, tiles etc., set horizontally in a wall or on a roof.

Cove
A concave moulding at the junction of a ceiling and a wall.

Coved ceiling
A ceiling rounded at its edges to merge into the head of the wall.

Cover fillet/Moulding
A moulding fixed over a joist on a flush surface.

Cowl
A slotted metal cap fitted to a chimney pot to improve the draught in the flue.

Cramp
A U-shaped piece of metal used for holding stone slabs together; a metal strap fixed to a frame and built into a wall as a fixing for the frame.

Creasing course
Usually a double layer of tiles laid under a brick-on-edge coping and projecting proud of the wall face, to provide weathering to the top of a wall.

Cricket
A small pitched roof running between the main roof slope and the back of a chimney stack.

Crockets
Carved features on the inclined sides of spires or pinnacles.

Crossing
The intersection of the nave, chancel and transepts of a church in the form of a cross.

Crucks
Large curved timbers used in pairs as the basic framing of medieval houses and extending from the ground to the apex of the roof.

Crypt
The vaulted area beneath a church, often used as a burial place.

Cupola
A domed ceiling on top of a building; sometimes an open-sided feature capped with a dome.

Curb/Kerb
An upstand of wood or stone etc. to form an edging to an opening or round a fireplace.

Curb rafter
A rafter of the upper flatter slope of a **Mansard roof.**

Curtain walling
Lightweight glass and panel cladding fixed between the floors of modern multi-storey buildings.

Curtilage
The total land area attached to a dwelling house.

Dado
The panelling or decorative covering applied to the lower part of an internal wall, above the skirting.

Dado rail/Chair rail
The moulding along the top of a dado or in a corresponding position where there is no dado.

Dais
A raised platform or floor as at the end of a medieval hall.

Damper
An adjustable plate across a flue or ventilator to enable the flow of gases or air to be regulated.

Damp-proof course (DPC)
A layer of impervious material placed in a wall as a barrier to the movement of dampness.

Damp-proof membrane (DPM)
A layer of impervious material incorporated in a solid floor as a barrier to dampness rising by capillary action.

Deadening/Pugging
Material placed beneath floors, usually on **Pugging boards** or sound boards set between the joists to provide sound-proofing.

Deck
The term sometimes used for a roof deck, i.e. the boards or sheet material which support the weather-proofing layers of a flat roof.

Dentils
A row of small rectangular blocks forming part of the bed mould of a cornice.

Dinging
A rough single coat of sand and cement stucco applied to a wall and sometimes marked to represent the joints of masonry.

Dog-legged stair
A stair of two flights between two storeys with the outer string of each flight carried in the same newel on the connecting rectangular landing so that there is no stair well.

Dome
A hemispherical roof, circular or polygonal in plan.

Door case
The frame and linings within which a door is fitted.

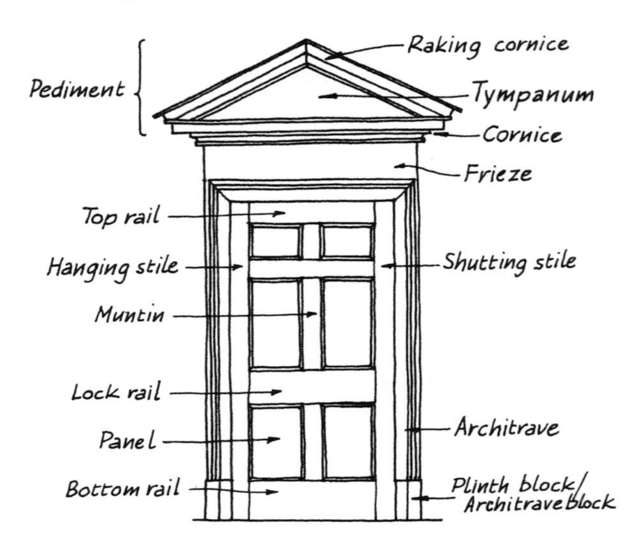

Door frame
The frame composed of two jambs and a head in which a door is hung.

Door jamb/Doorpost
The upright member of a door frame.

Door lining
The wooden lining to the reveal of a doorway.

Door sill/Threshold
The horizontal member at the bottom of an outside doorway.

Door stop
The projecting strip on a door frame against which the door closes.

Dormer
A vertical window projecting from the slope of a roof and having its own roof: it gives light to rooms wholly or partially within the roof.

Dowel
A cylindrical pin of wood or metal used for fixing one piece of material to another. Often inserted through a mortice and tenon joint to prevent the tenon withdrawing from the mortice.

Downpipe/Fallpipe/Rainwater pipe
The vertical pipe to convey rainwater collected in the gutters of a roof to lower levels or the ground.

Dragon beam
A horizontal timber set in the angle at the intersection of the wall plates at the corner of a building and into which the hip rafter is framed. Its inner end is carried in the dragon tie.

Dragon tie
See **Angle tie**.

Drip
A groove under the overhanging edge of a sill or other feature to stop rainwater running back to the face of the wall; a joint formed as a step between two sheets of metal covering a flat roof or gutter across the fall and so designed as to permit thermal movement.

Dripstone
A moulding projecting over a window or doorway to throw off rainwater.

Dry area
A narrow covered area between the external wall of a basement and a retaining wall to higher ground level, intended to keep the basement wall dry.

Dubbing out
The term applied to the filling of hollows or irregularities in a wall before the main coat of plaster is applied.

Duct
A chase or casing through which pipes are routed.

Dwang
Scots for the strutting between floor joists.

Eaves
The overhanging edge of the bottom of the slope of a roof.

Elbow
A right-angled joint in a pipe.

Embrasure
A splayed opening in a wall as in battlements.

Engineering bricks
Dense bricks of extreme hardness and uniform size used for heavy loadings and sometimes as a damp-proof course.

Entablature
The horizontal superstructure carried on the columns in classical architecture and composed of the architrave, frieze and cornice.

Entresol/Mezzanine
A low intermediate storey formed between the main storey levels.

Expansion strip
A resilient insulating material used in discontinuous construction to fill a joint as between a partition and the basic structure or between different materials.

Eyebrow dormer
A vertical window projecting slightly from a roof so that the main roof slope sweeps up and over it in the shape of an eyebrow.

Fabric
The carcase of a building without finishings etc.

Facing bricks
Bricks of suitable colour, texture and weathering quality for use in the external face of a wall.

Faience
Glazed terracotta or coloured pottery used for decorative panels, string courses etc. on the external face of buildings, *q.v.* **Terracotta**.

Fair-faced brickwork
Well-finished exposed brickwork.

Fallpipe/Rainwater pipe
See **Downpipe**.

False ceiling
A ceiling constructed on joists separate from the underside of a floor or beneath an existing ceiling.

Fanlight
A fixed light within a door frame above the transom, originally semicircular but now of any shape.

Fa(s)cia
A wide board fixed to the face of the rafter feet or joists at the eaves or to the head of a wall. May carry the gutter; the wide board over a shop front: a flat face in an entablature.

Fillet
A small square-sectioned wood moulding; a thin strip fixed between two angled surfaces; see also **Cement fillet.**

Finial
Ornamental vertical projection at the apex of a gable or on a pinnacle or newel.

Firewall
The projection of the party wall above the roof line in terraced properties.

Firring/Furring
Thin tapered pieces of wood fixed to the top of the joists of a flat roof to create a fall, or fixed to the tops of floor joists or to rafters to level up and overcome irregularities.

Fixing blocks/Plugs
Blocks or slips of wood set into a wall to provide fixings for joinery items; breeze blocks or soft bricks used for a similar purpose.

Flag/Flagstone
A large flat paving stone.

Flank wall
The external side wall of a building.

Flashing
A strip of material such as lead, copper, or felt used to weather the angle between a roof and an abutting wall or chimney stack to prevent the penetration of moisture. Often used to cover a **Soaker,** *q.v.* The edge of the vertical face is turned and wedged into the mortar joints and, when following a slope, is cut in steps as a **Stepped flashing**.

Flaunching
The cement mortar around the pots and over the top of a chimney stack to provide weathering.

Fleche
A slender spire usually of fairly light construction set on the ridge of a roof.

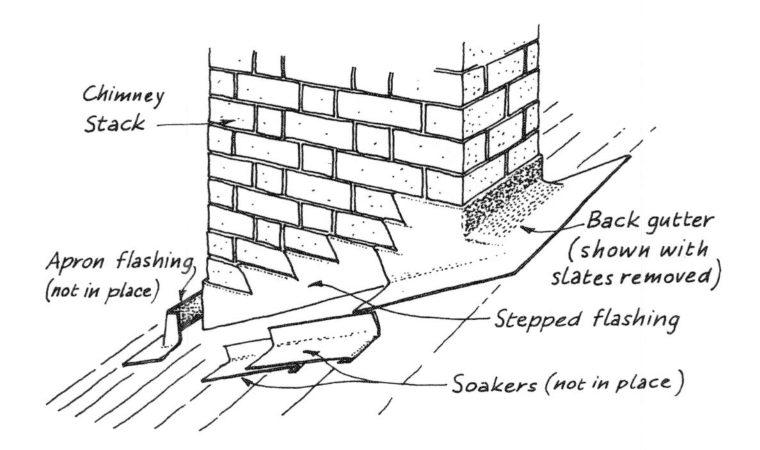

Fletton
A light-pink-coloured common brick made from the shale of the Peterborough district and widely used for internal work and backings.

Flier/Flyer
A step, in a flight of stairs, in which the long edges are parallel: *q.v.* **Winder,** in which the long edges are not parallel.

Flight
A series of steps without change of direction and unbroken by a landing.

Flitched beam
A beam made up of two or more timbers with vertical flat steel plates sandwiched between them and the whole secured by bolts.

Floor
Some of the component parts are shown in the diagram.

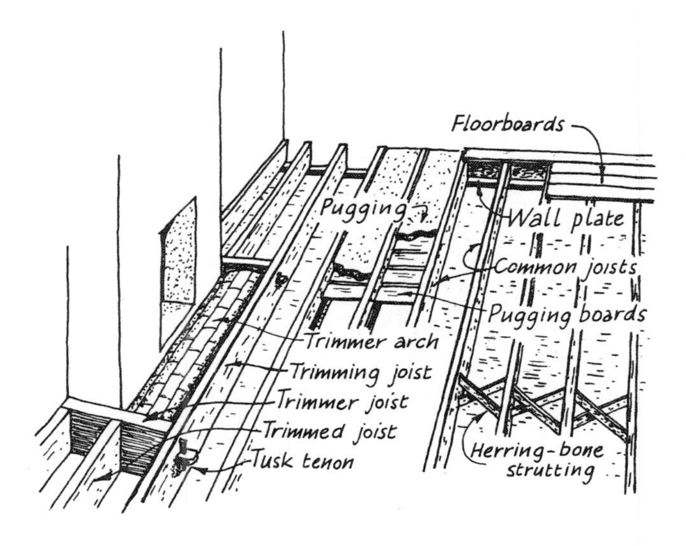

Floor joist
One of a series of wooden beams which support the floor surface and to which the floorboards are nailed.

Flue
The passage up a chimney through which the smoke and gases are drawn.

Flute
A vertical groove usually half-rounded in section.

Flying buttress
A buttress arched over the aisle of a church.

Folding casements
A pair of casements having rebated meeting stiles and no mullion in the frame between them.

Folding doors
A pair of doors having rebated meeting stiles as with French or casement doors; a door of two or more leaves hinged together.

Font
The vessel containing the baptismal water in a church. It is usually mounted on a pedestal and often has a wooden **Font cover**.

Footings
The wide courses of brickwork or masonry at the base of a wall, designed to spread the load of the wall over a greater area of the foundation.

Foundation
The stable base formed in the ground on which a building is erected.

Frame
The skeleton of a wooden-framed building, partition, door etc.

French door/French window
See **Casement door**.

Frieze
That part of a wall which is above the picture rail; the middle member of a classical entablature.

Gable
The triangular wall at the end of a pitched roof.

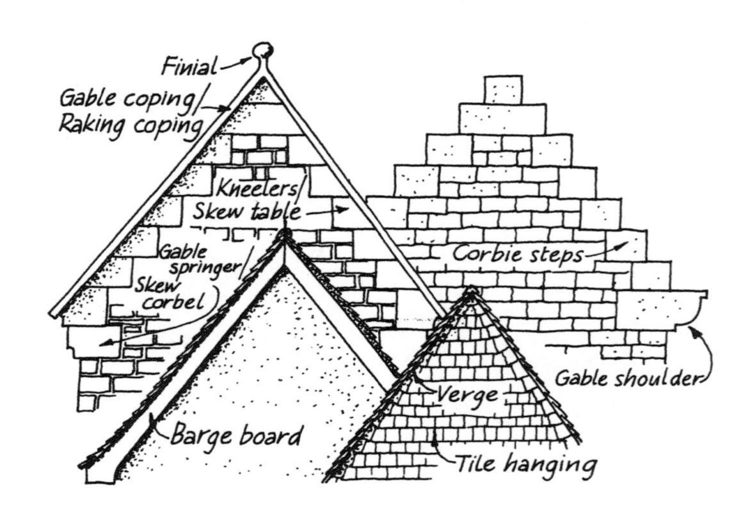

Gable coping
The coping to a gable wall which rises above the roof slopes.

Gable shoulder
The projection formed by the gable springer at the foot of a gable coping.

Gable springer/Skew corbel
The stone which overhangs at the foot of a gable coping.

Gablet
A small gable formed in the upper part of a hipped end of a roof. This is sometimes known as a **Gambrel roof**.

Gallery
An upper floor extended over part of another floor or a hall; a wide corridor connecting parts of a large house and used for exhibiting works of art.

Gambrel roof
A part-hipped roof having a small gable above the hip.

Gargoyle
A projecting water spout, often carved as a grotesque figure.

Garret
A room formed immediately under the roof and with sloping ceilings.

Gauged arch
An arch built with bricks which have been cut and rubbed to a taper to ensure a close fit over the curve of the arch.

Gazebo
A small garden building placed to command a view.

Geometrical stairs
A stair in which the outer string is curved round the wall so that the stairs rise continuously through a turn of 180° without an intermediate landing or newel post.

Glazing bar/A stragal (Scots)
A thin rebated wood or metal bar which divides a large window into smaller lights.

Glue block
See **Angle block**.

Going
The term applied to the horizontal dimension between the risers or the edges of the nosings of two adjacent stairs. This is the going of a tread; the going of a flight is the sum of the going of the treads.

Gravel board
The horizontal board fixed on edge at the bottom of a close-boarded fence.

Grille
A grating to a ventilation hole; a decorative screen.

Groin
The curved line formed by the junction of two surfaces of a vault.

Ground plate/Sole plate
The horizontal timber at the bottom of a timber-framed building.

Grounds
Pieces of wood fixed to a wall to give a level fixing to joinery items such as skirtings.

Grout
Thin mortar used as a filling; in some areas, a mortar applied over the slates of a roof.

Gulley
An opening, usually with a kerb around it, connected to a drain and receiving the discharge from rain or waste-water pipes.

Gutter
A channel or trough fixed at the eaves or behind a parapet wall into which rainwater drains and is conveyed to the fallpipes or other points of discharge.

Gutter bearers
The horizontal timbers to which the gutter boards of a parapet or valley gutter are fixed.

Gutter boards
The boards, fixed to the gutter bearers and which support the lead or other waterproof lining of a parapet or valley gutter.

Half-hipped roof/Hipped gable/Jerkin-head roof
A roof which is hipped at the upper part of its end but has a part gable below the hip.

Half-space landing
A landing joining two flights of stairs between two storeys, being level and extending across the width of both flights.

Half-span roof/Lean-to roof
See **Catslide roof**.

Half-timbered building
A timber-framed building in which the spaces between the exposed timbers are filled with brickwork, plaster etc.

Hammer beam
The short horizontal cantilevered beam of a hammer-beam roof at the springing level of the arch brace of the truss. It is often carved.

Hammer post
The post positioned at the side of a hammer-beam roof, with its foot on the hammer beam.

Hand
The term applied to the side to which a component is related, i.e., left or right.

Handed
The term applied to components matching as if by mirror-image.

Handrail
The rail fixed at waist height at the side of a stair or over the balustrade of a balcony.

Hanging post
The post to which the hinges of a door or gate are fixed.

Hanging stile
The stile of a door or window to which the hinges are fixed.

Hardcore
Broken brick and concrete rubble used to provide a base for site concrete or concrete sub-floors.

Harling
Scots for **Roughcast**.

Haunching
The concrete splayed up round the sides of a stoneware drain pipe to give protection against damage or settlement.

Head
The horizontal member at the top of a window or door frame.

Header
A brick laid across a wall so that it is end-on to the wall face.

Hearting
The filling of the centre of a wall between the facings, as with a rubble-filled masonry wall.

Heel strap
A U-shaped steel strap placed over the back of the principal rafter of a roof truss and bolted to the tie beam near to the wall plate.

Hench
The narrow side of a chimney stack.

Herring-bone strutting
Small-sectioned pieces of wood placed crosswise between the tops and bottoms of adjacent joists to stiffen them in the centre of their span, *q.v.* **Dwang** (Scots) and **Solid bridging**.

Hip/Piend (Scots)
The external angle formed where the slope of a pitched roof is returned along the side to form a hipped roof.

Hip hook/Hip iron/Piend strap (Scots)
A wrought-iron stay fixed at the foot of a hip to prevent slipping of the hip tiles.

Hipped gable/Jerkin-head roof
See **Half-hipped roof**.

Hipped roof
A roof which is sloped at its ends as well as on the sides.

Hip rafter
The rafter which forms the hip of a roof and to which the jack rafters are fixed.

Hip tile
The half-round or cap-shaped tile which covers the junction of the roof surfaces on a hip.

Hollow wall
See **Cavity wall**.

Honeycombed wall
A wall of half-brick thickness built with stretchers with a space between each brick on each course. This type of wall is used for sleeper walls under a timber floor to allow ventilation.

Hood
A canopy over a door or window to give some protection against rain.

Hopper head
A box or funnel-shaped head to a rainwater pipe into which water is discharged from a gutter or outlet.

Hopper light
A light which is hinged at the bottom to open inwards and which has triangular fixed lights or **Hoppers** at the sides to prevent draughts.

Horn
An extension of the stile beyond the rail as on some window sashes; the extension of the head over the stile of a window or door frame to give protection in transit.

Hurter
A block of stone or concrete placed against the coin of a wall to prevent damage by passing traffic.

Hyperbolic paraboloid roof
A double curved self-supporting shell roof.

Infilling
The brick nogging used to fill the spaces between the timbers in timber-framed buildings.

Ingle-nook
A recess with the fireplace within it and often a seat on each side of it.

Ingo/Ingoing
Scots for a Reveal.

Internal dormer
A vertical door or window recessed into a roof so that its head is under the main roof slope and its cheeks are within the roof. The flat surface in front of it is often covered with lead.

Intrados
The underside of an arch.

Jack rafters
Short rafters spanning between the wall plate and the hip or valley rafter of a roof.

Jack rib
A curved jack rafter as used for a small domed roof.

Jamb
The vertical side of an opening through the full thickness of a wall; the upright side pieces of a door or window frame.

Jamb lining
The wooden facing to the jamb of an opening.

Jerkin-head roof/Hipped gable
See **Half-hipped roof**.

Jettied construction
The projection of the upper part(s) of a building outwards over the lower parts as in timber-framed medieval buildings.

Joggled
Shaped with an indentation or a projection, or shouldered as with a king post where it receives the principal rafters and struts.

Joist
One of a series of timbers placed horizontally on edge and to which the boarding of a floor or flat roof is fixed.

Joist hanger
A steel strap or shoe in which the end of a joist is supported.

Kerb
See **Curb**.

Kerfing
A means of bending a piece of wood by making a series of saw cuts across the grain on the back face.

Keystone
The top central stone of an arch.

King post
The vertical post which hangs from the apex of a king post truss and carries the tie beam at its foot. It is usually joggled to receive the struts and principal rafters.

King post truss
A traditional wooden truss composed of a tie beam and two principal rafters with a king post connecting them below the apex and a pair of struts between the lower part of the king post and the principal rafters.

Kite winder
The middle one of the three winders of a stair when it turns through 90°. It is symmetrically shaped like a kite.

Knee
An angle bend in a pipe; the convex bend in a handrail.

Kneeler/Skew table
The stones used in a gable coping, bedded horizontally but with the upper edge sloped to follow the line of gable.

Laced valley
The intersection between two roof slopes which is weathered by the use of tile-and-a-half tiles, so obviating the need for a normal valley. A sharp angle is obtained as opposed to the curved angle of a **Swept valley**, *q.v.*

Lancet
A high narrow window with a sharply pointed arch at its head.

Landing
A platform at the termination of a flight of stairs.

Lantern
A raised structure on a roof with glazing round the sides to provide extra light and ventilation to the room below.

Lap
The length of the part of a slate or tile which is covered by two others.

Lath and plaster
The term applied to the traditional form of plastering to ceilings or stud walls.

Laths
Thin sawn or split pieces of wood nailed across joists or studs to form a base for plastering. Lath now also applies to plasterboard or metal lathing.

Lattice window
Usually casement-type windows in which the light is made up of a number of small pieces of glass, mostly rectangular or diamond in shape, which are bedded in metal cames.

Lay light
A horizontal light set in a ceiling, usually below a skylight in the roof above so as to allow daylight to reach dark areas below.

Leaded light
A light in which the small rectangular diamond-shaped pieces of glass are bedded in lead cames.

Lead flat
A flat roof covered with lead.

Leaf
One skin of a cavity wall; one door of a pair of doors or one casement of a pair of casements.

Lean-to roof/Half-span roof
See **Catslide roof**.

Lectern
A reading desk in a church.

Ledged and braced door
An unframed door of battens or boards fixed to two or three horizontal boards or **Ledges** and with diagonal braces between the ledges.

Lift shaft
The vertical well through a building in which the lift and its counterweight travels.

Liggers
Sticks of hazel or willow used in thatching and held down by spars at the ridge.

Light
One window as bounded by the mullions and transoms and sometimes itself divided into several panes.

Light well
See **Air shaft**.

Lining
The board or facing to an opening.

Link dormer
A large dormer which may join one part of a roof to another or be built round a chimney stack and which may have windows in the sides.

Lintel/Lintol
The beam spanning the opening of a window or doorway. It may be wood, concrete, stone or steel.

Lip
A strip of solid wood applied to the edges of a flush door.

Lobby
A small ante-room or inner vestibule.

Lock rail
The rail in a door which carries the lock. Where there are three rails, it is the middle one.

Lock stile/Shutting stile/Slamming stile
The stile on the side of the door which carries the lock or fastener.

Loft
A storage space in a roof; the gallery in a church containing the organ or a crucifix, *q.v.* **Rood loft**.

Loggia
A covered, open-sided, arcaded apartment or passage, usually on the ground floor.

Louvre
An opening for ventilation screened with horizontal sloping slats.

Lug sill
A sill, the ends of which are built into the jambs.

Lunette
A vertical window in a vaulted ceiling.

Mansard roof
A roof with two pitches to each slope, the lower pitch being much steeper than the upper one.

Mantelpiece
The ornamental surround to a fireplace.

Mantel shelf
A shelf fixed to the chimney breast over a fireplace opening.

Masonry
Construction in stone; the craft of building in stone.

Matchboards/Matching
Thin boards with tongued-and-grooved or rebated edges, sometimes incorporating edge beading or V-shaped edges, used for linings etc.

Mat sinking/Mat well
A section of floor just within an external doorway where the level of the surface is sunk slightly so as to contain the doormat.

Meeting rail
The rails which meet when a sash window is closed, i.e. the bottom rail of a top sash and the top rail of the bottom sash.

Meeting stile
The middle stile of sliding or folding doors or casements.

Merlons
The upstanding parts of battlements between the embrasures or notches.

Metal lathing
Expanded metal used as background for plastering.

Mezzanine
See **Entresol**.

Moisture barrier
A damp-proof course or membrane to prevent the passage of moisture.

Mortar
Cementing material composed of an aggregate such as sand, crushed clinker or ash mixed with Portland cement and/or lime.

Mortice (Mortise) and tenon joint
Commonly used for two timbers to be joined at right angles to each other. A rectangular slot (the mortice) cut in one timber houses a flat tongue (the tenon) cut in the other, and a treenail is often fixed through the joint to prevent withdrawal of the tenon.

Mosaic
Small chips or cubes of marble, glass etc., cemented to the surface of walls or floors in patterns. When the surface is polished, it is called **Terrazzo**.

Moulding
A strip of wood or other material shaped in section and used to cover joints, emphasise features or throw off rain. A **Stuck moulding** is shaped from solid wood: a **Planted moulding** has the shaped pieces applied to it.

Mullion/Munnion
The vertical member dividing the width of a window opening or dividing the opening between a door and a sidelight in a door frame.

Muntin
An intermediate vertical member in panelling, framed between two rails.

Narthex
A shallow arcaded porch extended across the full width of the west end of a church.

Nave
The main body of a church between the west end and the chancel.

Needle
A short length of wood or steel passed through a wall to support its upper part during underpinning. It is itself supported on dead shores, i.e. vertical shores.

Newel/Newel post
The vertical post which houses the outer strings and handrails of a stair and supports them at a corner.

Nib
A small projection from a tile which enables the tile to be hung from a batten; a term sometimes applied to a thin pier on a wall.

Niche
A shallow recess in a wall, usually to contain a vase or statue.

Nogging
A short horizontal timber placed between the studs of a partition, to stiffen it.

North-light roof
A pitched roof, usually with a steeper pitch on the north side, and with lights in the north slope. Used to provide daylight to factories, studios etc.

Nosing/Bottle nosing (Scots)
The half-round edge of a stair or window board etc.

Notch
A groove cut into a piece of wood to enable it to fit over another.

Offset
A ledge formed in a wall where it changes thickness and sometimes providing the location for the wall plate of a floor; the double bend in a pipe so that it continues parallel to its original direction.

Ogee
The shape formed by a double curve which is concave below and convex in its upper part, e.g. an **Ogee gutter**.

Open eaves
The projection at the eaves of a roof where the rafter feet are not concealed by soffit and fascia boards.

Open floor
A floor of which the joists are exposed below.

Opening light
The opening part of a window.

Open roof
A roof which has no ceiling and is therefore exposed below, as in many churches.

Oriel window
A window which projects from an upper storey and which is usually supported on corbels, *q.v.* **Bay window**.

Outer string
The string of a wooden stair which is furthest from the wall, *q.v.* **Wall string**.

Outlet
A point of drainage such as the aperture or chute in a parapet wall through which water drains into the hopper head of a rainwater pipe.

Overflow
A device incorporated into water storage or flushing cisterns, baths etc. so that water from overfilling is carried away, often through a pipe projecting through an outside wall so that attention is drawn to the occurrence.

Oversailing course
A projecting course of brickwork as with brick corbelling or the overhanging courses at the head of a chimney stack.

Oversite
The term sometimes applied to the surface of the ground exposed under the floor.

Oversite concrete
The layer of concrete placed over the ground under the floor of a building.

Padstone
A stone or block of concrete placed under the bearing end of a beam or girder to spread the load on a wall.

Pane
A single piece of glass to fill a light or to fit between the glazing bars or cames of a divided light.

Panel
The unit of wood or other material fitted into the grooves or rebates of a frame. It is often a flat surface area but sometimes shaped with bevelled edges or in other ways.

Pantile
A shaped tile with a double curve across its width from concave on one side to convex on the other so that it overlaps the tile adjoining it on the side.

Parapet
A low wall along the edge of a roof, balcony or bridge to act as a guard.

Parapet gutter
A gutter formed along the back of a parapet to a roof.

Parge
Lime mortar used for lining the inside of a flue.

Parget(t)ing
The decorative plaster facing applied to the outside walls of old timber-framed buildings, notably in parts of East Anglia.

Parquet floor
A surface of wooden blocks or of regular-sized thin pieces laid to a pattern over a sub-floor; narrow hardwood strips laid tight across a sub-floor known as **Parquet strip**.

Parting bead/Parting strip
The narrow vertical piece of wood which separates the two sliding sashes in a window.

Partition
A light non-load-bearing wall to separate rooms or areas.

Party fence
A wall or fence separating the gardens or grounds of two adjoining properties and shared by them.

Party wall
The wall separating two adjoining buildings and common to them.

Pavement light
A light composed of small squares of thick glass in an iron or concrete frame and set in a pavement to give light to an area or vault below.

Pavilion roof/Pyramid roof
A roof of which each side is a hipped slope equal to each of the others.

Pebble dash
Render or plaster applied to the external face of a wall and onto which small pebbles are thrown before it is dry, to stick and give texture.

Pediment
The triangular feature above a classical entablature or the smaller similar features over doors and windows of buildings of the Renaissance period.

Peg
A rough oak dowel or a metal peg pushed through the hole in a roof tile to provide the means of hanging it on a batten.

Pelmet
A decorative feature fixed to the **Pelmet board** above a window to hide the curtain rails or runners which are fitted to that board.

Pendant
A suspended carved ornament of wood or stone.

Penthouse
A small room or outhouse, or in present times a flat built on the flat roof of a building with walking space around it.

Penthouse roof/Pen roof
A single-slope roof differing from a lean-to roof in that it covers and protects the higher wall.

Pergola
A framework of beams carried on posts and open to the sky. Originally a garden feature, the term is now also applied to the screens on top of high-rise buildings to mask the lift motor and tank rooms etc.

Perpends
The vertical joints in brickwork or masonry.

Pew
In a church a fixed bench with a back sometimes enclosed with a door from the aisle.

Picture rail/Picture moulding
A wooden moulding placed horizontally along the upper part of a wall in a room, originally as a rail from which pictures could be hung.

Pieced timber
A timber from which a damaged section has been removed and a new piece fitted in a dovetail shape.

Piend
Scots for a **Hip**.

Pier
Load-bearing brickwork or masonry supporting an arch or as a projection from a wall of similar material to strengthen that wall or support a beam.

Pilaster
A rectangular pier projecting from a wall for either structural or decorative purposes. Usually fluted and with a cap and base.

Pillar
A vertical column to carry a load independently of the walls. The term is now usually confined to iron or stone.

Pinnacle
A small ornamental turret topped with a pyramid or cone and used to terminate features such as buttresses or towers, especially in Gothic architecture.

Piscina
A stone basin in an arched niche, usually in the south wall of the sanctuary of a church and used by the priest for rinsing the chalice.

Pitch
The angle of inclination from the horizontal of a roof or stair.

Plain tile
The commonly used nominally flat roofing tile of burnt clay or concrete.

Plaster
A material applied to walls in plastic form and which hardens to make a finish. The term covers materials such as gypsum, lime and Portland cement.

Plaster blowing/Pitting
The term used when plaster loses its adhesion to a wall.

Plasterboard
A sheet of gypsum plaster faced on both sides with stout paper and which, when nailed directly to the ceiling joists or wall studs, obviates the need for using wood laths.

Plaster dab
A small lump of gypsum plaster stuck to a wall and with others used as a means of fixing tiles or marble facings.

Plate
A horizontal timber used to distribute a load over a wider area as on a wall top or to carry floor joists; Scots for a broad thin board.

Platform roof
Scots for a flat roof.

Plinth
The projecting base of a column, pedestal or building. In many cases the projection is shallow. Often the plinth to a building is formed by a band of cement render.

Plinth block
See **Architrave block.**

Plugs
See **Fixing blocks/Plugs.**

Pocket
An opening in a wall to take a beam or for other housing purposes.

Pointing
The finishing of the mortar to the joints of brickwork or masonry to give good weathering to the wall face and provide a neat appearance.

Pole plate
A beam supported at its ends on the tie beams of the trusses and supporting the feet of the common rafters of a roof.

Poppy head
The ornamental carving on the end of a pew or choir stall in a church.

Porch
A roofed projecting structure to give protection against the weather to an entrance.

Porte-cochère
An entrance for carriages through a building; a projecting porch over an entrance large enough to allow people to alight from a carriage under cover.

Portico
An imposing porch open or partially enclosed but with a roof.

Post
The principal vertical timber of a wooden-framed structure.

Pot floor
A floor of hollow clay blocks usually supported on steel angles.

Presbytery
The part of a church occupied by the priests; the house occupied by a Roman catholic priest.

Principal rafters
The rafters which form part of the trusses in a roof and on which the purlins are carried.

Pugging
See **Deadening.**

Pulpit
The enclosed platform with reading desk and seat, which is raised above floor level and used by the priest for the delivery of sermons in a church.

Puncheon
A short post in the middle of a truss or a trussed partition.

Purlin
A horizontal timber carried on the principal rafters in a roof and supporting the common rafters. There may be one or more purlins to each roof slope.

Purlin roof
A roof used for some small buildings in which the purlins are housed in the end walls.

Putlog holes
Small holes left in brickwork to carry the **Putlogs** or horizontal members of the builder's scaffold.

Pyramid roof
See **Pavilion roof**.

Quadrant moulding
A quarter-round moulding used for trimming a joint in a right angle.

Quarry tile
A small burnt clay square tile used for flooring. It is hard and dense but unglazed.

Quarter-space landing
A small square landing, usually at the top of a flight and having its sides the same length as the width of the stair.

Queen post truss

A roof truss having two **Queen posts** one each side of the centre but no central **King post**.

Quoin
See **Coin**.

Rabbet/Rebate/Check (Scots)
A recess cut in the long side of a piece of wood to form a step in section.

Rafter
A sloping timber spanning between the eaves and ridge of a roof. See **Common rafter** and **Principal rafter**.

Rafter filling/Wind filling
See **Beam filling**.

Rail
A horizontal member in a door, framed between the stiles.

Rainwater goods
Rainwater gutters and downpipes etc.

Rainwater pipe/Downpipe
See **Fall pipe**.

Rake
An angle of inclination to the vertical.

Raking coping
A coping on a gable where the coping stones are laid on an inclined bed following the slope of the gable. It is sometimes finished with a **Raking cornice**.

Random rubble
Masonry construction utilising stone of irregular size and shape which is not laid to courses.

Rebate/Check (Scots)
See **Rabbet**.

Re-entrant angle
An internal angle or corner such as is formed where the flank wall of an extension meets the face of the main wall from which it projects, *q.v.* **Salient angle**.

Refectory
The dining hall in a monastic or scholastic building.

Register
A ventilation grille fitted with a damper so that the air flow may be controlled.

Relieving arch
An arch built over a wood or stone lintel to relieve the load over the opening.

Rendering
Facing a wall with cement mortar. In external work, rendering is covering a wall with this material and leaving a plain finish. In internal work, rendering usually applies to the first **Render coat** of a plastering process. In common usage. **Rendering** implies a sand-and-cement mix.

Reredos
A screen decorated by painting or carving and covering the wall behind the altar in a church.

Respond
A half-column terminating an arcade.

Retaining wall/Revetment
A wall built to hold back ground which is at a higher level and resist the lateral pressure of it.

Retro-choir
The part of a choir which is behind the high altar.

Return
A change in direction in a walls as at a corner or into a recess.

Reveal/Ingo (Scots)/Ingoing (Scots)
The part of the external jamb of a door or window which is not covered by the frame; the side of a fireplace or opening which is exposed.

Revetment
See **Retaining wall**.

Rhone/Rone
Scots for an eaves gutter.

Rib
A projecting moulding used to separate the segments or panels of vaulting or a ceiling.

Ridge
The line of the apex of a pitched roof.

Ridge board
The horizontal board set on edge along the apex of a pitched roof and against which the tops of the rafters are set.

Ridge tile
A tile for covering the ridge of a roof: commonly of half-round or angular section.

Ringing chamber
The room in the lower part of a church tower where the ringers stand when ringing the bells.

Riser/Breast (Scots)
The upright face of a step in a stair.

Rising main
The extension of the main inlet electricity cable or gas or water supply pipe, vertically through one or more storeys.

Riven laths
Laths made by splitting wood lengthwise rather than by sawing.

Rodding eye
See **Cleaning eye**.

Roll
A form of joint employed in sheet-metal roofing, sometimes formed as a hollow roll or sometimes as a solid roll over a wooden former.

Rolled steel joist (RSJ)
An I-sectioned steel beam.

Rone/Rhone
Scots for an eaves gutter.

Rood screen
A wooden or carved stone screen separating the choir from the nave in a church and sometimes topped with a **Rood loft**; a gallery in which a crucifix or other images are placed.

Roof-light
A skylight or lantern in a roof.

Roofs
The main types and some of the principal components are shown in the diagram.

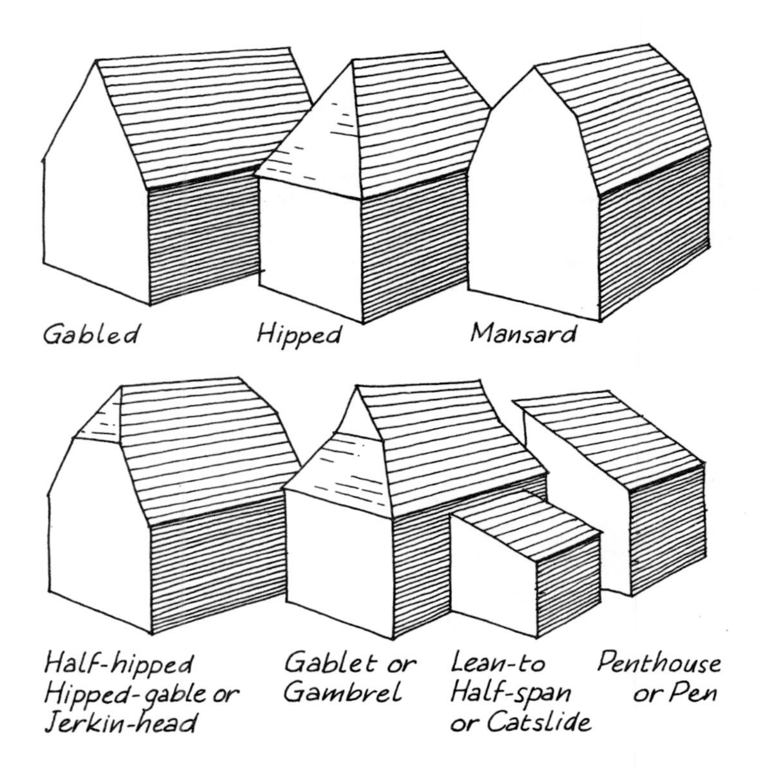

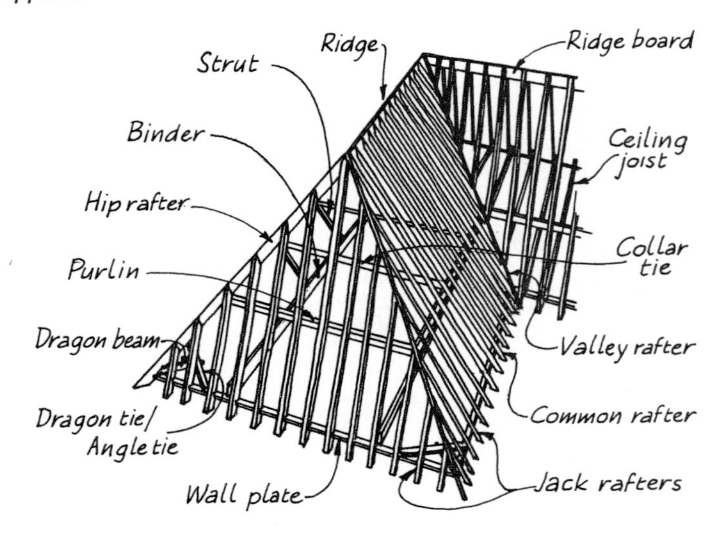

Roughcast/Harling (Scots)
A textured render finish for external walls in which gravel or stone chips are mixed with the mortar rather than thrown onto it as with pebble dash.

Rubble
Masonry stones which are left rough so that the joints between them are coarser than in ashlar work; debris such as that composed of broken brick, concrete, plaster etc.

Runners
Long strips of withies or osiers laid in horizontal bands along thatch and held down with spars.

Rusticated ashlar
A squared building stone the face of which is left rough and so stands proud of the joints.

Sacristy
A place where the vestments and vessels are kept in a church, *q.v.* **Vestry**.

Saddle
A square of lead or other flexible metal dressed under the tiles or slates to weather a roof junction such as that where the ridge of a projecting section meets the main slope of a roof.

Salient angle
The external angle made by a projecting corner and thus the opposite of a **Re-entrant angle**, *q.v.*

Salon/Saloon
A drawing room or state apartment in a large house or palace.

Sanctuary
The part of a church round the altar.

Sarking
Boarding fixed over the rafters of a roof to carry the tiles or slates (especially in Scotland); **Sarking felt** is the bituminous felt sometimes laid over the rafters and under the tile battens.

Sash/Case (Scots)
The sliding light of a **Sash window**.

Sash door
A door of which the upper part is glazed.

Scallops
Short lengths of osier (withies) used to secure thatch at the verge of a roof, being held down by spars.

Scarfed joint
A method of joining or repairing pieces of wood lengthwise whereby the ends of the joining surfaces are cut to fit on bevelled faces.

Scissors truss
A truss comprised of the two principal rafters and two additional members. The latter span between the wall plate and a position part way down the opposite rafter so that they cross in the middle like scissors. In some church roofs, each pair of common rafters is trussed in this way.

Scotia
A concave moulding such as that commonly used under the nosings on a stair.

Screed
A layer of cement mortar between 25 and 65 mm thick laid as a floor finish.

Screen
Used to separate parts of some churches. It is usually panelled in the lower part, with a carved tracery above.

Secret gutter
A gutter which is almost concealed from view by overhanging tiles: a gutter passing through a roof space to take water from a central valley gutter to an outside fallpipe.

Secret-nailed flooring
Flooring in which the boards or strips are nailed through the tongue on a side rebate so that the nails are invisible on the floor surface.

Shingles
Thin rectangular pieces of wood used like tiles for cladding a roof or for vertical hanging on walls. Usually of western red cedar, but in the past they have been of oak.

Shiplap
Weather boarding with a rebate formed on both long edges so that when the boards are laid they overlap at the rebates.

Shoe
A short angled section of pipe fixed to the bottom of a fallpipe to direct the outflow away from a wall.

Shore
A prop for giving temporary support to a wall or a building.

Shuttering
Structure to support concrete when it is placed until it is set.

Shutters
Covers for protecting windows. Most commonly of wood and hinged: can be fitted externally to fold back against the wall when open, or internally to fold back into boxing at the side of the windows.

Shutting stile/Slamming stile
See **Lock stile**.

Side Gutter
A gutter formed at the intersection of the side of a dormer or chimney, with the slope of a roof.

Sidelight
A glazed light at the side of a door frame.

Sill
See **Cill**.

Sillboard
Scots for **Window board**.

Single-hung window
A sash window of which only one sash is made to move.

Single-lap tiles
Shaped roofing tiles such as pantiles or interlocking tiles, which overlap only the tiles of the course immediately below.

Single roof
A roof supported only on the common rafters, being without trusses or purlins.

Sinking
A recess in a surface as with a mat well in a floor.

Skeeling
A sloping ceiling as fixed under the rafters in attic rooms.

Skew corbel
See **Gable springer**.

Skew flashing
A flashing between a gable coping and the roof shape.

Skew table
See **Kneeler**.

Skirting/Skirting board/Base plate (Scots)
The vertical board fixed along the base of internal walls to protect the plaster from kicks. Sometimes a skirting is run in plaster or formed with tiles.

Skylight
A glazed light on a roof and in the same plane as the roof. It may open or be fixed.

Slamming stile/Shutting stile
See **Lock stile**.

Slat
A thin strip of wood as in a louvre.

Slate
Very hard material formed in strata from clay and carbonaceous matter under great natural pressure. It is cut and laminated to make **Slates** for roofing.

Sleeper wall
A dwarf wall for supporting a timber ground floor. It carries a **Sleeper plate** which in turn supports the joists. A sleeper wall should be honeycombed to allow ventilation of the sub-floor space.

Slip sill
A sill which fits between the jambs of an opening and is not built into the jambs.

Slovens
Mortar and pieces of brick which have dropped into the cavity of a wall during construction.

Slurry
A fluid mix of cement mortar.

Snow boards
Narrow strips of wood fixed with gaps between them to bearers and placed over parapet or valley gutters to allow melting snow to drain away without choking the gutter; Scots for **Snow guards**.

Snow guards
Wooden boards or strips of wire mesh fixed on edge above the eaves to prevent masses of snow from sliding off the roof.

Soaker
A rectangular piece of flexible metal, usually lead, about 150 mm wide and of a length governed by the size of the tiles or slates, which is bent to fit the angle between a roof slope and a chimney or other vertical abutment. It is placed under the tiles or slates with its vertical face against the wall where it is covered by a **Flashing**, *q.v.*

Soffit(e)
The undersurface of any recess or feature or of a stair but not a ceiling.

Soil pipe/Soil stack
A fallpipe for carrying sewage from sanitary fittings on upper storeys to **Soil drains** in the ground. It is extended above roof level to provide a vent (**Soil vent pipe**).

Soldier
A short vertical ground fixed to the wall to take a built-up skirting.

Sole piece
A short piece of wood placed across the wall top over the wall plate(s) in an ashlar area, in the manner of a short length of a beam and carrying the foot of the common rafter at its outer end and the foot of the ashlar piece or ashlar post at its inner end.

Sole plate
See **Ground plate**.

Solid bridging/Solid strutting
Short pieces of board fixed between the joists of a floor or flat roof to stiffen the structure, *q.v.* **Herring-bone strutting** and **Dwang** (Scots).

Sound boarding
Short boards laid on fillets fixed to the sides of joists under a floor and carrying pugging for sound deadening.

Sounding board
A wooden canopy over the pulpit of a church.

Spalling
The eruption or disintegration of plaster or render as a result of dampness or the action of hygroscopic salts.

Spandrel/Spandril
A roughly triangular space or panel.

Spandrel steps
Solid steps made in triangular section to give a smooth soffit to the stair.

Span roof
A simple pitched roof.

Spars
Pieces of split hazel or willow about 600 mm long, bent double and driven into thatch over runners, liggers or scallops; common rafters.

Spigot and socket joint
A pipe joint in which the end of one pipe of normal diameter, the **Spigot**, fits into the enlarged end, the **Socket**, of another pipe. The joint is sealed by caulking.

Spire
A tall conical or pyramidal structure rising above a tower.

Splay
A bevel formed across the whole face of a surface.

Splice
A joint made between two halved timbers and secured with bolts or plates.

Springer/Springing
The bottom stone of an arch, being laid on the **Springing line**, where the arch begins to curve.

Sprocket piece
See **Cocking piece**.

Square-edged boards
Floorboards with flat long edges as opposed to tongued-and-grooved edges.

Squint coin
A corner of a wall which is at an acute or obtuse angle.

Stable door
A door divided into two horizontally with each piece separately hinged and fastened.

Stack
A chimney stack; a rainwater downpipe; a soil pipe.

Stair(s)
A series of steps including handrails, balustrades and landings where present, and giving access between storeys.

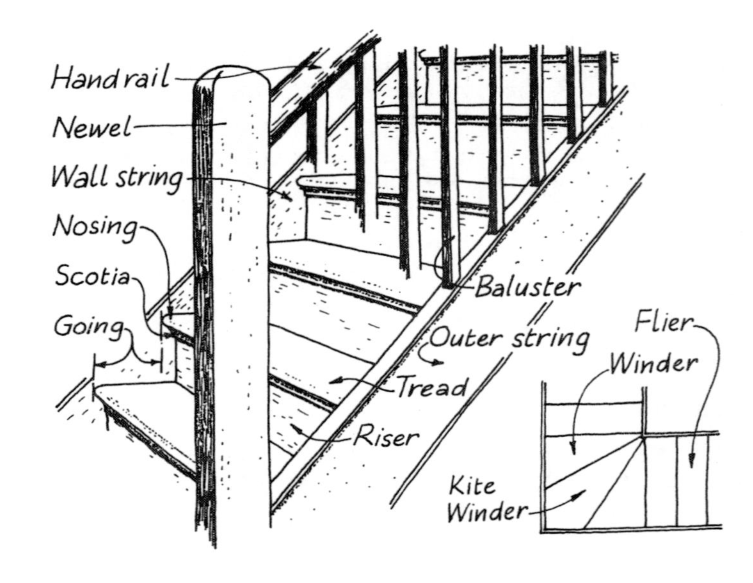

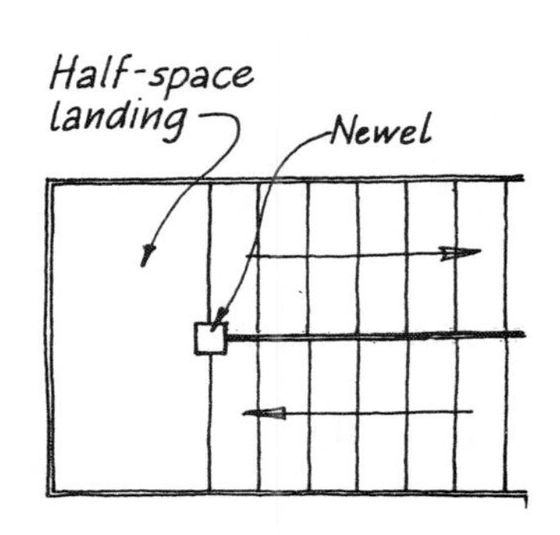

DOG-LEGGED STAIR

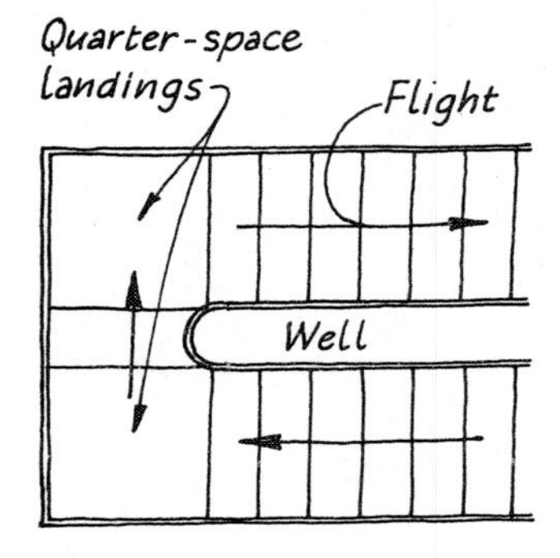

GEOMETRICAL STAIR

Staircase
Originally the space in which a stair is constructed. In present-day usage it is the term applied to a stair as described above.

Stairwell
The horizontal distance or space between the flights of a stair.

Stall board
The shelf or base to the display space of a shop window, including its framing.

Stall riser
The vertical facing on the outside of a shop between the pavement and the base of the shop window.

Stanchion
A column or pillar of steel or cast iron.

Standard
The upright pole of a scaffold.

Stay
A rod or bar used to hold an object in position, as with a **Casement** stay which holds a window open.

Steeple
A tower surmounted with a spire.

Step
A tread and riser comprising one unit of a stair.

Stile
The outer upright member of a frame, especially of a door or window frame.

Stock brick
The brick most commonly available in a district.

Stone
Various natural rocks which are cut and used for building material.

Stone slate
Thin flat pieces of stone used for roofing.

Stoneware
Hard salt-glazed ceramic material from which drain pipes, channels, gulleys etc. are made.

Storey
The part of a building between each floor level and the floor above it.

Strapping
Grounds fixed to a wall to provide an even base for a lath-and-plaster or plasterboard finish.

Stretcher
A brick laid with its long face parallel to the face of a wall.

String
A stout sloping board on each side of a wooden stair into which the treads and risers are housed.

String course
A decorative or slightly projecting horizontal band of brickwork or stone in the external face of a wall.

Strip flooring
Narrow strips of tongued-and-grooved wood of carefully controlled size often fixed over a sub-floor by secret nailing after tight cramping. The completed surface is usually sanded and polished, *q.v.* **Parquet floor**.

Strut
A supporting member in a structure and one which is placed in compression at an angle; see **Braces**.

Stucco
Smooth plaster or mortar facing on the outside of a wall, commonly painted.

Stud
A vertical member in a **Stud partition**, i.e. a partition which is framed in wood.

Sub-basement
A storey set below the main basement of a building.

Sub-floor
The structural base of a floor which is then covered with a screed or other floor finish.

Sub-floor space
The space beneath a floor, as between a ground floor and the oversite.

Sub-sill
An additional sill fixed to the outside of the sill of a window.

Surround
The object or feature placed around something to protect or decorate it, e.g. a gulley surround or a fireplace surround.

Suspended ceiling
A ceiling supported on its own framework rather than on the underside of a floor.

Suspended floor
A floor which is suspended at its sides but not in the middle.

Swan neck
An S bend in a pipe, most usually seen in a rainwater pipe where it sets back to the wall face from a projecting eaves gutter.

Sway
A long length of hazel or willow used for securing thatch.

Swept valley
A roof valley where the tiles, slates or shingles are cut and laid to give continuity round a valley and so dispense with the need for a valley gutter of lead or other material, *q.v.* **Laced valley**.

Tank
A vessel for the storage of fluid.

Tanking
A system of forming a continuous water-proof lining, usually in asphalt, round the walls and floor of a basement as a barrier to rising and penetrating dampness.

Terracotta
Clay material moulded and burnt and used for features such as cornices, vases etc. Can be used with or without a glazed finish, *q.v.* **Faience**.

Terrazzo
Small pieces of marble or stone set in cement usually to patterns, ground smooth and polished. Used for flooring, *q.v.* **Mosaic**.

Thatch
A roof covering of reeds or straw.

Thermoplastic tiles
Flooring tiles made from thermoplastic resins, asbestos fibre or similar materials which are fixed with adhesive to a rigid sub-floor.

Threshold
See **Door sill**.

Throat
The groove cut in the underside of a sill to stop water running back onto the wall; the contracted part of a flue above the fireplace.

Tie beam
The horizontal member of a roof truss which acts as a tie across the roof holding the feet of the principal rafters on each side.

Tie rod
An iron or steel rod placed across a building with the ends passing through the opposite walls and clamping them by means of plates or X bars, to restrain outward movement of the walls.

Tile battens
Small-sectioned battens fixed horizontally across the rafters of a roof or across a wall to form ledges on which tiles or slates are hung.

Tile creasing
See **Creasing course**.

Tile hanging/Weather tiling
The use of tiles hung vertically on a wall as a cladding.

Tile pegs/Tile pins
Oak pegs pushed through the holes in tiles to provide the means of hanging them on battens.

Tilting fillet
A length of wood of triangular section nailed along the bottom of the rafters or roof boarding to tilt the bottom edge of the first course of slates or tiles to ensure better bedding. Also used at the bottom of tile hanging to tilt the bottom tiles out for better weathering.

Timber connectors
Metal fastenings in various forms including toothed plates, and rings which are used to increase the shear strength of timber at joints in roof trusses or framed construction.

Tongued-and-grooved boards
Boards used for flooring and lining and having a groove cut in one long edge and a matching tongue on the other so that they interlock with one another when laid.

Torching
Pointing with haired mortar to the underside of the head of slates where they bed on the battens of a roof. Full torching entails the covering of the whole of the underside of the slates exposed between the battens.

Tracery
Pierced decorative pattern work in wood or stone, employed in the heads of Gothic windows or in screens of that period.

Transept
Part of a church set at right angles to the nave so that where the plan of the building is in the form of a cross, it forms an arm of the cross.

Transom/Transome
A horizontal member dividing the lights of a window or a door from the fanlight above it.

Trap
A term applied to a short length of floorboard often fixed with screws, to give ready access to electrical or plumbing connections under a floor; the access point to a roof space with a door or moveable cover, hence **Trap door**.

Tray
A sheet, usually of flexible material, placed in a cavity wall over an opening to prevent any rainwater which may have penetrated the outer skin from soaking into the head of the opening.

Tread
The horizontal part of a step.

Treenail/Trenail
A hardwood pin driven across a mortice and tenon joint through a pre-bored hole to secure the joint.

Triforium
A gallery without natural light over the aisle of a large church with an arcaded front opening onto the nave.

Trimmed joist
A joist which is supported at least at one end by a trimmer or trimming joist.

Trimmer arch
An arch formed within the fireplace area of an upper floor to carry the hearth. It spans from the chimney back to the trimming joist.

Trimmer joist/Bridle (Scots)
A joist which encloses a short side of a rectangular hole in a wood floor. The end of the trimmer joist is carried in the trimming joist.

Trimming joist
A joist running parallel to the common joists but of heavier section and forming the long side of a rectangular hole in a wood floor.

Trunking
A wooden or metal pipe for ventilation.

Truss
A strong frame of wood (or metal or concrete nowadays) to form the principal support of a roof. Some typical examples are shown in the diagram.

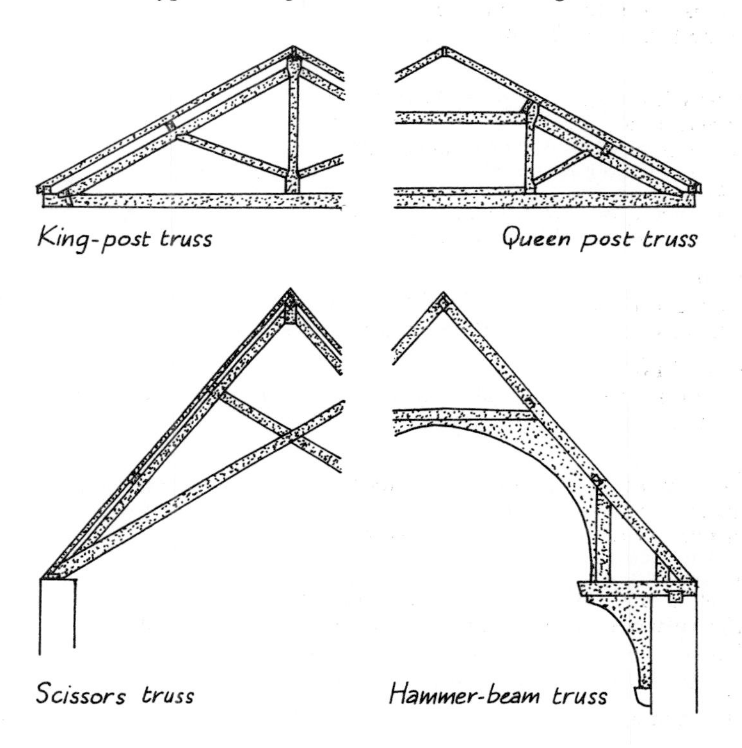

Trussed partition
A partition constructed with strong bracing to enable it to carry a load in addition to its own weight.

Trussed rafter roof
A roof in which each pair of rafters is tied and braced.

Tumbling courses
The raking of sloping courses of brickwork intersecting with the horizontal courses on a raking buttress or a gable end.

Turret
A small tower projecting from the main part of a building or tower. In some churches a turret, housing the head of a spiral stair, projects from the corner of the tower roof.

Tusk tenon
A form of tenon usually employed at the end of a trimmer joint. It passes right through the mortice in the trimming joist and is itself morticed and secured with a wooden peg.

Tympanum
The triangular vertical face of a pediment where it is contained by the horizontal and raking cornices.

Undercloak
The additional course of tiles or slates installed at the eaves under the lowest course of the main tiles.

Undercroft
A crypt.

Underpinning
The building of new foundations and brickwork underneath existing work to stabilise it when movement or weakness has occurred.

Universal beam
A rolled steel joist (RSJ).

Upstand
The turned-up part of felt or flexible metal roofing against a wall. It is not turned into the wall but is covered by a flashing.

Valley
The angle formed at the junction of two roof slopes either as the opposite of a hip or between two parallel pitched roofs.

Valley board
A board laid under a valley and following its line to provide support for the valley gutter.

Valley gutter
The gutter formed in the angle of a valley.

Valley rafter
The rafter which is positioned under a valley formed at the intersection of two roof slopes in the form which is the opposite of a hip.

Vapour barrier
A membrane incorporated in a ceiling or wall to prevent warm moist air passing through and reaching colder surfaces beyond the insulation layer where the moisture would condense.

Vault
An arched roof of brickwork or masonry over a building or part of a building or a cellar.

Vee roof
A roof composed of two lean-to sections which meet in a central valley.

Vent
The outlet point or pipe of a ventilation system.

Veranda(h)
A covered terrace, open on one side, along the side of a house.

Verge
The edge of a sloping roof which overhangs the gable wall.

Vestibule
The ante-room to an important room or hall; an entrance hall.

Vestry
A room usually in or attached to the east end of a church, used for the keeping of vestments, registers etc. In addition to the vestry used by the priest, there is often a separate choir vestry, *q.v.* **Sacristy**.

Voussoirs
The wedge-shaped stones or bricks of an arch.

Wag(g)on roof
See **Barrel roof**.

Wainscot
A wood-panelled dado.

Wall anchor
A steel strap fixed to the end of a joist and built into the wall to give some support to the wall through the joists.

Wall plate
A timber bedded horizontally on or in a wall to carry floor joists or the roof timbers.

Wall post
The vertical timber set against the wall and usually supported on a corbel under the beam of some roof trusses. Sometimes Called a **Wall piece**.

Wall string
The stair string which is against a wall and thus the opposite of the **Outer string**, *q.v.*

Wall tie
A metal tie built into the two leaves of a wall across a cavity to stabilise the two leaves.

Waste pipe
The pipe which carries waste water away from a sink, bath, basin etc.

Water bar
A galvanised steel bar set on edge into the sill or threshold of an opening or doorway to prevent rainwater driving across it.

Water-waste preventer (WWP)
The term sometimes used for the cistern for flushing a water closet.

Wattle and daub
A plaster of clay, cow dung and chopped straw applied over wattles or thick sticks wedged between the timbers of timber-framed buildings and thus providing an infilling for the walls.

Weather board
A horizontal board set at an angle on the bottom of a door to throw off rain and prevent it driving underneath.

Weather-boarding
See **Clapboard**.

Weathering
A surface covering or a means of giving protection from rain.

Weather tiling
See **Tile hanging**.

Weep hole
A small hole left for drainage as in the outer brickwork of a cavity wall in front of a tray over a lintel or in a wood sill to allow the escape of condensed moisture.

Well
An open space passing through one or more floors, such as a **Light well**.

Wicket gate
A small door, for use by people on foot, set in a large door, as in a warehouse.

Wind brace
A wide straight or curved board set at an angle between the trusses and the purlins or wall plates in a roof slope to give lateral stability to a roof.

Winders
Triangular or kite-shaped steps giving a change of direction in a stair.

Wind filling
See **Beam filling**.

Window
A glazed framed opening in a wall to give light to the interior of a building.

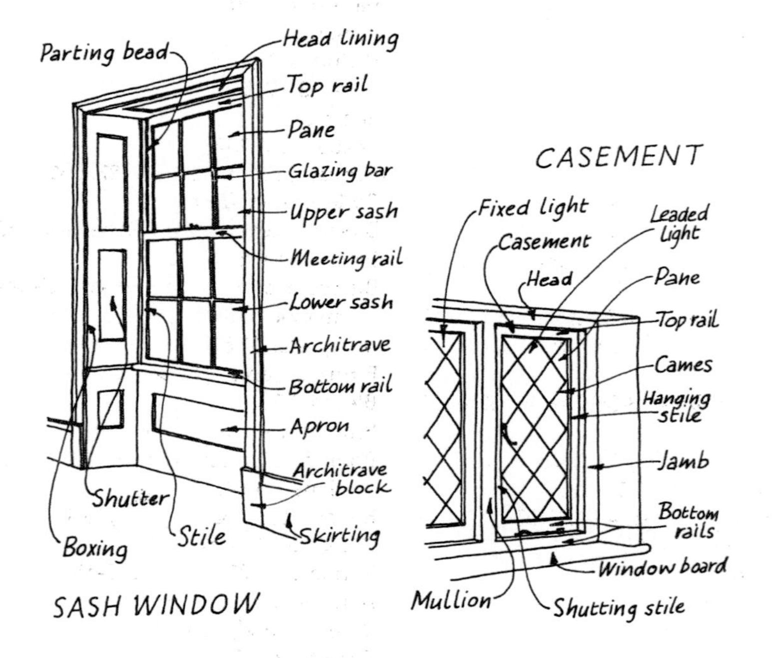

Window board/Sill board (Scots)
The wooden sill on the inside of a window.

Wing
Part of a building which projects from the main body of the structure. Usually applied to large buildings, e.g. the east and west wings.

Wood-block flooring
Rectangular blocks of wood set in patterns in mastic or pitch on a solid sub-floor. The blocks are often tongued-and-grooved.

Index

NOTE:
For wood-borers, insects, etc. *see also* Appendix 1, pages 191-244.
For wood rots, moulds, lichens, etc. *see also* Appendix 2, pages 245-252.
For structural terms, etc. *see also* Appendix 7, pages 293-336.